Advances in Intelligent Systems and Computing

Volume 296

Series editor

Janusz Kacprzyk, Polish Academy of Sciences, Warsaw, Poland
e-mail: kacprzyk@ibspan.waw.pl

For further volumes:
http://www.springer.com/series/11156

About this Series

The series "Advances in Intelligent Systems and Computing" contains publications on theory, applications, and design methods of Intelligent Systems and Intelligent Computing. Virtually all disciplines such as engineering, natural sciences, computer and information science, ICT, economics, business, e-commerce, environment, healthcare, life science are covered. The list of topics spans all the areas of modern intelligent systems and computing.

The publications within "Advances in Intelligent Systems and Computing" are primarily textbooks and proceedings of important conferences, symposia and congresses. They cover significant recent developments in the field, both of a foundational and applicable character. An important characteristic feature of the series is the short publication time and world-wide distribution. This permits a rapid and broad dissemination of research results.

Advisory Board

Gordan Jezic · Mario Kusek · Ignac Lovrek
Robert J. Howlett · Lakhmi C. Jain
Editors

Agent and Multi-Agent Systems: Technologies and Applications

Proceedings of the 8th International
Conference KES-AMSTA 2014
Chania, Greece, June 2014

 Springer

Editors
Gordan Jezic
University of Zagreb
Faculty of Electrical Engineering and
    Computing
Department of Telecommunications
Croatia

Mario Kusek
University of Zagreb
Faculty of Electrical Engineering and
    Computing
Department of Telecommunications
Croatia

Ignac Lovrek
University of Zagreb
Faculty of Electrical Engineering and
    Computing
Department of Telecommunications
Croatia

Robert J. Howlett
KES International
Shoreham-by-sea
United Kingdom

Lakhmi C. Jain
Faculty of Education, Science, Technology
    and Mathematics
University of Canberra
ACT 2601
Australia

ISSN 2194-5357   ISSN 2194-5365   (electronic)
ISBN 978-3-319-07649-2   ISBN 978-3-319-07650-8   (eBook)
DOI 10.1007/978-3-319-07650-8
Springer Cham Heidelberg New York Dordrecht London

Library of Congress Control Number: 2014940147

Printed on acid-free paper

Springer is part of Springer Science+Business Media (www.springer.com)

Preface

This volume contains the proceedings of the 8th KES Conference on Agent and Multi-Agent Systems – Technologies and Applications (KES-AMSTA 2014) held in Chania on the island of Crete in Greece, between June 18 and 20, 2014. The conference was organized by KES International, its focus group on agent and multi-agent systems and University of Zagreb, Faculty of Electrical Engineering and Computing. The KES-AMSTA conference is a subseries of the KES conference series.

Following the successes of previous KES Conferences on Agent and Multi-Agent Systems – Technologies and Applications, held in Hue, Vietnam (KES-AMSTA 2013), Dubrovnik, Croatia (KES-AMSTA 2012), Manchester, UK (KES-AMSTA 2011), Gdynia, Poland (KES-AMSTA 2010), Uppsala, Sweden (KES-AMSTA 2009) Incheon, Korea (KES-AMSTA 2008) and Wroclaw, Poland (KES- AMSTA 2007), the conference featured the usual keynote talks, oral presentations and invited sessions closely aligned to the established themes of the conference.

The aim of the conference was to provide an internationally respected forum for scientific research in the technologies and applications of agent and multi-agent systems. This field is concerned with the development and evaluation of sophisticated, AI-based problem-solving and control architectures for both single-agent and multi-agent systems. Current topics of research in the field include (amongst others) agent-oriented software engineering, BDI (beliefs, desires and intentions) agents, agent co-operation, co-ordination, negotiation, organization and communication, distributed problem solving, specification of agent communication languages, formalization of ontologies and conversational agents. Special attention is paid on the feature topics: learning paradigms, agent-based modeling and simulation, self-organizing multi-agent systems, digital economy, and advances in networked virtual enterprises.

The conference attracted a substantial number of researchers and practitioners from all over the world who submitted their papers for three main tracks covering the methodology and applications of agent and multi-agent systems, and two invited sessions on specific topics within the field. Submissions came from 21 countries. Each paper was peer reviewed by at least two members of the International Programme Committee and International Reviewer Board. 30 papers were selected for oral presentation and publication in the volume of the KES-AMSTA 2014 proceedings.

The Programme Committee defined the following main tracks: Modeling and logic agents, Knowledge based agent systems, and Cognitive and cooperative multi-agent systems. In addition to the main tracks of the conference there were the following invited sessions: Agent-based Modeling and Simulation, and Learning Paradigms and Applications: Agent-based Approach.

Accepted and presented papers highlight new trends and challenges in agent and multi-agent research. We hope that these results will be of value to the research community working in the fields of artificial intelligence, collective computational intelligence, robotics, dialogue systems and, in particular, agent and multi-agent systems, technologies and applications.

We would like to express our thanks to the keynote speaker, Prof. Nikos Tsourveloudis from Technical University of Crete Chania, Greece, for his interesting and informative talk of a world-class standard.

The Chairs' special thanks go to special session organizers, Dr. Roman Šperka, Silesian University in Opava, Czech Republic, Prof. Mirjana Ivanović, University of Novi Sad, Serbia, Prof. Costin Badica, University of Craiova, Romania, and Prof. Zoran Budimac, University of Novi Sad, Serbia for their excellent work.

Thanks are due to the Programme Co-chairs, all Programme and Reviewer Committee members and all the additional reviewers for their valuable efforts in the review process, which helped us to guarantee the highest quality of selected papers for the conference.

Our special thanks go to Springer and Prof. Janusz Kacprzyk, Systems Research Institute, Polish Academy of Sciences, Warsaw, Poland for publishing the proceedings in Advances in Intelligent Systems and Computing series.

We cordially thank all of the authors for their valuable contributions and all of the other participants in this conference. The conference would not be possible without their support.

April 2014

Gordan Jezic
Mario Kusek
Ignac Lovrek
Robert J. Howlett
Lakhmi C. Jain

Organization

KES-AMSTA 2014 Conference Organization

KES-AMSTA 2014 was organized by KES International – Innovation in Knowledge-Based and Intelligent Engineering Systems.

Honorary Chairs

I. Lovrek	University of Zagreb, Croatia
L.C. Jain	University of South Australia

Conference Chair

G. Jezic	University of Zagreb, Croatia

Executive Chair

R.J. Howlett	University of Bournemouth, UK

Programme Chair

M. Kusek	University of Zagreb, Croatia

Publicity Chair

I. Bojic	University of Zagreb, Croatia

Keynote Speakers

Prof. Nikos Tsourveloudis
Technical University of Crete, Greece
Bio-inspired Robots: Learning from Nature

International Program Committee

Asst. Prof. Ahmad Taher Azar	Benha University, Egypt.
Dr. Messaouda Azzouzi	University of Djelfa, Algeria
Dr. Marina Bagic Babac	University of Zagreb, Croatia
Prof. Costin Badica	University of Craiova, Romania
Dr. Iva Bojic	University of Zagreb, Croatia
Dr. Dariusz Barbucha	Gdynia Maritime University, Poland
Prof. Maria Bielikova	Slovak University of Technology, Bratislava
Prof. Andrej Brodnik	University of Ljubljana and University of Primorska
Zoran Budimac	University of Novi Sad, Serbia
Assoc. Prof. Dr. Frantisek Capkovic	Institute of Informatics, Slovak Academy of Sciences, Slovakia
Dr. Jessica Chen-Burger	University of Edinburgh, Scotland, UK
Prof. Barbara Dunin-Kęplicz	University of Warsaw, Poland
Dr. Trong Hai Duong	Vietnam National University HCM, Vietnam
Dr. Jose Luis Fernandez-Marquez	University of Geneva, Switzerland
Dr. Konrad Fuks	Poznan University of Economics, Poland
Dr. Arnulfo Alanis Garza	Instituto Tecnologico de Tijuana, México
Prof. Anne Håkansson	KTH Royal Institute of Technology, Stockholm, Sweden
Prof. Chihab Hanachi	University of Toulouse 1 Capitole - IRIT Laboratory, France
Dr. Ronald Hartung	The Design Knowledge Company, USA
Dr. Quang Hoang	Hue University, Vietnam
Prof. Robert J. Howlett	Bournemouth University, UK
Prof. Mirjana Ivanovic	University of Novi Sad, Faculty of Sciences, Serbia
Assoc. Prof. Jason J. Jung	Yeungnam University, Korea
Prof. Dr. Dragan Jevtić	University of Zagreb, Croatia
Dr. Arkadiusz Kawa	Poznan University of Economics, Poland
Dr. Adrianna Kozierkiewicz-Hetmanska	Wroclaw University of Technology, Poland
Dr. Konrad Kułakowski	AGH University of Science and Technology, Poland
Prof. Kazuhiro Kuwabara	Ritsumeikan University, Japan
Dr. Marin Lujak	University Rey Juan Carlos, Madrid, Spain
Prof. Jaeho Lee	The University of Seoul, Korea
Assoc. Prof. Hanh H. Hoang	Hue University, Vietnam
Prof. Tzung-Pei Hong	National University of Kaohsiung, Taiwan
Dr. Manuel Mazzara	Polytechnic of Milan, Italy & Newcastle University, UK

Dr. Daniel Moldt	University of Hamburg, Department of Informatics, Germany
Dr. Marin Orlic	Ericsson Nikola Tesla, Croatia
Assist. Prof. Vedran Podobnik	University of Zagreb, Croatia
Prof. Radu-Emil Precup	Politehnica University of Timisoara, Romania
Dr. Rajesh Reghunadhan	Central University of Bihar, India
Prof. Dr. Vladimir Rybakov	Manchester Metropolitan University, UK
Dr. Adam Sędziwy	AGH University of Science and Technology, Cracow, Poland
Prof. Giovanna Di Marzo Serugendo	University of Geneva, Switzerland
Assist. Prof. Roman Šperka	Silesian University in Opava, Czech Republic
Prof. Andrzej Szalas	University of Warsaw, Poland and University of Linkoping, Sweden
Dr. Wojciech Thomas	Wroclaw University of Technology, Poland
Dr. Bogdan Trawinski	Wroclaw University of Technology, Poland
Dr. Krunoslav Trzec	Ericsson Nikola Tesla, Croatia
Dr. Taketoshi Ushiama	Kyushu University, Japan
Dr. Bay Vo	Ton Duc Thang University, Ho Chi Minh, Vietnam
Dr. Toyohide Watanabe	Nagoya Industrial Science Research Institute, Japan
Dr. Mahdi Zargayouna	IFSTTAR, France
Prof. Arkady Zaslavsky	Commonwealth Science and Industrial Research Organisation (CSIRO), Australia
Prof. Wen-Ran Zhang	Georgia Southern University, USA
Prof. Krzysztof Zielinski	AGH University of Technology, Krakow, Poland

Workshop and Invited Session Chairs

Learning Paradigms and Applications: Agent-Based Approach

Prof. Mirjana Ivanović	University of Novi Sad, Serbia
Prof. Costin Badica	University of Craiova, Romania
Zoran Budimac	University of Novi Sad, Serbia
Prof. Lakhmi Jain	University of South Australia, Australia

Agent-Based Modeling and Simulation

Dr. Roman Šperka	Silesian University in Opava, Czech Republic

Contents

Modeling and Logic Agents

IS: Agent-Based Modeling and Simulation

IS: Learning Paradigms and Applications: Agent Based Approach

Bio-inspired Robots: Learning from Nature

Nikos Tsourveloudis

Technical University of Crete, Greece
nikost@dpem.tuc.gr

Abstract. The fundamental motivation behind the development of bio-inspired multi-robot teams is the ability of living organisms to successfully cope and provide good solutions to almost all robotic related problems. Navigation, material handling and sensors, machine learning are only some of the research areas benefited from examining and adopting methodologies, techniques or mimicking behaviors proved sustainable and successful for animals and humans.

The talk will follow the bio-inspired paradigm of hunting mammals in land (wolves) and the sea (dolphins), intending to make this knowledge applicable to the coordination problem of heterogeneous robotic teams. The objective will be to present, define and discuss the required level of inference capabilities needed for robotic navigation and coordination purposes. Emphasis will be given on the fact that humans and animals decide and conclude about unknown features of their world under constraints of limited time, knowledge, and computational capacity. And despite their "bounded rationality" (or cognitive limitations) tend to built and use domain specific heuristics that allow for fast problem solving (and task specific successful behaviors). Robots and agents may be benefited from this fact.

G. Jezic et al. (eds.), *Agent and Multi-Agent Systems: Technologies and Applications,*
Advances in Intelligent Systems and Computing 296,
DOI: 10.1007/978-3-319-07650-8_1, © Springer International Publishing Switzerland 2014

Conflicts Resolution in Heterogenous Multiagent Environments Inspired by Social Sciences*

Bartosz Ziembiński

Institute of Computer Science, Polish Academy of Sciences, Warsaw
b.ziembinski@phd.ipipan.waw.pl

Abstract. Conflict, which is an inherent part of multiagent environments, is also a natural element of any social structure. Therefore, it might be useful to research social sciences in order to find out what is the state of the art concerning conflicts and then try to investigate how it can be utilized in multiagent settings. A synergy of the two fields may also lead to significant insights about conflicts for social sciences.

In our approach we focus on Thomas and Kilmann's classifications of conflict resolution strategies. Following them, we design the behaviour semantics of five different styles of dealing with conflict and obtain a new method of conflict resolution in heterogenous multiagent environments where agents differ among themselves either physically (e.g. do not have the same sensors) or concerning their roles (e.g. their goals are different). Then we conduct series of simulations in order to understand the nature of modeled strategies. Investigation lets us answer the questions about existence of a dominant strategy, influence of proportions of agents of various types and influence of number of conflicts in a population on the performance of distinct strategies. Finally, we are able to find the best circumstances for each strategy in which it can be adopted.

Keywords: Conflict Resolution, Multiagent Systems, Social Simulation.

1 A New Perspective on Conflict Resolution in Heterogenous Multiagent Environments

Conflict is a natural consequence of agents' social behaviour in multiagent environments. Given they cooperate [1], they can easily be involved in a competition over shared resources. Given they compete with each other [2], conflict is a direct extension of their rivalry. Thus, as a frequent phenomenon in multiagent environments, it has been given some attention by researchers in the area. Some of the investigations are application dedicated. For instance, authors are trying to resolve conflicts concerning Air Traffic Management (ATM), willing to allow the possibility of *free flight*, in which aircraft choose their own optimal routes, altitudes, and velocities, as opposed to being restricted to planned

---

* The paper is cofounded by the European Union from resources of the European Social Fund. Project PO KL "Information technologies: Research and their interdisciplinary applications".

G. Jezic et al. (eds.), *Agent and Multi-Agent Systems: Technologies and Applications*,
Advances in Intelligent Systems and Computing 296,
DOI: 10.1007/978-3-319-07650-8_2, © Springer International Publishing Switzerland 2014

jetways [3, 4]. Another example is utilizing multiagent systems in project management, where conflicts among scheduling activities and distributing resources are inevitable [5]. However, majority of the literature presents more general look on the topic. Authors in [6] try to coordinate the multiagent system by capturing social constraints with norms. However, at times these norms may conflict with one another. Therefore, they present mechanisms for detection and resolution of such normative conflicts. Another approach is to resolve normative conflicts online [7]. The work of [8] postulates that following fixed behavioral rules can be limiting in performance. Authors utilize learning techniques in order to provide agents with adaptability and flexibility. Different way to resolve conflicts among agents is through negotiation [9–12]. In these approaches agents are trying to present stronger arguments than their opponents using some sort of negotiation protocol.

On the other hand, conflict is not only yet another issue in multiagent systems. In fact, it can be found everywhere where a social behaviour is present: among humans, animals, insects etc. Therefore, it might be useful to research social sciences in context of conflicts to find out their possible applications in agency. This idea has been given some attention [13]. For instance, authors of [14] try to import sociological insights (mainly from the theory of autopoietic social systems and the pragmatist theories of symbolic interaction) into Distributed Artificial Intelligence. However, their work lacks an attempt of translating sociological knowledge to some kind of mathematical or computational formalism. In [15] the sociological debate concerning the *micro-macro-link* finds its counterpart in the investigation of internal and external conflicts of agents.

In our approach, we are focusing on one of the classifications of conflict resolution strategies proposed by Thomas and Kilmann [16]. We try to investigate if agents cognitions based on the types mentioned in the classification can provide a reasonable results in multiagent settings. We also try to find out if approach where each agent has some kind of attitude and behaviour characteristics can lead to better performance in some domain of problems. Just like in the book [17] where authors investigate both individual cognitive modeling as well as complex agents' interactions, and however synergy of the fields has not been sufficiently developed, they believe that the interaction of the two may be more significant than either alone.

In this paper we address the problem of conflict resolution in multiagent environments where agents are not homogenous. They can differ among themselves either physically (e.g. do not have the same sensors) or concerning their roles (e.g. their goals are different). In such a situation it can be useful to differentiate styles of conflict resolution for different types of agents. Therefore, we propose a new method of conflict resolution in multiagent environments consisting of five strategies: *Competitive*, *Collaborative*, *Compromising*, *Accommodating* and *Avoiding*. Then we analyse those strategies and provide a description of the best circumstances to utilize them, as well as the types of agents that can most benefit from using them. Results contribute both to conflict resolution in agency, as well as in social sciences.

The rest of the paper is structured as follows. Section 2 introduces *The Game of Resources Collecting* that constitutes an enviroment for agents in which they can resolve conflicts. The game itself is a simple model of the problem of resources distribution. Section 3 presents the behaviour semantics of five different strategies of conflict resolution, while Section 4 evaluates them describing results of conducted simulations. We end with conclusions in Section 5.

2   The Game of Collecting Resources

In order to find out if agents with the behaviour semantics proposed by us will be able to reasonably resolve conflicts emerging among them, firstly we need to clarify how the multiagent environment and conflicts in this environment are defined. We address the problem of conflict resolving in a heterogenous multiagent environment and we also want to somehow connect conflict resolving strategies with some classical problems occurring in distributed systems and therefore with some real-world applications. Example of such problem, that can be easily encountered in conflictogenous environments, is the problem of resources distribution. It is an issue of splitting possibly divisible resources among agents. The problem itself has many links to real-world applications. It can be a group of heterogenous agents, each of them having a different set of sensors, being involved in a mission to explore an unknown terrain, as typically they have to split the terrain among themselves (e.g. according to their sensors capabilities). During their negotiations conflicts can easily appear. Different examples of resources distribution can be: processes competing for the CPU time, clients competing for the time of the server, agents taking part in electronic auctions etc.

As a simple model of resources distribution we propose *The Game of Resources Collecting* that gives a possibility to investigate how each strategy proposed in our method works and to evaluate their efficiency. It is a game in which agents compete with each other for resources. In order to formalise the game and agents environment, we will first introduce some basic notions and then we will describe the rules of the game.

The Game of Collecting Resources can be defined as a tuple $<\mathcal{A}, \mathcal{R}, \gamma>$, where:

- $\mathcal{A} = \{1, 2, ..., n\}$ is the set of agents,
- $\mathcal{R} = \{R_1, R_2, ..., R_m\}$ is the set of sets of resources, where m is a number of rounds in the game. Each of sets R_i for $i = 1, 2, ..., m$ consists of natural numbers that correspond to the resources that are available in the round i,
- $\gamma \in [0, 1]$ is the *conflict factor*. It is a real number which determines how big is the part of agents involved in conflicts in every round.

The rules of the game:

1. There are n agents that take part in the game. They create the set $\mathcal{A} = \{1, 2, ..., n\}$.
2. The game consists of specified number of m rounds.
3. In each round $i \in 1, 2, ..., m$ resources from the set $R_i \in \mathcal{R}$ are split among agents.

4. Each of the sets $R_i \in \mathcal{R}$ consists of n natural numbers $r_1, r_2, ..., r_n$ representing resources that are drawn from the exponential distribution.

5. During each round specified number of agents $k = \gamma n$ is involved in conflicts. Agents taking part in encounters are randomly chosen (in each round).

6. Conflicts are bilateral. In each round k randomly chosen agents are matched in pairs. Then, in this $\frac{k}{2}$ pairs agents are competing for resources using their own startegies of conflict resolution.

7. Agents, that in given round are involved in conflicts, have a chance to compete for larger resources from the pool. More formally, without loss of generality assume that resources $r_k \in R_i$ are numbered in such a way that $r_1 \geq r_2 \geq r_3 \geq ... \geq r_n$. Then, each of $\frac{k}{2}$ conflicting pairs will compete for resoures $r_1, r_2, ..., r_{k/2}$.

8. Resources are divisible. Agents that are competing for resource $r \in R_i$ can split it into parts r_1 and r_2, such that $r_1 + r_2 \leq r$ and $r_1, r_2 \geq 0$.

9. The biggest resources, that are left after conflict resolutions, are distributed among other $n - k$ agents which were not involved in conflict situations.

10. Number of agents taking part in successive rounds may decrease, because some agents may drop out off the game. Such agents can keep their gained resources and are taken into consideration in the final result of the game, but cannot participate in next rounds.

11. Agent, that will gain the largest amount of resources during the whole game, wins.

Thus, the game in a simple way models a situation in which agents have to split some resources (terrain, time of CPU, items in an auction) among themselves. Of course, some of them are more attractive (in the game - bigger resources), therefore agents compete for them. Less attractive resources can be acquired without a rivalry. Only limitation of this model is that conflicts are bilateral, therefore it is not as general as one could expect. However, still a lot of cases can be modeled with this approach. The model with multilateral conflicts can be a topic of further research.

3   Behaviour Semantics

After introducing the environment, now we want to focus on possible actions of agents. Thomas and Kilmann described five styles of reacting to conflict situations based on their assertiveness and cooperativeness level: competition, collaboration, compromise, accommodation and avoidance [16] (see Fig. 1). In our research we wanted to create a semantics of these behaviours which would be useful in multiagent environments i.e. agents' strategies of conflict resolution. To describe it, we used a Prolog-like language based on first order logic. It consists of predicate symbols and operators. From their connections we obtain logical rules. Some of the most general predicate symbols used to describe agents' behaviour semantics are as follows:

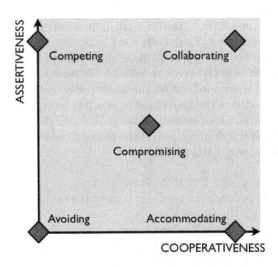

Fig. 1. Thomas and Kilmann's styles of dealing with conflict based on their level of assertiveness and cooperativeness

1. *Conflict*: $\mathcal{A} \times \mathcal{A} \times R \to \{0,1\}$ is a predicate symbol that signals a conflict between agents. $Conflict(i,j,r)$ denotes that agents i and j are in conflict concerning resource r. The predicate symbol is symmetric, which means that $Conflict(i,j,r) \Leftrightarrow Conflict(j,i,r)$.
2. *Competitive, Collaborative, Compromising, Accommodating, Avoiding*: $\mathcal{A} \to \{0,1\}$ are predicate symbols denoting type of a given agent.
3. *Gets*: $\mathcal{A} \times R \to \{0,1\}$ is a predicate symbol denoting how much resources is gained by a given agent.
4. *GetsRandom*: $\mathcal{A} \times \mathcal{A} \times R \times (\mathcal{A} \to \{0,1\}) \to \{0,1\}$ is a predicate symbol that randomly chooses which of the agents will receive the resource r and to which of them the predicate symbol, that is the last argument, will be applied.

The rest of the predicate symbols, which were associated with a particular type of reaction to conflict, will be introduced during description of the particular types of agents.

3.1 Reactions to Conflicts

Competitive Style. Thomas and Kilmann wrote about this type of reaction that individual, which resolves conflicts in this manner, is highly assertive and self confident, he knows what he wants and knows that he has a power to achieve it. He is also not willing to cooperate. Therefore, because we wanted to be as close to the original description as possible, we decided to make *Competitive* agents a type that always want to grab all resources for itself. Thus, it comes as a winner from every encounter with an agent of the other type (receives whole resource r, and the opponent gets 0 reward). However, problem occurs when two *Competitve* agents meet. They both want to possess the whole resource and they both do not want to give up. In such situation, stronger agent (the one that collected more resources so far) is rewarded with the resource and the other agent receives nothing and is dropped out off the game (however, it keeps the collected resources and is included in the final classification). In case of both agents having collected the same amount of resources so far, it is randomly chosen which of them is rewarded and which drops out off the game. *Competitive* style semantics described in a formal way looks as follows:

```
Gets(i, r), Gets(j, 0), OutOfTheGame(j) :-
Conflict(i, j, r), Competitive(i), Competitive(j), HasMore(i, j)

GetsRandom(i, j, r, OutOfTheGame) :-
Conflict(i, j, r), Competitive(i), Competitive(j),
-(HasMore(i, j) | hasMore(j, i))

Gets(i, r), Gets(j, 0) :-
Conflict(i, j, r), Competitive(i), -Competitive(j)
```

where:

5. *OutOfTheGame*: $\mathcal{A} \to \{0,1\}$ is a predicate symbol informing about an agent falling out of game. It is included in the final classification though, with the amount of resources that it has collected so far.
6. *HasMore*: $\mathcal{A} \times \mathcal{A} \to \{0,1\}$ is a predicate symbol denoting that so far the first agent collected more resources than the second one.

Collaborative Style. According to previously mentioned paper [16], individuals, that are collaborative, try to satisfy all parties that are involved in the conflict. They try to come with an optimal solution that will be right for everyone. They are also highly assertive. Model of such behaviour in our multiagent environment was obtained by splitting resources in a proportional way. Say that the agent i has r_i resources and the agent j has r_j resources. They are competing for the resource r. Then, the agent i will receive $r \cdot \frac{r_i}{r_i+r_j}$ and the agent j will receive $r \cdot \frac{r_j}{r_i+r_j}$. Of course, it is only a situation in which two *Collaborative* agents meet. When a *Collaborative* agent meets a *Competitve* agent, the second one takes the whole resource and leaves his opponent with nothing. When a *Collaborative* agent meets *Compromising* or *Accommodating* agent, notice that he is more assertive than them. Therefore his opponents must adopt his way of thinking. Thus, the resource is splitted proportionally. The case with *Avoiding* agent will be discussed later. The *Collaborative* strategy, written in a formal way, looks as follows:

```
GetsProportional(i, j, r), GetsProportional(j, i, r) :-
Conflict(i, j, r), Collaborative(i),
(Collaborative(j) | Compromising(j) | Accommodating(j))
```

where:

7. *GetsProportional*: $\mathcal{A} \times \mathcal{A} \times R \rightarrow \{0,1\}$ is a predicate symbol that split the resource among agents in a proportional way. *GetsProportional*(i,j,r) means that agent i receives $r \cdot \frac{r_i}{r_i+r_j}$ resources (of course if $0 = r_i = r_j$ then both agents will receive $\frac{r}{2}$ resources).

Compromising Style. Compromise is a method of conflict resolution where both parties are trying to come up with a solution, which will at least partially satisfy them [16]. In this sense, the approach is similiar to collaboration, but the difference is that in compromise it is expected from both sides to abandon some of their demands in order to propose a solution right for everyone. Model of such behaviour in our multiagent environment is not that easy to achieve, because main things, that agents can resign from, are resources. If both agents will not take some of their resources, it does not make much sense. Therefore, in our semantics abandoning some of agents' demands has more of a temporal aspect. *Compromising* agent will want to sacrifice the resource, if it will consider it not worth fighting for, but it will also state, that in the future it may request the return of a favor. To go more into the detail, the *Compromising* agent tries to hand over the resource to its opponent in order to mitigate the conflict, but for each of this kind of favors it receives a *ticket*. In the future, during different encounter, it can request to take the whole resource for itself giving back the *ticket*. Thus, tactics of a *Compromising* agent looks as follows. Let us assume that during the round t, it meets another agent of a type different than *Competitive* and *Collaborating* (because they are more assertive and will force their own solution of a conflict) and *Avoiding* (because they avoid conflicts at all). They have r_t resources to split among themselves. Then the *Compromising* agent will check if r_t is bigger than his average gain during previous rounds, i.e. if $r_t > \frac{\sum_{i=1}^{t-1} r_i}{t-1}$. If it is not the case, then it will hand over the resource to its opponent and get a *ticket*. If the equality holds and the agent has a ticket, then it takes whole resource for itself and throws away the ticket. If it does not have a ticket, it must hand over the resource to its opponent. Another issue is when two *Compromising* agents meets and both of them want to grab the resource or both want to hand it over to the other one. Then, it is randomly chosen which one of them gets the resource and which one gets the *ticket*. In a formal way, we can write:

```
Gets(i, 0), Gets(j, r), DropsTicket(j) :-
Conflict(i, j, r), Accommodating(i), Compromising(j),
IsBiggerThanAverage(r, j), HasTicket(j)

Gets(i, r), takesTicket(j) :-
```

```
Conflict(i, j, r), Accommodating(i), Compromising(j),
(-isBiggerThanAverage(R, B) | -hasTicket(B))

Gets(i, r), DropsTicket(i), TakesTicket(j) :-
Conflict(i, j, r), Compromising(i), Compromising(j),
( (IsBiggerThanAverage(r, i), -IsBiggerThanAverage(r, j)) |
(IsBiggerThanAverage(r, i), HasTicket(i),
IsBiggerThanAverage(r, j), -HasTicket(j)) )

GetsRandom(i, j, r, TakeTicket) :-
Conflict(i, j, r), Compromising(i), Compromising(j),
( (IsBiggerThanAverage(r, i), HasTicket(i),
IsBiggerThanAverage(r, j), HasTicket(j)) |
(IsBiggerThanAverage(r, i), -HasTicket(i),
IsBiggerThanAverage(r, j), -HasTicket(j)) |
(-IsBiggerThanAverage(r, i), -IsBiggerThanAverage(r, j)) )
```

where:

8. *IsBiggerThanAverage*: $R \times \mathcal{A} \to \{0, 1\}$ is a predicate symbol, which tells if the resource possible to be collected in this round is bigger than an average gain in previous rounds. Say we are in round t and resources possible to collect in this round are r_t. Then, the predicate symbols is true when $r_t > \frac{\sum_{i=1}^{t-1} r_i}{t-1}$.

9. *hasTicket*: $\mathcal{A} \to \{0, 1\}$ is a predicate symbols, which tells if an agent has one or more free *tickets*.

10. *takeTicket*: $\mathcal{A} \to \{0, 1\}$ is a predicate symbol denoting that an agent receives another *ticket*.

11. *dropTicket*: $\mathcal{A} \to \{0, 1\}$ is a predicate symbol denoting that if an agent has more than a zero *tickets*, then he is throwing one away.

Accommodating Style. According to Thomas and Kilmann's words, accommodating is a style of dealing with conflicts where an invidual is willing to accept the opponent demands even though they may be contrary to his own interests. The individual representing this type of reactions is highly cooperative but not assertive. In our model, an agent that represents *Accommodating* style is adopting the solution proposed by the opponent. If it meets *Collaborative* agent, they will collaborate, i.e they will split resources proportionally. If he meets *Compromising* agent they will compromise etc. The only issue is what happens when two *Accommodating* agents meet. Then, they both want to accept the opponent demands. In our semantics they both receive half of the resource. In terms of formal description, this approach looks as follows:

```
Gets(i, r/2), Gets(j, r/2) :-
Conflict(i, j, r), Accommodating(i), Accommodating(j)
```

Avoiding Style. Individuals representing this style try to avoid conflicts at all costs. Thomas and Kilmann wrote that in the name of this goal, they are ready to accept default solutions, even if they will not be the best options at the moment. In our model we used this characteristic and made *Avoiding* agents not being involved in conflicts at all and giving all the resource to the opponent. Instead of normal fights for resources, they have their own way of getting them. Namely, they gather the resources that are left after the encounters between agents and have the priority to collect the bigger resources before the agents that in a given round were not involved in conflicts at all. Only exception is when two agents of type *Avoiding* meet. Then, it is randomly chosen which one of them will get the resource and which one will collect one of the resources left after all encounters. These semantics written in a formal way:

```
GetsFromRest(i), Gets(j, r) :-
Conflict(i, j, r), Avoiding(i), -Avoiding(j)
```

```
GetsRandom(i, j, r, GetsFromRest) :-
Conflict(i, j, r), Avoiding(i), Avoiding(j)
```

where:

12. *GetsFromRest*: $\mathcal{A} \to \{0,1\}$ is a predicate symbol denoting that the agent receives the resource from the pool $r_{k/2+1}, r_{k/2+2}$, etc.

4 Simulations

In order to see how given conflict resolving strategies interact with each other, we have undertaken simulations. We wanted to find out:

- If there is a dominant strategy.
- If the effectiveness of a given strategy is dependent on the number of other agents using it. If the effectiveness of a given strategy is dependent on the number of agents using other strategies. If yes, then in what way?
- How is the effectiveness of a given strategy dependent on a number of conflicts emerging in a population?
- If every strategy is worth adopting in some circumstances, or if some strategies are not worth using at all.

Multiagent simulations, that were supposed to answer those questions, consisted of series of experiments. Each experiment involved 1000 games (agents played *The Game of Collecting Resources* 1000 times). Each game consisted of 10 rounds and in each game there were 100 agents participating. Among these 100 agents there were agents of different types. Agents of each type were utilizing their own strategies in order to resolve conflicts. Experiments differed between themselves by the numbers of agents of each type and by the value of the conflict factor. These were the *degrees of freedom* of conducted experiments. For instance, a sequence (5, 15, 20, 25, 35) denoted that in a given experiment, in each game there were 5 *Competitve*, 15 *Collaborative*, 20 *Compromising*, 25 *Accommodating* and 35 *Avoiding* agents. The conflict factor $\gamma = 0.2$ denoted that in a given experiment, in every game 20% of agents were involved in conflicts during every round.

$\lambda = 2$ was the parameter of exponential distribution from which resources were drawn.

In order to track agents' performance we introduced two statistics. First of them, Q statistic, describes what is the expected number of times for an agent of type $<T>$ to be classified on places from intervals $[1, 20]$, $[21, 40]$, $[41, 60]$, $[61, 80]$, $[81, 100]$ in the final classification during 1000 games:

$$\begin{smallmatrix}[p+1,p+20]\\<T>\end{smallmatrix}Q = \frac{\sum_{i=1}^{1000} \begin{smallmatrix}[p+1,p+20]\\<T>\end{smallmatrix}q_i}{N_{<T>}}, \text{ where:}$$

- $\begin{smallmatrix}[p+1,p+20]\\<T>\end{smallmatrix}q_i$ denotes the number of agents of type $<T>$ taking places from interval $[p + 1, p + 20]$ in the final classification during round i. $<T> \in \{Competitive, Collaborative, Compromising, Accommodating, Avoiding\}$, $p \in \{0, 20, 40, 60, 80\}$
- $N_{<T>}$ denotes the number of agents of type $<T>$ in the experiment.

The second statistic, R statistic, describes what is the expected amount of resources that an agent of type $<T>$ gains during 1000 games.

$$_{<T>}R = \frac{\sum_{i=1}^{1000} {}_{<T>}r_i}{N_{<T>}}, \text{ where:}$$

- $_{<T>}r_i$ denotes the amount of resources gained by agents of type $<T>$ during round i.
- $N_{<T>}$ has the same meaning as above.

4.1 Results of Simulations

Is there a dominant strategy? There is no dominant strategy. In an average case, where all types of agents are represented in the same proportion, i.e. (20, 20, 20, 20, 20) and among different conflict factors representing less, average and more conflictogenous environments, i.e. (0.2, 0.5, 0.8) *Competitive* strategy proved to be the best one. In all cases it had the biggest expected payoff (R statistic) and the biggest expected number of agents classified on places from the interval $[1, 20]$. Also, the bigger the value of the conflict factor was, the better was the performance of the strategy. Thus, we considered it as a strong candidate for a dominant strategy. However, it turned out that with the bigger part of the population being *Competitive* agents, it is harder for them to achieve good results. It is because they bump into each other and eliminate themselves from the game. For instance, in a population (40, 15, 15, 15) and $\gamma = 0.5$ they still have the biggest expected payoff, but they win only slightly. $R = [491235, 459534, 439420, 478698, 446263]$ (the order is the same as in the proportions of the population: *Competitive*, *Collaborative*, *Compromising*,

Accommodating and *Avoiding*). In a population (52, 12, 12, 12, 12) they take fourth place concerning expected payoff, and in a population (60, 10, 10, 10, 10) they are the last one. The bigger is the ratio of *Competitive* agents in a population, the worst is their performance. *Competitive* strategy is also very risky. Typically, taking into account the Q statistic, there is a lot of agents of this type taking places in the interval $[1, 20]$, but in the interval $[81, 100]$ as well. Thus, if there is a lot of agents of this type in a given game, one of them will probably win the whole contest, but lots of them will occupy the last positions in the final classification.

Is the effectiveness of a given strategy dependent on the number of other agents using it. Is the effectiveness of a given strategy dependent on the number of agents using other strategies. If yes, then in what way? In this subsection all experiments were conducted with a fixed value of $\gamma = 0.5$ which corresponds to an average conflictogenous situation. Different values of the factor will be discussed in the next chapter.

The case of *Competitive* type was described above. It is sensitive to the number of agents of the same type and not really responsive to the number of agents of different types. *Collaborative* agents in an average case ((20, 20, 20, 20, 20)) are on a third place concerning R statistic. If there will be more agents of this type ((10, 60, 10, 10, 10) or (5, 80, 5, 5, 5)), they will take fourth place. Situation changes a little bit when the conflict factor has a bigger value (*Collaborative* agents are able to maintain third place). The case will be described in the next subsection. For agents of the *Compromising* type the value of conflict factor equal to 0.5 is the worst case scenario. They always occupy fifth place as far as expected payoff is concerned, no matter if there is a lot of them, or if they form a small part of the population. If we take a look at *Accommodating* agents, it is better for them if their number in the society is low ((24, 24, 24, 4, 24), (20, 20, 20, 20, 20)). Then, they hold a strong second place concerning R statistic. For a ratio (10, 10, 10, 60, 10) and higher they loose their second place in favor of *Avoiding* agents and are third. For *Avoiding* agents it is also important not to constitute a big part of the population. They hold strong second place for (24, 24, 24, 24, 4), third place for (22, 22, 22, 22, 12) and fourth place for an average case (20, 20, 20, 20, 20). In case of bigger number of agents of this type in the society (40% and more) they occupy fifth place.

It is worth noticing that all types of agents in a semi-conflictogenous environment ($\gamma = 0.5$) were not that much sensible to numbers (and therefore proportions) of agents of different types. Their performance depended mostly on the number of agents of their type. Another fact was that for almost all types of agents it was better if the number of individuals of the same type was smaller (did not really matter for the *Compromising* strategy). Thus, it was good to come up with an original strategy for a given game.

How is the effectiveness of a given strategy dependent on a number of conflicts emerging in a population? Simulations have shown that the conflict factor is the most significant for *Avoiding* and *Compromising* styles, thus for the strategies that have low both levels of assertiveness and cooperativeness. For *Avoiding* agents it is better if there is not a lot of conflicts in the population (value of γ factor in [0, 0.2]) and for *Compromising* agents it is better if there are many of them. If the value of γ is high (γ in [0.8, 1]) then it is also better for *Compromising* agents to increase their number in the population. For $\gamma = 0.9$ they occupy fourth place concerning R statistic in configuration (20, 20, 20, 20, 20), third place if there is 44% of them or more. For both high and low values of γ it is better for *Avoiding* agents to keep their number small. For the other types the conflict factor is not that important, but it is rather better for them if it has bigger value. In this case *Competitive* type can expect a bigger average payoff. Styles with high cooperativeness level, i.e. *Collaborative* and *Accommodating*, will have a little bit smaller payoffs (2-4%), but in case of increasing their numbers in population, they can count on keeping their third and second place, respectively, in the classification.

Is every strategy worth adopting in some circumstances, or are some strategies not worth using at all? The worst strategy to adopt seems to be to compromise. It was on the last place in the classification of expected payoffs in many cases, taking third place in the best scenario. Remember though, that this strategy is picky: wants to take bigger resources in exchange for giving away the small ones. Thus, in a case when the bigger resources would have additional value or a feature to themselves, which would be significant for this type of agents, this type of strategy would also be worth taking into consideration. *Competitive* style seems to be efficient, but is risky (many agents in the top but also in the bottom of the classification). Thus is not worth using for agents not willing to bear the risk. *Avoiding* is a very good strategy if the value of the conflict factor is low or/and if the number of such agents in the population is small. Accommodation and Collaboration are safer strategies (will not win, but take a fairly good place in the classification). Their performance is satisfying in most scenarios. Mostly they take second and third place concerning expected payoff. Taking into consideration the Q statistic, agents of these types are most likely to fall into intervals [21, 40] or [41, 60], some of them into [61, 80]. Thus, these styles are not leading to victory in every encounter, but are also not likely to fail on every occasion.

5   Conclusion

We have modeled five different styles of handling a conflict. Those strategies are different in many ways: some of them are more risky, other are safer, they are suitable to adopt in distinct circumstances. Their usage also depends on what an agent actually wants to achieve. In this way, differentiating agents' behaviour can be useful in situations, where agents are not homogenous either physically (e.g. they do not have the same sensors) or concerning their roles (e.g. their goals are different). In such situations it is better not to search for *optimal* strategies, but for *maximizing* ones. In game theory maximal players try to exploit perceived weaknesses in their opponent's way of playing [18]. Maximizing strategy in our framework would stand for a style that best corresponds with current goals of a given agent (which may be completely different from goals of agents of different types) and the current state of a population, i.e. number of conflicts and proportions of agents representing different styles.

In order to understand the nature of modeled strategies we have conducted series of experiments. We have answered the questions about existence of a dominant strategy and came to the conclusion that there is no strategy that will win in all circumstances. In fact, the best way to choose a strategy for a new coming player is to choose a most original one. That is a style which is currently not commonly used in the population. We have also answered the questions about influence of proportions of agents of various types and influence of number of conflicts in a population on the performance of distinct styles. We have investigated what are the best circumstances for each strategy in which it can be adopted. However, we are also aware of the fact that our empirical study has not exhausted all possible cases and in order to fully understand the behaviour of said dynamical environment it would be useful to come up with a mathematical or computational model of it. Such investigation would allow to identify the underlying mechanisms of interactions between agents and may be a good topic for further research.

References

[1] Dunin-Keplicz, B., Verbrugge, R.: Teamwork in Multi-Agent Systems: A Formal Approach. Wiley and Sons (July 2010)

[2] Scheutz, M.: Surviving in a Hostile Multi-agent Environment: How Simple Affective States Can Aid in the Competition for Resources. In: Hamilton, H.J. (ed.) Canadian AI 2000. LNCS (LNAI), vol. 1822, pp. 389–399. Springer, Heidelberg (2000)

[3] Tomlin, C., Pappas, G.J., Sastry, S.: Conflict Resolution for Air Traffic Management: A Study in Multiagent Hybrid Systems. IEEE Transactions on Automatics Control 43(4), 509–521 (1998)

[4] Wollkind, S., Valasek, J., Ioerger, T.R.: Automated Conflict Resolution for Air Traffic Management Using Cooperative Multiagent Negotiation. In: Proc. AIAA Guidance, Navigation, and Control Conference and Exhibit, August 16-19, Providence, Rhode Island (2004)

[5] Yan, Y., Kuphal, T., Bode, J.: Application of Multiagent Systems in Project Management. Int. J. Production Economics 68, 185–197 (2000)

[6] Vasconcelos, W.W., Kollingbaum, M.J., Norman, T.J.: Normative conflict resolution in multi-agent systems. Autonomous Agents and Multi-Agent Systems 19(2), 124–152 (2009)

[7] Gaertner, D., Garcia-Camino, A., Noriega, P., Rodriguez-Aguilar, J.-A., Vasconcelos, W.: Distributed Norm Management in Regulated Multi-Agent Systems. In: Proc. AAMAS 2007, Honolulu, Hawaii, USA, May 14-18 (2007)

[8] Haynes, T., Lau, K., Sen, S.: Learning Cases to Compliment Rules for Conflict Resolution in Multiagent Systems. AAAI Technical Report SS-96-01 (1996)

[9] Sillince, J.A.A.: Multi-agent conflict resolution: a computational framework for an intelligent argumentation program. Knowledge-Based Systems 7(2), 75–90 (1994)

[10] Kakehi, R., Tokoro, M.: A negotiation protocol for conflict resolution in multi-agent environments. In: Proc. International Conference on Intelligent and Cooperative Information Systems, Rotterdam, The Netherlands (1993)

[11] Sycara, K.: Resolving Goal Conflicts via Negotiation. In: Proc. AAAI 1988 (1988)

[12] Berker, I., Brown, D.C.: Conflicts and Negotiation in Single Function Agent Based Design Systems. Concurrent Engineering 4(1), 17–33 (1996)

[13] Tessier, C., Chaudron, L., Muller, H.-J.: Conflicting Agents: Conflict Management in Multi-Agent Systems. Springer (2001)

[14] Malsch, T., Weiss, G.: Conflicts in social theory and multi-agent systems. In: Conflicting Agents: Conflict Management in Multi-Agent Systems. Springer (2001)

[15] Hannebauer, M.: Their problems are my problems. In: Conflicting Agents: Conflict Management in Multi-Agent Systems. Springer (2001)

[16] Thomas, K.W.: Conflict and conflict management: Reflections and update. Journal of Organizational Behavior 13(3), 265–274 (1992)

[17] Sun, R.: Cognition and Multi-Agent Interaction: From Cognitive Modeling to Social Simulation. Cambridge University Press (2006)

[18] West, R.L., Lebiere, C., Bothell, D.J.: Cognitive architectures, game playing, and human evolution. In: Cognition and Multi-Agent Interaction: From Cognitive Modeling to Social Simulation. Cambridge University Press (2006)

Expanding the Control Scope of Cooperative Multiple Robots

Ko Shibata[1,*], Munehiro Takimoto[2], and Yasushi Kambayashi[1]

[1] Department of Computer and Information Engineering, Nippon Institute of Technology, 4-1 Gakuendai, Miyashiro-machi, Minamisaitama-gun, Saitama, 345-8501 Japan
c1085262@cstu.nit.ac.jp, yasushi@nit.ac.jp
[2] Department of Information Sciences, Tokyo University of Science, 2641 Yamazaki, Noda 278-8510, Japan
mune@cs.is.noda.tus.ac.jp

Abstract. When we attempt to construct a multi-robot system in out-door setting, it is natural to provide Mobile Ad Hoc Network (MANET) to form cooperative work. When a robot accidentally moves out from the communication range, it becomes impossible to communicate with one another. In order to mitigate this situation, we employ visible ray to widen the control scope so that the stray robot can move into the radio communication range. In this paper, we propose a method for controlling multi-robot system. That is a combination of visible light and radio communication. Robots are controlled by mobile software agents that convey task information to each robot through MANET. The controlling software agent intermittently visits each robot to check the status of the robot. When the agent finds a missing robot, it uses a beacon light to signal the aerial robot to hover above the current ground robot, and send beacon to the missing robot to come back into the range. The contribution of this research is to expand the scope of communication range of robots. It should provide effective use for the robot system with limited resource in the field.

Keywords: Multi-agent, Multi-robot, ad-hoc network, Light signal, Aerial-robot.

1 Introduction

In the last decade, we have witnessed the advent of multi-robot systems. A multi-robot system consists of a large number of homogeneous robots that have limited capacity but, when combined into a group, they can generate more complex behaviors [1]. In multi-robot systems, robots communicate with each other to achieve cooperative behaviors. There are three major advantages of multi-robot systems over single robot systems [2, 3]. The first is parallelism; a task can be achieved by autonomous and asynchronous robots in a system. The second is robustness; it is realized through redundancy. The system can have more robots than required for a certain task. The third is scalability; a robot can be added to or removed from the system easily.

---

* Corresponding author.

G. Jezic et al. (eds.), *Agent and Multi-Agent Systems: Technologies and Applications*,
Advances in Intelligent Systems and Computing 296,
DOI: 10.1007/978-3-319-07650-8_3, © Springer International Publishing Switzerland 2014

On the other hand, excessive interactions among agents in the multi-agent system may cause problems in the multiple robot environments. In order to mitigate the problems of excessive communication, mobile agent methodologies have been developed for distributed environments. Since a mobile agent can bring the necessary functionalities with it and perform its tasks autonomously, it can reduce the necessity for interaction with other sites. In the minimal case, a mobile agent requires that the connection is established only when it performs migration [4]. We have implemented several multi-robot systems that search and recollect arbitrary targets without redundant movements [5-9]. We have designed and implemented multi-agent systems that control the robot systems. A control system based on multiple software agents can control robots efficiently [5, 8]. Multi-agent systems introduced modularity, reconfigurability and extensibility to control systems, which had been traditionally monolithic. It has made easier the development of control systems on distributed environments such as multi-robot systems.

In such systems, we have we assumed wireless LAN is available as the communication environment [5-12]. In other words, software agents move from a robot to another robot through wireless LAN with TCP/IP connection. They are good for indoor experiments. When we attempt to extend our multi-robot system to more realistic environments, i.e. out-door setting, using wireless LAN is not an ideal choice. It is particularly not a good choice for rescue missions in disaster area. Upon the above observation, it is natural to provide the multi-robot systems Mobile Ad Hoc Network (MANET) environment to form cooperative work. MANET is a computer network that is dynamically formed by autonomous mobile nodes [13]. Such mobile nodes are connected through wireless links without relying on any central controller or established infrastructure. The participating mobile nodes can freely and dynamically self-organize into arbitrary and temporary network topologies. Therefore, MANET can handle many problems with the systems in the distributed environment very well.

As for wireless device equipment in the robot, ZigBee and Bluetooth are popular, and their effective wireless ranges are approximately 100m radius. When a robot accidentally move beyond the communication range, it becomes impossible to communicate with each other. In order to mitigate this situation, we employ visible ray to widen the control scope so that the stray robot can move into the radio communication range. Transmitting information by way of vision, such as flag and light, has been used from ancient human history. We employ an aerial robot that emits visible light as the beacon to send signals to make the robots come back into the range. The beacon provides flexibility and adaptability to the robot system. It also contributes to the energy saving, because the robots need not to use strong radio wave. If the robots are given some covert tasks, we can simply use infrared rays to achieve the goal.

In this paper, we propose a method for controlling multi-robot systems. That is a combination of visible light and radio communication. Robots are controlled by mobile software agents that convey task information to each robot through MANET. The controlling software agent intermittently visits each robot to check the status of the robot. When the agent finds a missing robot, it uses a beacon light to signal the aerial robot to hover above the current ground robot, and send a beacon to the missing robot to come back into wireless range. The contribution of this research is to expand the scope of communication range of robots. It should provide effective use for the robot system with limited resource in the field.

The structure of the balance of this paper is as follows. In the second section, we describe the background. The third section describes the proposed method. In fourth section, we demonstrate the effectiveness of the proposed system, and describe the results of the experiments with a simulator. Finally, we conclude our discussion with future works in the fifth section.

2   Background

We have investigated the search problem for multi-robot system using the mobile software agents [5-9]. The robot system is designed to search a target cooperatively. If the system makes the entire robots search, the system can achieve the goal fast. However, robots that could not find the target would result in the unnecessary consumption of energy. The robots are not connected to the plug; they work by batteries. We have succeeded to reduce the energy consumption as much as possible [5-9]. However, when a robot moves too far from other robots or moves behind a large obstacle, it may fail to communicate with its colleagues by using radio wave. The system needs a technique for reconstructing the communication environment for agent migration.

Yokoyama et al added an aerial robot to the multi-Robot system [12]. Agents facilitate the control of the aerial robot in a three-dimensional environment. Aerial robot hovers over the robot where the mobile software agent currently resides in this system. Adding an aerial robot certainly widen the control scope of the group of multiple robots.

There are other researches of multi-robot systems that use mobile software agents. Those include collecting robot using Ant Colony clustering [10]. The study uses mobile agents to reduce the communication cost. A research of alignment of the gathered robots uses mobile agents to reduce the cost of moving the actual robots [11].

In those researches, the steady communication is assumed. However, the environment where the information infrastructure is not available should be considered. We investigate the cooperative multi-robot system in such environment. We employ an aerial robot as a beacon to signal stray robots to re-assemble.

3   The Proposed Method

Our proposed method uses visible ray as a beacon to guide missing robots. This method can recollect any stray robots that are outside of the wireless communication range. We employ an aerial robot to emit the guidance signal light. A missing robot can re-connect to the other robots when it comes back to the communication range of other connected robots. Even though many robots go beyond the communication range, they can be reconnected by repeating this procedure.

In our multi-robot system, we have two types of robots. One is aerial robot and the other is ground robot. The ground robots are scattered on the field. Ground robots are executing a given task, e.g. searching for an object. The aerial robot is given the role of the lighthouse that guides the ground robots. The ground robot has an omnidirectional camera in addition to the equipment necessary for its own work, so that it can sense the visible ray from any directions. The ground robot has a light too. The ground robot uses this light to communicate with the aerial robot. When a ground robot wants to establish a connection to a certain other ground robot, it uses the light as a signal that calls the aerial robot. When the aerial robot senses the request light from one of the ground robots, it moves to and hovers above the lighting ground robot. Then the aerial robot puts on its own signal light to send a signal to the other ground robot to come close. The reason why we employ an aerial robot is that flying robots can send visible light signal farther than radio signal used in MANET.

3.1   Constructing Network of Robots

Each robot is equipped with a wireless device, i.e. ZigBee. The entire ground robots construct a MANET by using the weak radio wave. The reason why we employ ZigBee as the wireless device is that it consumes less energy than Bluetooth. In addition, weak radio is also desirable for small-scale experiments. In our assumed communication system, the ground robots act as the mobile nodes in the traditional MANET experiments. A robot joins to the network automatically when it moves into the inside of the communication range. When a participating robot moves out of the network range, it is disconnected.

The aerial robot has the key role. As shown in Fig. 1, it monitors the ground robots. When it finds a request light from the ground robot as shown in Fig. 2, it moves to and hover above the requesting ground robot, and puts on its own signal light. The ground robot that is out of the communication range finds the light of the aerial robot and moves toward the aerial robot until it is in the communication range. If the newly participated robot is not the robot the original request robot wants to communicate, the newly participated robot put on the light to make the aerial robot come and hover. This procedure repeats until the right destination robot is found and the requested communication is established.

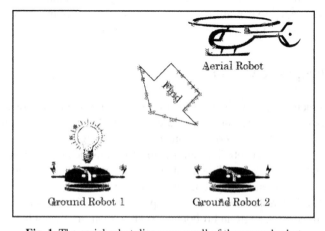

Fig. 1. The aerial robot discovers a call of the ground robot

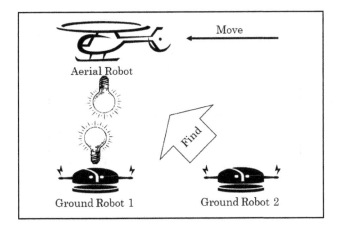

Fig. 2. The aerial robot moves and turns on the light. A missing ground robot can notice the light of the aerial robot.

3.2   System Flow

This section explains the flow of our method by using an example. The scenario is to re-assemble scattered robots and to re-construct one MANET. Fig. 3 shows the start state. Each dot represents one robot, and each circle represents the wireless communication range. There are six robots namely Robot 1 (R1) through Robot 6 (R6). R1 and R2 create one network using ad-hoc communication because they are in the overlapping area of their wireless communication ranges. The line that connects the two robots represents this communication network.

Supposing R1 wants to establish an ad hoc connection to R5 so that it can send a mobile agent to R5. R1 turns on the request signal to call nearby unconnected robots. In order to convey the calling signal to other robots, the aerial robot moves to and hover above the R1 and turns on its own signal light. The dark colored dot of R1 denotes this situation in Fig. 3.

Robots (R3 through R6) are monitoring surrounding environment using their omnidirectional cameras to notice signal from the aerial robot. The robots that notice the aerial signal light move toward the signal. When a robot obtains the connection to R1, it stops moving. R3 and R4 sense the signal light, and move to connect to R1 as shown in Fig. 4. Since R3 had the connection to R5, and that is the robot R1 desires to establish a connection, R3 turns on its light signal to call the aerial robot to repeat the procedure. As the aerial robot move to and hover above R3, R5 notices the aerial signal and comes close to re-connect to R3. Then the connection from R1 through R5 is established to form the desired MANET. Fig. 5 represents this situation. In this figure, we assume that R6 is too far to recognize the light signal.

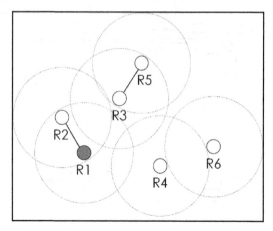

Fig. 3. Initial state

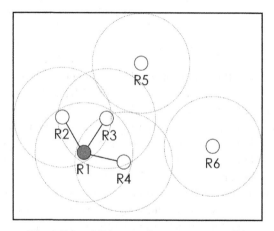

Fig. 4. R3 and R4 come close to connect to R1

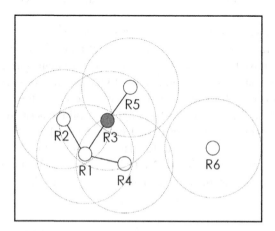

Fig. 5. R3 issues a signal to call R5, and connect

4   The Experiment with a Simulator

In order to demonstrate the effectiveness of our multi-robot system, we have implemented the system and conducted experiments using a simulator. In the present experiment, we investigate two relationships.

First topic is the relationship between the sensibilities of ground robots that notice the light signal from the aerial robot and the success rate of connection. It is natural that the more sensible the sensor, the easier they connect, but we have numerically measured.

Second topic is the relationship between the number of ground robots and the success rate. Since we want to measure the success rate to connect from an arbitrary robot to an arbitrary robot, we made the simulator to try to connect all the robots in the field starting from a randomly selected robot. The randomly selected robot in the field puts on the request signal to connect. Other robots in the view range move toward the signal until they can be in the wireless range to join the MANET to which the initial robot belongs. Other robots can notice the signal but are not connected with wireless. In this section, we report the results of the experiments and discuss about the observations.

Fig. 6 and 7 show a typical result of the experiments. The dots denote robots. Fig. 6 shows the initial state. The field is 50×50, and we put fifty robots. Each robot has view range fifteen. The view range means the robot can notice the light up to fifteen-pixel distance. The lines denote wireless connections. The robots that are in the overlapping area of their wireless ranges can make their own local ad-hoc network. The goal is making them into one network starting from an arbitrary robot. The dark colored dot denotes the starting robot that initially request connection. The aerial robot is supposed to be above it. Upon receiving the signal from the aerial robot, the ground robots in the view range start to move closer until they get into the connection range. Fig. 7 shows the MANET that the robot forms. The dark colored dot denotes the last robot that joins to the network. The aerial robot is supposed to be hovering above it. As shown in Fig. 7, almost all robots are connected but two of them are not. They are too far away from the aerial robot and cannot sense the signal light. If we increase the sensibility to the aerial signal of the robots, they can participate into the network connection. In actual system, photo sensors can catch subtle light. Therefore, we can expect that we can connect all the robots in relatively small field such as we use for experiments. The main problem may be how to achieve stable hovering of the aerial robot over the right spot.

The graph of Fig. 8 shows the result of simulations, executed while changing the setting of the total number of robots and the recognition distance to the light. Common settings of each simulation are that network range is five pixels and the field size is 50×50. The horizontal axis of the graph shows the recognition distance to the light and the vertical axis shows success rate. Success rate is the percentage of robots that connected to a request robot. Each line of the graph denotes the total number of robots. The success rate is the average of all the executions.

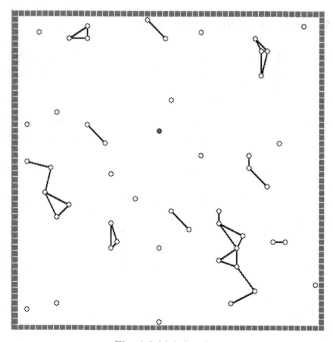

Fig. 6. Initial situation

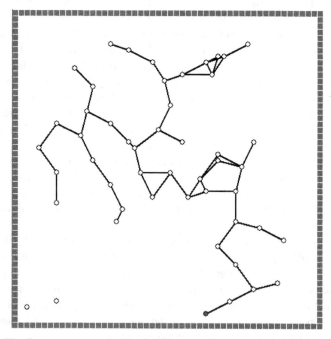

Fig. 7. Formed network. Forty-eight out of fifty robots are connected.

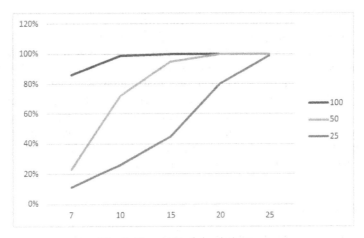

Fig. 8. The result of simulations

In Fig. 8, we can observe that the success rate and the recognition distance to the light are proportional in all lines. If the recognition distance is short, the success rate increases as the number of robots increases. In addition, if the total number of robots is small, the success rate increases as the recognition distance to the light increases, because the influence of the recognition distance and the total number of robots are related to the field. The recognition distance is related to the range of fields that can be covered by the light signal. The total number of robots is the density of the robot in the field. In our method, the range that can be covered by the light signal is expanded artificially by network connection of many robots. The range is getting longer as the recognition distance to the light is getting longer. Percentage of the covered field by the range affects success rate. Therefore, we can interpret the results that increasing the recognition distance to the light that affects direct range will affect improvement of the success rate. In addition, the range expands by an increase of connected robots and an increase of the total number of robots. Therefore, it affects the success rate.

5   Conclusion and Future Works

We have proposed a method to construct MANET using multiple mobile robots. The usefulness of cooperative multi-robot system is well recognized. Most of them relay on the steady communication infrastructure, and such assumption is not realistic in out-door environments. Therefore, we propose a method that employs visible light as well as wireless to provide communication for multi-robot system. A robot that wants to start to communicate with a certain other robot emits a light signal to call an aerial robot. Upon receiving the request, the aerial robot hovers above the ground robot and emits a light signal to make nearby robot come close to establish an ad hoc communication. Employing the aerial robot solves many problems when we try to form a MANET using small multiple robots. Because the aerial robot provides bird-view to find the missing ground robot. In addition, the height of aerial robot solves intrinsic wireless communication such as hidden terminal problem. It is easy to detect wireless.

Therefore, it is not desirable to use powerful wireless in some applications. In addition, some environments prohibit the use of powerful wireless, because it interferes with other electronic devices. We believe our multi-robot system forming ad hoc network is practical enough. We can confirm that our basic design is right by constructing a simulator. We are implementing an actual prototype system by using iRobot creates and AR Drone. There are several mechanical problems. Small aerial robots cannot carry even portable computers. Therefore, the current implementation makes the ground robot control the aerial robot, i.e. AR Drone. The usual web camera easily lose target. We have to overcome such problems before achieving stable hovering. Those are naturally the next step of our research conduct.

References

1. Parker, L.E.: Distributed intelligence: overview of the field and its application in multi-robot systems. Journal of Physical Agents 2(1), 5–14 (2008)
2. Peter, S., Manuela, V.: Multiagent systems: A survey from a machine learning perspective. Autonomous Robots 8(3), 345–383 (2000)
3. Yasuda, T., Ohkura, K.: Autonomous role assignment in a homogeneous multi-robot system. Journal of Robotics and Mechatronics 17(5), 596–604 (2005)
4. Hulaas, G., Binder, A.V.W.J.: Portable resource control in the j-seal2 mobile agent system. In: Proceedings of International Conference on Autonomous Agents, pp. 222–223 (2001)
5. Takimoto, M., Mizuno, M., Kurio, M., Kambayashi, Y.: Saving Energy Consumption of Multi-robots Using Higher-Order Mobile Agents. In: Nguyen, N.T., Grzech, A., Howlett, R.J., Jain, L.C. (eds.) KES-AMSTA 2007. LNCS (LNAI), vol. 4496, pp. 549–558. Springer, Heidelberg (2007)
6. Nagata, T., Takimoto, M., Kambayashi, Y.: Suppressing the Total Costs of Executing Tasks Using Mobile Agents. In: Hawaii International Conference on System Sciences, CD-ROM (2009)
7. Abe, T., Takimoto, M., Kambayashi, Y.: Searching Targets Using Mobile Agents in a Large Scale Multi-robot Environment. In: O'Shea, J., Nguyen, N.T., Crockett, K., Howlett, R.J., Jain, L.C. (eds.) KES-AMSTA 2011. LNCS, vol. 6682, pp. 211–220. Springer, Heidelberg (2011)
8. Shibuya, R., Takimoto, M., Kambayashi, Y.: Suppressing Energy Consumption of Transportation Robots using Mobile Agents. In: International Conference on Agents and Artificial Intelligence, vol. 1, pp. 219–224 (2013)
9. Nagata, T., Takimoto, M., Kambayashi, Y.: Cooperatively Searching Objects Based on Mobile Agents. In: Nguyen, N.T. (ed.) Transactions on Computational Collective Intelligence XI. LNCS, vol. 8065, pp. 119–136. Springer, Heidelberg (2013)
10. Sugiyama, S., Yamachi, H., Takimoto, M., Kambayashi, Y.: Aggregating Multiple Robots with Serialization. In: Pan, J.-S., Chen, S.-M., Nguyen, N.T. (eds.) ACIIDS 2012, Part I. LNCS, vol. 7196, pp. 177–186. Springer, Heidelberg (2012)
11. Satta, K., Takimoto, M., Kambayashi, Y.: Making Autonomous Robots Form Lines. In: Pan, J.-S., Chen, S.-M., Nguyen, N.T. (eds.) ACIIDS 2012, Part I. LNCS, vol. 7196, pp. 198–207. Springer, Heidelberg (2012)
12. Yokoyama, C., Takimoto, M., Kambayashi, Y.: Cooperative Control of Multi-Robot Using Mobile Agents in a Three-dimensional Environment. In: 2013 IEEE International Conference on Systems, Man, and Cybernetics, pp. 1115–1120 (2013)
13. Jamalipour, A., Ma, Y.: Intermittently Connected Moblie Ad Hoc Networks. Springer, Heidelberg (2011)

An Agent Cognitive Model for Visual Attention and Response to Novelty

Cynthia Ávila-Contreras, Ory Medina, Karina Jaime, and Félix Ramos

Department of Computer Science, Cinvestav Unidad Guadalajara, Zapopan, Mexico
{cavila,omedina,ajaime,framos}@gdl.cinvestav.mx

Abstract. Cognitive virtual agents are useful in human behavior simulation. We present a biologically inspired cognitive model for visual attention that takes into account the occurrence of novel stimulus, and it deals with the habituation to novelty. Our approach relies on the identification of cerebral areas involved in attention, semantic memory and non-associative learning; the processes related to each of them and the hypothetical information generated in each step. The model described in this paper is capable to be integrated in a cognitive architecture to interact with other cognitive functions.

Keywords: Cognitive Agent Models, Perception, Visual Attention, Novelty Handling.

1 Introduction

The virtual agents, with human-like behavior, are a current topic for research. The visual attention and the response to novelty are some of the desired features in these kind of virtual agents. These human cognitive processes are being studied from different sciences: the psychology, the neurophysilogy, and the neuroscience, among others [1].

The visual attention is the ability to select an object or location from others for further processing. The visual attention has two factors for the stimulus selection: bottom-up and top-down [2]. The bottom-up is based on physical features of the stimulus, such as intensity, hue color or line's orientation [3]. The features stimulate the retina and primary visual areas . A region is called *salient* if its features are sufficiently different with respect to its surrounding. The top-down attention is based on the current goals, expectations or knowledge. The relevant information is taken from the working memory [2].

Novelty is a factor that captures involuntary attention [4]. The process of novelty detection compares the perceived stimuli with the information stored in memory. If a novel stimulus is detected in the environment, the brain releases a charge of Acetylcholine (ACh). The charge alerts the attentional system and the memory system, This alert is known as the orienting response.

Habituation is the decrement in the behavioral response to a repeated, non-threatening stimulus [5]. If a series of novel stimuli represent no threat, the orienting response, and thus the novelty, will eventually habituate [6,7].

Borisyuk and Kazanovich [8] proposed an oscillatory model for object selection, it is based on attention. It selects only the objects in the attentional focus and it is capable

G. Jezic et al. (eds.), *Agent and Multi-Agent Systems: Technologies and Applications*,
Advances in Intelligent Systems and Computing 296,
DOI: 10.1007/978-3-319-07650-8_4, © Springer International Publishing Switzerland 2014

of detecting novel objects. If an item is outside of the attentional focus, the model does not switch the attention like the human does[4]. It does not take into consideration the habituation to novelty itself. Vikram et al. [9] habituate their saliency maps [10], to obtain the change in the regions to attend, with an habituated model proposed by Marsland [11]. The habituation happens after the top-down influences, which creates two issues. First, it causes a time overhead after the processing. Second and more important, if the gaze moves, the habituation is lost.

Our purpose in this paper is to develop a cognitive architecture model for visual attention and response to novelty based on the current knowledge of processes involved in the brain. The model will be included in a complete architecture. The architecture has memory system, attention system, motor system, among others.

This document is structured as follows: section 2 shows the model and explains the visual processing stage, novelty handling stage, and the attention selection stage; section 3 describes the model applied to a behavioral neuroscience experiment; and the section 4 discusses our results.

2   Model of Visual Attention and Response to Novelty

In order to endow virtual agents with human-like behavior, the question of how do we perform such functions arises. Biological sciences, like neurobiology or neurophysiology, provide information about which brain areas activate during cognitive processes. The model presented in this paper integrates the brain structures involved in visual attention and novelty detection, with special emphasis in their role and the possible data treated in each of them.

Fig. 1 depicts a diagram of the Visual Attention-Novelty Model, its components and interactions between them. Next subsection describes in detail this model.

2.1   Components

A thorough review of neuroscientific evidence led to the determination of main components of the diagram in Fig. 1 representing a brain structure. The idea is that their functionality could be abstracted and implemented as flat algorithms to process the inputs, and then forward the processed data to the next components. In Table 1 the brain structures and their distinguished processes – related with the proposed model – are summarized.

Table 1. Description of brain structures–components

Brain structure	Related functions
Dorsolateral Prefrontal Cortex (dlPFC)	It is closely related to working memory and it plays a role in giving attentional sets among different task demands [12]. Its activity is dependent of task-relevant information, providing top-down control to other areas to prepare for forthcoming events.
Pulvinar (PUL)	It mediates communication between cortical areas [13]. It is a center piece for indirect visual communication and is part of a pathway for attentional top-down signals to cortical areas [14].

Table 1. (*continued*)

Brain structure	Related functions
Lateral Geniculate Nucleus (LGN)	It receives inputs directly from the retina and projects to striate cortex (STC)[2]. During visuo-spatial task, input from retina is modulated in LGN in favor of the cued region [15]. This process is related to the activity in thalamic reticular nucleus (TRN).
Thalamus Reticular Nucleus (TRN).	It modulates the activity of LGN. TRN receives excitatory inputs from the LGN and STC. In contrast, the output projections to LGN are inhibitory [15]. Neurons in the TRN shows the opposite pattern of activation than LGN: directing spatial attention to a stimulus decreases the firing rate of correspondent TRN neurons
Striate Cortex (STC).	STC – or primary cortex (V1) – has a retinotopic map of the visual scene.It is suggested that maps representing saliency for a single basic feature (i.e. orientation, direction, color or spatial frequency) [16] are created in parallel in this area by bottom-up mechanisms [13]; and then combined in an overall saliency map representing the locus of attention.
Extrastriate Cortex (ESTC).	Cells in ESTC have several feedback connections to improve discrimination of figures from their background [16]
Inferior Temporal Cortex (ITC).	It is crucial for visual object identification [17]. It is the main visual input for the perirhinal cortex [18]
Temporal Parietal Junction (TPJ).	It responds to salient, infrequent, or rare stimulus. Together with the orbitofrontal cortex (OFC), it works as a circuit breaker of ongoing cognitive activity when a behaviorally relevant, novel or unexpected stimuli is detected [19].
Intraparietal Sulcus (IPS).	Studies have shown that human IPS could contain or be the putative homologue of the monkey lateral intraparietal area (LIP) [19]. In monkey LIP exists a topographical representation of attentional weights, as a priority map, and its neurons are strongly modulated by spatial attention, taking in account stimulus-driven saliency and task-relevant information, with or without ocular movement. In humans, such functions has also been observed in part of IPS. [20,21].
Frontal Eye Fields (FEF).	FEF are structures intrinsically related with voluntary control of ocular movement (saccades) [22]. But also FEF is involved in attentional mechanisms, like target selection [19]. In spatial selective attention, FEF manages target-oriented information, such as top-down signals about the relevant region to attend [20].
Superior Parietal Lobule (SPL).	In contrast with IPS, SPL is more closely related to the modification of spatial coordinates linked to attentional priorities [21]. In [23] is showed that SPL could be the source of top-down attentional biasing signal to increase responses for the location to attend.
Superior Colicullus (SC).	It is involved in spatial attention, mostly in voluntary saccadic eye movements and overtly shifts in both gaze and attention [13]. The path for ocular movement goes from FEF to SC and towards brainstem circuitry.
Perirhinal Cortex (PC).	It serves as a direct input for the EC. It has recently been related to recognition of singular objects [24] [7].

Table 1. (*continued*)

Brain structure	Related functions
Entorhinal Cortex (EC).	It is the sensory input for CA3, CA1 and the DG [24]. It's also the structure in charge of eliciting the orienting response when contextual novelty is found [7].
Dentate Gyrus (DG).	It is the area responsible of creating new neural representations for novel stimulus [25].
CA3.	It has an important role in storing new, fast memories [25], it's also the area in charge of retrieving fast representations of the objects stored in memory. At the same time, it forms predictions of the upcoming stimuli (context) [7] [26].
CA1.	Besides being a relay station to cortical regions from the CA3-DG network, it works as a comparator between the external stimuli, received from the EC and the predictions (context) generated by CA3 [25] [7].
Subicullum (SB).	It is the output structure of the CA3-CA1 network, without it, the hippocampus would not be able to communicate with cortical regions [27].
Nucleus Basalis of Meynert (NBM).	It is the nucleus of the cholinergic system, and the main producer of ACh in the brain. It is also the area responsible for generating the orienting response [28].
Orbitofrontal Cortex (OFC).	It is believed to be critically involved in a form of integrative memory, it links personal experiences with external cues. Due to the extensive connections with the medial temporal lobe, and subcortical regions involved in motivation, it may have an important role in the assembling and monitoring of relevant experiences [29].
Semantic Memory	This a special node. It sends semantic information about the objects in the stimuli and its functionality follows the ideas of Martin [30].

2.2   Processing Stages

The flow of information in the model is divided in the three previously mentioned stages. The numbering indicates the sequential order of execution through the whole system, as indicated in Fig. 1.

A. Visual Processing. In this part of the processing, the initial image is segmented like a grid and treated as values with an (x, y) location. Coordinates are preserved through the processing even if the data is transformed in each component.

0. dlPFC sends elements of *attentional-set* to PUL, TRN and TPJ. The *attentional-set* has top-down information, it could be *relevant region*, *relevant color*, or *relevant shape* to attend.
1. Retina sends n *visual* values to SC and LGN. These *visual* values have the hue color and intensity of the (i, j) segment of the image that represents.
   PUL relays *relevant region attentional-set* elements to TRN and *relevant color attentional-set* elements to STC
   TRN sends *spacial modulation* signals to LGN. A low value of *spacial modulation*

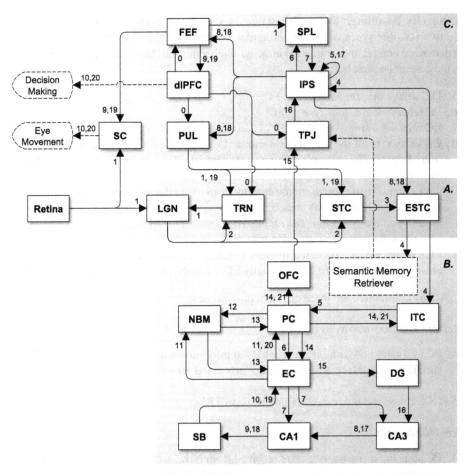

Fig. 1. Model of visual attention and response to novelty. Processing stages: A. Visual processing, B. Novelty handling and C. Attentional selection.

   is calculated if it (x, y) location is in *relevant region attentional-set* received from PUL, and high value otherwise.

2. LGN increases the *visual* values if their correspondent *spacial modulation* has a low value.
   LGN combines k *visual* values to send *simplified visual* data to STC.

3. STC increases *visual* values if its color is similar to the *relevant color attentional-set* element received from PUL
   STC measures how different is in hue *color* each of the k *simplified visual* data comparing it with the rest. It does similar actions with *intensity* and *orientation* features. This processes creates *saliency by features* values for ESTC

4. ESTC relays *saliency by features* values received from STC to TPJ and IPS.
   ESTC extracts figures from their background. This figures are send as *singular types* to ITC.

B. Novelty Handling. During this stage, the stimulus is evaluated for novelty. There are two possible types of novelty: singular novelty, when a stimulus has never been experienced before; and contextual novelty, when a stimulus has never been experienced in the current context before.

5. ITC relays the *singular type* to PC.
6. PC relays the *singular type* to EC
7. EC relays the *singular type* to CA3 and CA1
8. CA3 recovers all the information similar to the *singular type*, along with a prediction of the possible stimuli coming next (context). This information is sent to CA1 as a *processed singular type*.
9. CA1 compares the *singular type* coming from EC to the prediction in the *processed singular type*, if it doesn't match, a novelty signal is attached to the *processed singular type*, finally it sends the *processed singular type* to SB.
10. SB relays the *processed singular type* to EC.
11. EC evaluates if the *processed singular type* has a novel signal, if it does, EC requests an *orienting response* to NBM. Finally EC sends the *processed singular type* to PC.

At this point, there are three different possibilities:

Memory Found (No Novelty)

14. PC evaluates how similar is the retrieved information to the original stimulus, if the similarity trespasses a threshold, the *processed singular type* is sent to OFC and ITC.
15. OFC relays the *processed singular type* to TPJ

Semantic Novelty

14. PC adds the *orienting response* received from NBM to the *processed singular type* and then sends the *processed singular type* to OFC and ITC.
15. OFC relays the *processed singular type* to TPJ.

Singular Novelty

12. PC evaluates how similar is the retrieved information to the original stimulus, if the similarity does not trespass a threshold, PC requests an *orienting response* to NBM.
13. NBM generates an *orienting response* which intensity depends on the time and intensity of the previous *orienting response*, then sends the generated *orienting response* to PC and EC.
14. PC adds the *orienting response* to the *processed singular type* and sends it to both, OFC and EC
15. OFC relays the *processed singular type* to TPJ
    EC relays the *processed singular type* to DG.
16. DG encodes the original stimulus in the *processed singular type* and generates a (*new neural representation*), then sends the *new neural representation* to CA3.
17. CA3 stores the *new neural representation*, then recovers all the information similar to it, generating a *processed singular type* that is sent to CA1.

18. CA1 compares the *singular type* coming from EC to the prediction in the *processed singular type*, if it doesn't match, a novelty signal is attached to the *processed singular type*, finally it sends the *processed singular type* to SB.
19. SB relays the *processed singular type* to EC.
20. EC evaluates if the *processed singular type* has a novel signal, it requests an *orienting response* to NBM and sends the *processed singular type* to PC.
21. PC relays the *processed singular type* to OFC and ITC.

C. Attentional Selection. This stage involves the sources of top-down influences and operations to select a stimulus among others. Deployment of attention – shift of gaze and decision-making – is reached when selection is made.

0. dlPFC sends the *ocular movement allowance* signal to FEF.
   dlPFC sends *relevant shape attentional-set* elements to TPJ.
1. FEF relays *ocular movement allowance* to SPL
5. IPS lineally combines the *color, orientation* and *intensity saliency by features* values for each (x, y) location. It first normalize the values in a same scale and then it do linear sums and average. The results are *unified saliency features* values .
6. IPS sends *unified saliency features* to SPL.
7. SPL increases values of *unified saliency features* if the are inside the *relevant region attentional-set* element received from FEF.
8. IPS calculates the region to attend taken the highest values of *unified saliency features* received from SPL, only if there is no other information received from TPJ to take in account. These *global saliency* values are sent to FEF, PUL and ESTC.
9. FEF sends *global saliency* to dlPFC.
   FEF computes the *ocular movement* signal based on *ocular movement allowance*. It codifies the locus of attention.
10. dlPFC sends the *attended region* to other cognitive processes, i.e. decision-making or working memory
    SC sends *executive eye movement* commands to brainstem circuitry only if *ocular movement* signal received from FEF is strong enogh to activate it.
16. TPJ breaks out the ongoing information in IPS if the *ns* received from OFC is strong enough.
    TPJ computes if *processed singular type* are behaviorally or biologically relevant. It compares the categories of *relevant shape attentional-set* element received from dlPFC with categories of *processed singular type* received from Semantic Memory and assigns a *singular type relevance* value that is sent to IPS.
17. IPS recalculates *unified saliency features* values received from SPL combining them with *singular type relevance* from TPJ. During combination, *singular type relevance* values are less significant than *unified saliency features*.
18. IPS determines the *attended region* taken the highest values of *unified saliency features*.The result is *global saliency* values sent to FEF, PUL and ESTC.
19. Same process as 9.
20. Same process as 10.

3   Case Study

The next step in our research process is to prove the model in a real case. The case study is based on the work of Yamaguchi et al. [6]. They develop a behavioral task where visual stimuli were displayed on a LCD. They used a bi-field visual-selective attention paradigm allowing examination of the response to attended versus unattended novel events. Visual stimuli consisted of three categories: standard, target and novel stimuli. Standard and target stimuli were triangle. The target triangle was rotated 10° clockwise, relative to the upright triangle standard stimuli. Novel stimuli were different images. The subject's task was to make a speeded button-press response on each target presentation only in the attended visual field, while stimuli in the opposite field were to be ignored.

Next, we describe the six sub-cases and how the model will respond to each one.

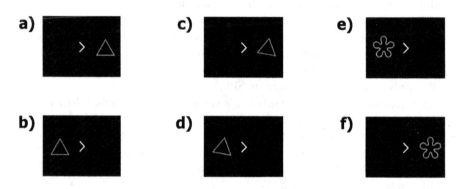

Fig. 2. Sub-cases' representation. **a)** Standard stimulus on attended visual field (AVF). **b)** Standard stimulus on non-attended visual field (NAVF). **c)** Target stimulus on AVF. **d)** Target stimulus on NAVF. **e** Novel stimulus on AVF. **f)** Novel stimulus on NAVF. These sub-cases are inspired on the work of Yamaguchi et al. [6].

In case **a.**, the standard stimulus is attended; there is no action. Case **b.**, the cue signal is attended, but not the standard stimulus; there is no action. Case **c.**, the target stimulus is attended; there is an action. Case **d.**, the target stimulus is not attended; there is no action. Case **e.** and case **f.**, if the system were habituated to novelty, the novel stimulus would not be attended, otherwise, it would be attended; there is no action.

The habituation to novelty will not be noticeable until tests with continuous stimuli are executed.

4   Discussion

Visual attention and response to novelty are important abilities that must be taken into account in the development of cognitive architectures for virtual agent. The virtual agent with our architecture will be able to filter the incoming stimuli by detecting goals and physical attributes. The model focuses on response and habituation to novelty. So far, there is no other agent cognitive architecture that performs these processes in a way the human does: our model simulates each of the brain areas involved in visual attention and novelty handling. Therefore, the result is a realistic behavior.

Based on the characteristics of the model, we selected a behavioral task as a case study. We will prove our model by using the case study inputs and comparing both outputs. If the outputs were similar, the model would be correct.

Each model component is developed as a node in a distributed system, which is the nucleus of our cognitive architecture. If the correct node processed the data in the correct sequence, the system execution would be successful.

We are assuming the vision system can detect objects in the scene; for that reason, we do not have the certainty that it works with cluttered scenes. The semantic memory and the episodic memory are not complete yet; therefore, some unknown processes can emerge in the future.

We believe that the fusion of the model with the cognitive architecture will get emerging function like emotional attention and priming.

Acknowledgment. We would like to thank CONACyT for its financial support.

References

1. Kandel, E.R., Schwartz, J.H., Jessell, T.M., Siegelbaum, S.A., Hudspeth, A. (eds.): Principles of neural science, vol. 4. McGraw-Hill, New York (2013)
2. Frintrop, S., Rome, E., Christensen, H.I.: Computational visual attention systems and their cognitive foundations: A survey. ACM Transactions on Applied Perception (TAP) 7(1), 6 (2010)
3. Diaz-Barriga, S., Torres, G., Ramos, F.: Feature-based saliency for a model of bottom-up visual attention. In: 11th IEEE International Conference on Cognitive Informatics and Cognitive Computing, pp. 399–406 (2012)
4. Sokolov, E., Vinogradova, O.: Neuronal mechanisms of the orienting reflex. L. Erlbaum Associates (1975)
5. Rankin, C.H., Abrahams, T., Barry, R.J., Bhatangar, S., Clayton, D.F., Colombo, J., Coppola, G., Geyer, M.A., Glanzman, D.L., Marsland, S., McSweeney, F.K., Wilson, D.A., Wu, C.F., Thompson, R.F.: Habituation revisited: An updated and revised description of the behavioral characteristics of habituation. Neurobiology of Learning and Memory 92, 135–138 (2009)
6. Yamaguchi, S., Hale, L.A., D'Esposito, M., Knight, R.T.: Rapid prefrontal-hippocampal habituation to novel events. The Journal of Neuroscience 24, 5356–5363 (2004)
7. Ranganath, C., Rainer, G., et al.: Neural mechanisms for detecting and remembering novel events. Nature Reviews Neuroscience 4(3), 193–202 (2003)
8. Borisyuk, R.M., Kazanovich, Y.B.: Oscillatory model of attetion-guided object selection and novelty detection. Neural Networks 17, 899–915 (2004)
9. Vikram, T., Tscherepanow, M., Wrede, B.: Integrating habituation into saliency maps. In: 2012 IEEE International Conference on Development and Learning and Epigenetic Robotics (ICDL), pp. 1–2 (2012)
10. Itti, L., Rees, G., Tsotsos, J.K.: Neurobiology of attention. Access Online via Elsevier (2005)
11. Marsland, S.: Using habituation in machine learning. Neurobiology of Learning and Memory 92, 260–266 (2009), Special Issue: Neurobiology of Habituation

36   C. Ávila-Contreras et al.

12. Banich, M.T., Milham, M.P., Atchley, R.A., Cohen, N.J., Webb, A., Wszalek, T., Kramer, A.F., Liang, Z.P., Barad, V., Gullett, D., Shah, C., Brown, C.: Prefrontal regions play a predominant role in imposing an attentional 'set': evidence from fmri. Cognitive Brain Research 10(1-2), 1–9 (2000)
13. Shipp, S.: The brain circuitry of attention. Trends in Cognitive Sciences 8(5), 223–230 (2004)
14. Saalmann, Y.B., Kastner, S.: Gain control in the visual thalamus during perception and cognition. Current Opinion in Neurobiology 19(4), 408–414 (2009)
15. Mayo, J.P.: Intrathalamic mechanisms of visual attention. Journal of Neurophysiology 101(3), 1123–1125 (2009)
16. VanRullen, R.: Visual saliency and spike timing in the ventral visual pathway. Journal of Physiology-Paris 97(2-3), 365–377 (2003), Neurogeometry and visual perception
17. Miyashita, Y.: Inferior temporal cortex: where visual perception meets memory. Annual Review of Neuroscience 16(1), 245–263 (1993)
18. Sigala, N., Logothetis, N.K.: Visual categorization shapes feature selectivity in the primate temporal cortex. Nature 415(6869), 318–320 (2002)
19. Corbetta, M., Shulman, G.L.: Control of goal-directed and stimulus-driven attention in the brain. Nature Reviews. Neuroscience 3(3), 201–215 (2002)
20. Geng, J.J., Mangun, G.R.: Anterior intraparietal sulcus is sensitive to bottom–up attention driven by stimulus salience. Journal of Cognitive Neuroscience 21(8), 1584–1601 (2009)
21. Molenberghs, P., Mesulam, M.M., Peeters, R., Vandenberghe, R.R.: Remapping attentional priorities: differential contribution of superior parietal lobule and intraparietal sulcus. Cerebral Cortex 17(11), 2703–2712 (2007)
22. Wardak, C., Ibos, G., Duhamel, J.R., Olivier, E.: Contribution of the monkey frontal eye field to covert visual attention. The Journal of Neuroscience 26(16), 4228–4235 (2006)
23. Yantis, S., Schwarzbach, J., Serences, J.T., Carlson, R.L., Steinmetz, M.A., Pekar, J.J., Courtney, S.M.: Transient neural activity in human parietal cortex during spatial attention shifts. Nature Neuroscience 5(10), 995–1002 (2002)
24. Aggleton, J.P., Brown, M.W.: Contrasting hippocampal and perirhinal cortex function using immediate early gene imaging. The Quarterly Journal of Experimental Psychology 58(3-4), 218–233 (2005)
25. Lee, I., Hunsaker, M.R., Kesner, R.P.: The role of hippocampal subregions in detecting spatial novelty. Behavioral Neuroscience 119, 145–153 (2005)
26. Daselaar, S.M., Fleck, M.S., Cabeza, R.: Triple dissociation in the medial temporal lobes: Recollection, familiarity, and novelty. Journal of Neurophysiology 96(4), 1902–1911 (2006)
27. Afifi, A.K., Bergman, R.A.: Functional Neuroanatomy. McGraw-Hill New York (1998)
28. Mesulam, M.M., Mufson, E.: Neural inputs into the nucleus basalis of the substantia innominata (ch4) in the rhesus monkey. Brain 107(1), 253–274 (1984)
29. Cavada, C., Compañy, T., Tejedor, J., Cruz-Rizzolo, R.J., Reinoso-Suarez, F.: The anatomical connections of the macaque monkey orbitofrontal cortex. a review. Cerebral Cortex 10(3), 220–242 (2000)
30. Martin, A.: Semantic memory. In: Squire, L.R. (ed.) Encyclopedia of Neuroscience. Academic Press (2009)

Self-composition of Services in Pervasive Systems: A Chemical-Inspired Approach

Francesco L. De Angelis, Jose Luis Fernandez-Marquez,
and Giovanna Di Marzo Serugendo

Institute of Information Service Science, University of Geneva,
Route de Drize 7, Carouge, Switzerland
{francesco.deangelis,joseluis.fernandez,giovanna.dimarzo}@unige.ch

Abstract. Service-oriented programming has dramatically changed the way software applications are developped, promoting reusability of code and easing the design of complex applications. Actual techniques for automatic composition of services present several limitations to be used in the context of future pervasive scenarios: (1) limited scalability due to centralised computations, (2) slow reactivity with respect to appearance and removal of services, and (3) no support for context-aware applications. In this paper we define a chemical-model and two chemically inspired approaches for self-composition of services operating in a pervasive system. We show how distributed shared data spaces can be exploited to design spontaneous and emergent compositions that deal with context information and a dynamic set of available services. This new approach, taking inspiration from chemical reactions, turns to be completely decentralised and self-adaptive to service appearance and disappearance.

Keywords: Self-composition, chemical-model, services, chemical reactions, context-awareness, dynamic environment.

1 Introduction

Next generation of socio-technical infrastructures will be characterised by the presence of complex networks of pervasive systems, composed of thousands of heterogeneous devices consuming and producing high-volumes of interdependent data. Open smart environment, such as smart-cities, represent an example of these future digital scenarios: by using wide area mobile ad-hoc networks (MANETs), data will be shared among applications placed on cars or running on several devices such as smartphones, tablets, public displays and sensors placed at the edges of the roads; moreover, all these devices will access traditional remote Web services. Smart-cities depict the emergence of new open-infrastructure pervasive systems, where scalability and dependability will be achieved by developping and adapting (at run-time) applications through compositions of customised services.

The static character of traditional approaches for composition of services, such as orchestration and choreography, has been recently challenged by so-called dynamic service composition approaches, involving semantic relations [1], or AI

G. Jezic et al. (eds.), *Agent and Multi-Agent Systems: Technologies and Applications*,   37
Advances in Intelligent Systems and Computing 296,
DOI: 10.1007/978-3-319-07650-8_5, © Springer International Publishing Switzerland 2014

(Artificial Intelligence) planning techniques to generate process automatically based on the specification of a problem [2]. All these approaches turn to be unfeasible for being adopted in future pervasive systems because of their restricted scalability due to the centralisation of the composition process, slow reactivity to sudden appearance or disappearance of services, and limited support for context-awareness applications.

To tackle the scalability issue, Banâtre and Priol [3] propose a new approach based on the Higher-Order Chemical Language [4]. That paper extends the Gamma programming model [5] in the context of service-based infrastructures. Self-compositions of services are automatically performed in a decentralised way by means of chemical reactions (i.e. local interactions) of virtual molecules containing services names and input-output data. Even though the computation is implicitly parallel and the reaction processes are autonomic (i.e. thus increasing the scalability), such an approach still lacks a way for dealing with a high volatility in the number, type and availability of services.

A recent chemical-inspired framework for pervasive systems, SAPERE [6], brings together novel concepts in order to accommodate dynamic arrival of entities in the system, and to provide a common treatment of data and functionalities. Services, such as sensors embedded in a mobile phone, Web services, wireless sensors connected by bluetooth or wi-fi, are mapped into an active (i.e. dynamically updated) shared space of data tuples. Even though the SAPERE framework allows different services to be alive in the same computational environment, i.e. sharing the same tuple space, the interactions among the different services in order to create higher level functionalities need to be designed in advance.

This paper presents two novel service self-composition approaches that: (1) take inspiration from the chemical reaction concept presented by Banâtre and Priol for self-composition of services, thus increasing the *scalability* compared with current centralised approaches; (2) use the m active tuple space concept provided by SAPERE, supporting services appearance and disappearance. This increases the *reactivity of the system*, and automatically introduces *contextual information* as part of the system. We have implemented our approach as an extension of the SAPERE framework, and preliminary results show the feasibility of our self-composition approaches.

This paper is organised as follows, Section 2 introduces the model that we use in Section 3 to present our approach for self-composition of services. Finally, Section 4 shows a case study involving self-composition of Web services.

2   Reference Model for Self-composition of Services

The model considered in this paper is an abstraction of the SAPERE active tuple space model [7, 6], a model for pervasive ecosystems inspired by chemical reactions. The system is composed of five entities: *tuples, active tuple space, chemical reactions, agents* and *services*, as shown in Figure 1(a).

Tuples are dynamically updated vectors of properties *(name=value)* used to describe and represent services, applications and contextual information in a

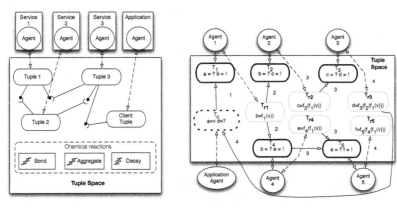

(a) Reference model   (b) Executing all services: generation of value d starting from a

Fig. 1. Active tuple space

common way, easing the integration of heterogeneous devices; e.g. *(id=1234, Temp=10C, position=20,30)* describes the information coming from a sensor situated at position (20,30).

Active Tuple Space is a container of tuples, located in a computing node, shared by services, sensors or applications represented by tuples injected into that tuple space.

Chemical reactions provide the system with an automatic way for tuples interactions by combining and updating them. Formally, they are functions used to modify, update and delete a subset of tuples. Chemical reactions are key entities of the active tuple space, i.e. tuples stored in a tuple space are subject to the action of the chemical reactions.

Services are entities producing data as result of a computation performed on data passed as input or data coming from external sources, such as, sensors. Services act as an external entity, thus, it is required to instantiate an agent that interfaces between the service and the tuple space. Services can be represented by using a precondition-postcondition notation. For example, a tuple $T_1 = (city =?, weather =!)$ represents a service producing an output (!) result of type *weather* (i.e., weather forecast) given an input (?) value of type *city*.

Agents are external active entities representing sensors, mobile phones or services associated with a tuple, which is the interface between services/applications and the tuple space. An agent interacts with the tuple space updating or deleting tuples, retrieving information from tuples, and receiving a notification each time an interaction is performed on its tuple by a chemical reaction.
In this paper we use the following chemical reactions [7, 6]:

Bond establishes a relationship between two distinct tuples containing the same property names. To request a bond chemical reactions, agents inserts a question

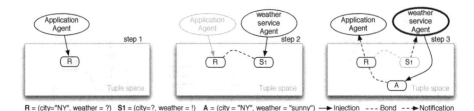

R = (city="NY", weather = ?)   S1 = (city=?, weather = !)   A = (city = "NY", weather = "sunny")   ──▶ Injection --- Bond --▶ Notification

Fig. 2. Self-composition principle

mark "?" as property value in one of the two tuples. Every time a bond is established or removed and every time a property value involved in a bond is updated an event of activation is delivered to the agent exhibiting the value "?".

Aggregate is in charge of merging together two or more tuples, producing a new one that contains synthesised data. This reaction is carried on by using aggregation specifications, *spec*, contained in the value of the property ($aggregate = spec$), as discussed in Section 3.4.

Decay reduces the relevance of information throughout time in order to adapt to dynamic environments, and to free resources by removing tuples from the space when relevance is below a determined threshold. If an agent wants its tuple to be removed after t_{rem} units of time, it adds to the tuple vector a property of this type: ($decay = t_{rem}$). When the property expires, the decay chemical reaction delivers a notification to the agent and proceeds with the removal of the tuple from the space.

Detailed information about the SAPERE model can be found in [8].

3   Self-composition Algorithms

In this section we describe two approaches used to compute self-compositions of services with single input and output parameters, as well as a generalisation to n parameters.

3.1   Self-composition Principle

Self-compositions of services are based on expected input and output types and spontaneously arise when the output type of a service matches the input type of another one. Matching occurs through the Bond chemical reaction. Figure 2 shows an Application Agent injecting a tuple $R = (city = \text{"NY"}, weather = ?)$ requesting the weather in New York (step 1). Upon the arrival of a Weather Service, its corresponding Agent injects tuple $S_1 = (city = ?, weather = !)$ representing the Weather Service. Tuples R and S_1 bond on property "city" (step 2). The bond activates the Weather Service Agent, that in turn invokes the Weather Service itself for computing the answer to the query. The result is then injected into the tuple space as an additional tuple $A = (city = \text{"NY"}, weather = \text{"sunny"})$. Tuples R and A now bond on the property *"weather"*. This activates the Application Agent that retrieves the answer to its original query (step 3).

3.2   Executing All Services

This approach, depicted in Figure 1(b), has two main steps: a service request followed by a series of bonds (i.e. chemical reactions), activating agents which invoke their respective services, ultimately leading to a self-composition result.

Services Appearance/Disappearance: when a service joins the system, its agent injects a tuple describing its expected input and output. For instance $Agent_1$ injects tuple $T_1 = (a =?, b =!)$ specifying that $Agent_1$ expects one input of type a and one output of type b. Similarly, when a service leaves the system, the agent removes the corresponding tuple. It is important to notice that arrival and departure of services can happen at any time and are independent of the self-composition process itself.

Service Request: an application encodes a service request by creating a tuple $T_c = (a = \text{``}v\text{''}, d =?)$ containing the initial input value v of type a and the question mark "?" for bonding with the expected output of type d.

Composition Execution: the query tuple T_c will bond with tuples injected by agents expecting as input a value of type a. In our example, T_c bonds with T_1, causing the invocation of $Agent_1$, that will inject T_{r_1} providing a value of type b. In turn T_{r_1} bonds with both T_2 and T_4, causing the activation of $Agent_2$ and $Agent_4$ and the execution of their respective services. This process carries on until eventually generating a tuple T_{r_3} with the requested type d, produced by $Agent_3$. T_{r_3} now bonds with the original query since it contains the expected output requested by T_c through the "?".

This approach presents several advantages: (1) the self-composition process is completely decentralised, thus favoring *scalability*, (2) there is no prior planning of services executions, i.e. there is no prior knowledge about the services available or not, and invocations of services are completely transparent with respect to the application. This helps coping with *dynamic appearance and removal of services*; (4) services may represent sensors injecting contextual information at specific rates, this data influences and participates to the self-composition process, thus providing *context-awareness* support for applications.

The main disadvantage of this approach is represented by the fact that it may invoke a large number of services, i.e. more than the number of services actually needed to produce the requested value.

3.3   Designing All Compositions

This second approach keeps the advantages of the first approach, while solving the disadvantages of the previous one by selecting the sequence of services to actually execute. In this second approach all services contribute in maintaining a graph of interdependencies (as reported in Figure 3) used to inform every agent about the set of all other agents whose input types, if satisfied, will activate that agent. This set is then used by application agents in order to select a sequence of services to invoke to produce a specific composition. This approach

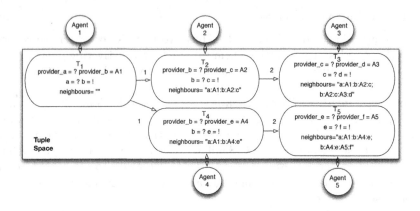

Fig. 3. An instance of the tuple space at the end of the discovery process

has three main steps. First a discovery process where all tuples bond with tuples of other agents matching either their input or their output types. The discovery process is independent of any actual query for service. It happens permanently during the life-time of the system and updates seamlessly when services appear or disappear. Second, an application agent injecting a query tuple starts a design process, where by a series of bonds, sequences of potential services compositions are spontaneously designed. It is important to notice that no service is actually executed. Finally, the application agent selects the best composition of services to execute, this is done according to its own criteria. The actual execution is triggered when the application agent actually injects the input value.

Discovery Process: Every agent S accepting input value of type x and producing output value of type y injects a tuple of the form: $(x = ?, y =!, provider\_x = ?, provider\_y = "S", neighbours = \varnothing)$. The first two properties are the common ones expressing the inputs and output types of the service; $provider\_x$ is used to discover services providing a result of type x, whereas $provider\_y$ contains the identifier of the current service ("S"). These two properties trigger bonds among agent tuples. Through a chain of reactions, this process creates a graph of interdependencies that propagates across the tuple space. This directed graph is coded in sequences $input_1 : Service_1 : output_1 : Service_2 : output_2$, stored in the property *neighbours*. Sequences are propagated across tuples to build the graph of interdependencies. Figure 3 reports the graph for the same set of services as the one reported in Section 3.2. When $Agent_3$ receives a notification (because of the property $provider\_c$ of T_2) it updates *neighbours* of T_3 by aggregating the (only) *neighbours* sequence read from T_2 ($a : A1 : b : A2 : c$) and the new sequence ($b : A2 : c : A3 : d$), representing the interaction between $Agent_3$ and $Agent_2$. The property *neighbours* of T_2 was previously generated because of the interaction ($provider\_b = A1$) in T_1 bonding with ($provider\_b =?$) in T_2 between $Agent_2$ and $Agent_1$.

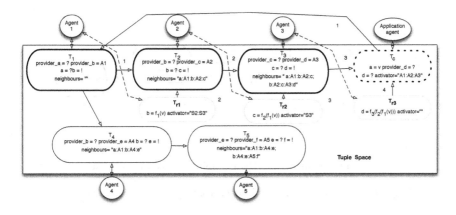

Fig. 4. An instance of the tuple space at the end of the composition execution

Design Process: This phase aims at designing one or more suitable service compositions based on the discovery process established above. As shown on Figure 4, an Application Agent starts the design process by injecting tuple T_c with the following properties ($a = v, provider\_d =?$). This tuple bonds with all tuples related to services providing an output of type d. The Application Agent then collects a set of sequences by accessing the properties *neighbours*. A set of services $S_1, \ldots, S_n$ is a candidate for the composition requested by the Application Agent if the postcondition of each service S_i is the precondition of S_{i+1} (i.e. a chain reaction propagates through the services from S_1 up to S_n) and the initial precondition and final postcondition are respectively a and d (i.e. the composition starts from an input of type a and produces a final value of type d). In this case the sequence $A1 : A2 : A3$ is provided.

Depending on its own criteria (e.g. Quality of Service), the Application Agent chooses the best service composition sequence (if many are generated).

Composition Execution: The Application Agent updates the tuple $T_c = (a = "v", d =?, activator = "A1 : A2 : A3")$ in the tuple space, starting the execution process (Figure 4). In this case, the property *activator* allows executing only those services involved in the selected composition.

Graph Maintenance: The graph of interdependencies is a flexible structure that automatically reacts to the creation and removal of services updating the bonds among tuples, as reported in Figure 5(a).

3.4   Generalisation for Services with $n > 1$ Input Parameters

The approaches presented so far assume that the input and output parameters of the services are single values. This is a strong limitation, so we extended the approaches to services accepting and generating vectors of one or more values. The approach of Section 3.2 is extended to cope with such cases by assigning a

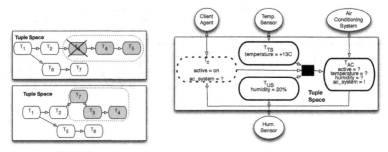

(a) Graph maintenance   (b) Combining several tuples of sensors

Fig. 5. Graph maintenance and generalisation to multiple parameters

Table 1. Summary of our Web services

Service name	Input	Output
W_1	GPS	city
W_2	GPS	country
W_3	city	weather
W_4	country	UTC
W_5	country	season

distinct property to each sub-value of the output. To generalise also the second approach, we need a mechanism to: (1) index tuples and; (2) merge together several indexed tuples. Thus, producing new tuples filled with parameters generated independently by distinct reactions. This mechanism is represented by the *aggregate* reaction rule, whose execution is depicted in Figure 5(b): a system is composed of a temperature sensor, a humidity sensor and an air-conditioning system (AC system), which starts refreshing / warming the air around depending on temperature, humidity and its current status (property *active = on/off*). In this case, a client agent injects a request to turn on the air-conditioning system; the three tuples containing the three input parameters are merged together by the *aggregate* chemical reaction, producing a new tuple (black square) that finally interacts with the AC system tuple, actually invoking the service. We formally proved that chain reactions involving tuple merging can be performed in our model [9].

4 Implementation

In order to show the feasibility of our approach and its potential for service self-composition in pervasive environments, we extended the SAPERE middleware[1] with the implementation of the two presented approaches for self-composition. The simulator providing the SAPERE middleware is called TheOne-SAPERE[2],

---

[1] https://bitbucket.org/gcastelli/saperemiddleware
[2] https://bitbucket.org/fdeangelis/theone-sapere-selfcomposition

Table 2. Active tuple space - Web service example

Tuple	Tuple Space
T_{w_1}	(GPS = ?, city = !)
T_{w_2}	(GPS = ? country = !)
T_{w_3}	(city = ?, weather = !)
T_{w_4}	(country = ?, UTC = !)
T_{w_5}	(country = ?, season = !)
T_{Req}	(GPS = "40.42,74.00", weather = ?, activator = "W_1:W_3")
A_{w_1}	(GPS = "40.42,74.00", city ="NY", activator ="W_3")
A_{w_3}	(city = "NY", weather = "Cloudy", activator = ∅)

it is an extension of The One simulator [10] where each simulated device is executing a real instance of the SAPERE middleware. The examples of Section 3 have all been implemented with TheOne-SAPERE simulator, and are available on-line2.

In this section, we report the additional case of five Web services injected within a single SAPERE node. Table 1 describes the used Web services. They all accept one input parameter and produce one output; they provide several information about a city, like weather forecast and Coordinated Universal Time.

Let us imagine that an application wants to gather the weather information for a given latitude and longitude coordinates set. Since no service provides such a result, the system automatically composes services *W1* and *W3*, i.e. from the GPS coordinates we obtain the city (*W1*), and from the city we obtain the weather forecast (*W3*). This example is implemented using our second self-composition approach. When the Web services appear in the system, the Web service agents inject tuples describing the services provided, i.e. tuples $T_{w_1}, T_{w_2}, T_{w_3}, T_{w_4}$ and T_{w_5} in Table 2. These tuples specify the input and output types for each service. An application agent injects a query tuple T_{Req}, requesting the weather for the provided GPS coordinates. The sequence of services that provides the desired result is progressively produced ($W_1 : W_3$). The composition execution starts with W_1 producing tuple A_{w_1}. This is followed by the execution of W_3 producing tuple A_{w_3} that contains the desired result. T_{Req} bonds with A_{w_3} and the application agent is notified with the desired information. Table 2 shows the final status of the shared tuple space after the self-composition process has finished. Notice that if the first approach would have been used, service W_2 would have executed and in cascade effect also services W_4 and W_5.

Implementation results shows that both approaches are able to answer queries as result of self-composition. Additionally, we observe that the system is able to deal with services removal and appearance on real time, because: (1) our self-composition approach does not need a previous design of the composition, (2) the status of services is dynamically updated in the active tuple space, allowing the system to react when a service appears or disappears.

5   Conclusion

Traditional approaches for composition of services turn to be inefficient in pervasive environments because of limited scalability, slow reactivity to frequent appearance or disappearance of services and no support for context-awareness.

To tackle these problems we combine different paradigms existing in the literature, (i.e. a chemical reaction approach for self-composition and a framework for pervasive computing), combining the scalability features of chemical approaches with the ability of dealing with dynamic and heterogeneous environment proper of pervasive system frameworks.

Experimental results show the feasibility of the approach, particularly for dynamic environments and context-aware applications. Future work include: (1) exhaustive analysis in term of resource consumption;(2) large scale experiments involving multiple computing nodes; and (3) identification of limitations.

Acknowledgment. This work has been supported by the EU-FP7-FET Proactive project SAPERE Self-aware Pervasive Service Ecosystems, under contract no.256873.

References

[1] Beek, M., Bucchiarone, A., Gnesi, S.: A survey on service composition approaches: From industrial standards to formal methods. In: Technical Report 2006TR-15, Istituto, pp. 15–20. IEEE CS Press (2006)
[2] Wu, Z., Ranabahu, A., Gomadam, K., Sheth, A., Miller, J.: Automatic composition of semantic web services using process and data mediation. In: Proc. of the 9th Intl. Conf. on Enterprise Information Systems, pp. 453–461 (2007)
[3] Banâtre, J.P., Priol, T.: Chemical programming of future service-oriented architectures. JSW 4(7), 738–746 (2009)
[4] Banâtre, J.P., Fradet, P., Radenac, Y.: Generalized multisets for chemical programming (2005)
[5] Banâtre, J.P., Métayer, D.L.: The gamma model and its discipline of programming. Sci. Comput. Program. 15(1), 55–77 (1990)
[6] Zambonelli, F., et al.: Self-aware pervasive service ecosystems. Procedia Computer Science 7, 197–199 (2011)
[7] Castelli, G., Mamei, M., Rosi, A., Zambonelli, F.: Pervasive middleware goes social: The sapere approach. In: Proceedings of the 2011 Fifth IEEE Conference on Self-Adaptive and Self-Organizing Systems Workshops, SASOW 2011, pp. 9–14 (2011)
[8] Montagna, S., Viroli, M., Fernandez-Marquez, J.L., Di Marzo Serugendo, G., Zambonelli, F.: Injecting self-organisation into pervasive service ecosystems. Mobile Networks and Applications 18(3), 398–412 (2013)
[9] De Angelis, F., Fernandez Marquez, J.L., Di Marzo Serugendo, G.: Self-composition of services with chemical reactions. 332/658, 650, ID: unige:32649 (2013)
[10] Keränen, A., Ott, J., Kärkkäinen, T.: The ONE Simulator for DTN Protocol Evaluation. In: SIMUTools 2009: Proceedings of the 2nd International Conference on Simulation Tools and Techniques. ICST, New York (2009)

A Trust-Based Approach to Clustering Agents on the Basis of Their Expertise

Francesco Buccafurri, Antonello Comi, Gianluca Lax, and Domenico Rosaci

University of Reggio Calabria, DIIES Department,
via Graziella, Feo di Vito, 89122 Reggio Calabria, Italy
{bucca,antonello.comi,lax,domenico.rosaci}@unirc.it

Abstract. The issue of detecting clusters of agents on the basis of their expertise in providing e-services is a crucial point in managing multi-agent systems, since generally a cost is associated with a given expertise level and user would have the possibility to select those agents that have the best quality of service compatibly with his budget. However, estimating agents' expertise in a multi-agent system is a hard task due to the generally large dimension of the system, so that only distributed approaches appear practicable, involving the participation of all the agents in the clustering task. In this paper, we propose to use the notion of *trust* as a basis for detecting clusters in a distributed manner. Our idea is based on the introduction of a simple trust model in a competitive multi-agent system, and on the assumption that the trust measure associated with each agent estimates the agent's expertise, provided that a sufficient number of competition steps is performed. Some preliminary tests we have performed on the ART platform show that our approach provides good results with a limited number of clusters, while the clustering capabilities worse when the number of clusters increases.

1 Introduction

In a multi-agent system, users request services to agents, which are capable of providing them with different quality of service (QoS). The QoS provided by an agent depends on his expertise in satisfying the type of request posed by the user, and different agents have different expertise for a given request type. The user would like to contact those agents having the best expertise or, in the case contacting an agent has a cost, those agents having the best expertise compatibly with the user's budget [7]. Unfortunately, agents' expertise is not usually known by the users. Also in the case an agent declares his expertise, it is not always possible to assess the reliability of this declaration, since the agent is not necessarily honest and, on the other hand, is not necessarily capable of accurately estimating his expertise by himself. Estimating agents' expertise in a multi-agent system is a hard task for the user. The most trivial approach is represented by periodically contacting all agents and evaluating in time the quality of their responses. However, this approach is not practically implementable in a large multi-agent system. A better solution would be to use some distributed

G. Jezic et al. (eds.), *Agent and Multi-Agent Systems: Technologies and Applications,*
Advances in Intelligent Systems and Computing 296,
DOI: 10.1007/978-3-319-07650-8_6, © Springer International Publishing Switzerland 2014

approach that involves the participation of the whole agent community in the estimation task, giving the possibility to each agent to express an evaluation of the expertise of the other agents. This is particularly indicated for those multi-agent systems that can be conceived as "social" communities, where each member can interact with each others in order to effectively perform its tasks. A relevant example in this context is represented by competitive agents, where each agent tries to maximize his income and minimize his outcome [8] during a competition with other agents. Over the last recent years, the introduction of trust-based approaches in competitive multi-agent systems (MAS) has been recognized as a promising solution to improve the effectiveness of these systems [15,18,9]. It is a matter of fact that nowadays MASs are more and more exploited in several application domains to provide e-services, as e-Commerce, e-Learning and so on. In such contexts, software agents are distributed in large-scale networks and interact to share resources with each other. Trust is essential to make social interactions as much fruitful as possible [13,6,17], but it can be also suitably used to estimate agents' expertise.

However, using a trust measure to punctually estimate the expertise of an agent would require a significant effort that is not necessary in the scenario depicted above, where the main issue is to provide the user with a coarse-grained partition in clusters of agents having similar expertise, such that each cluster represents a sort of expertise level. In the past, this problem has been faced in a multi-agent scenario using trust-based approaches, for example in [16] to form group of agents having similar communication capabilities, or to group agents based on the homogeneity of ideas of the group members [19] or based on inter-agent trust relationships [5].

Since often agents are abstracted through the services they provide, also research on web services is relevant in this context. For instance, in [11] a framework for establishing trust in service-oriented environments is presented, in which Web services share their experiences about the service providers with their peers through feedback ratings. The different ratings are aggregated to derive a service provider's reputation. This in turn is used to evaluate trust.

Other web service-based approaches using trust for classification and composition purposes are presented in [20,14]. These approaches are generally applied to build social communities of agents that desire to share resources or ideas about common subjects, as in the case of file sharing communities, thematic forums, scientific communities etc. However, none of the aforementioned approaches consider the problem of clustering agents based on agent expertise. In [2] a trust-based approach for partitioning the agents of a multi-agent system in two clusters of expertise is proposed, applying it to the problem of detecting compromised nodes in a SCADA system (thus the two clusters represent the compromised and the normal nodes, respectively).

Other techniques for detecting misbehaving nodes in sensor networks have been presented, for example, in [4], where an architecture to stimulate a correct routing behavior is proposed, or in [10], where the authors present an approach for mitigating routing misbehavior by detecting non-forwarding nodes and rating every path so those nodes are avoided when the routes are recalculated. Also the approach proposed in [3] introduces a secure routing protocol, called CONFIDANT, which makes misbehaviour less attractive for the nodes than proper routing. Another similar approach, which uses a reputation system, is presented in [12]. Differently from CONFIDANT, in this case the reputation system is not local, but global: some *reputation* information about the sensor nodes is transmitted all over the network, in order to warn all nodes about misbehaving nodes. In this approach, the compromised nodes are detected and isolated, since their reputation is rated as low.

In this paper, we propose to face the more general problem of detecting a given number of clusters based on the agent expertise. Our idea is based on the introduction of a simple trust model in a competitive multi-agent system, and on the assumption that the trust measure associated with each agent estimates the agent's expertise, if a sufficient number of competition steps is performed. In our approach, the computation of the trust values is distributed over the whole system, assigning to a central component only the task of computing the limits of the clusters. Some preliminary tests we have performed on the ART platform show that our approach provides good results with a limited number of clusters, while the clustering capabilities worse when the number of clusters increases.

The paper is organized as follows. In Section 2, we introduce the scenario we deal with as well as a sketch of our proposal. In Section 3, we describe the adopted trust model and our method for detecting clusters. Then, in Section 4, we present the experimental campaign we have performed to validate our proposal. Finally, in Section 5, we draw some conclusions and discuss our ongoing research.

2   An Overview of the Proposal

A multi-agent system is composed of a set of agents, such that each agent a can provide services of several types (e.g., in an e-Commerce environment, *selling a book, selling hardware and software*, etc.). Moreover, each agent has a different *expertise* in each service type T, due to his particular capability to provide that service. Indeed, it is possible that the agent can directly provide that type of service, so that he behaves as an expert in T, while it is also possible that he has to require the collaboration of other agents.

We suppose that each user U can send *service requests* to the agents. Assume that the agent A receives a service request. If A is not able to directly provide a response to this request, then he forwards the request to the agents considered the most suitable, based on a trust model. The number of agents to be contacted by A is a design parameter of the multi-agent system and can be suitably tuned to avoid an excessive traffic in the communication network. The selection of the most suitable agents is based on the effectiveness shown in the past by the agents in the service type T which the request falls into. During each temporal step, the agents of the security subsystem can interact with each others, in order to exchange information. The interactions among agents follow the protocol described below:

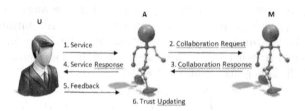

Fig. 1. The Multi-Agent System Scenario

- In order to provide a user U with a response to a request of a given type T, an agent A may decide to require the collaboration of another agent M.
- At the end of the step, the agent A receives a feedback from the user U. The feedback contains an evaluation of the quality of the response and consequently informs A about the quality of the contributions given by the contacted agents to provide the response. This way, A can use the feedback to update its internal *trust model* about the agents of the system.

The overall process that leads to provide the requesting agent with a response is logically decomposed into 6 phases, graphically represented in Figure 1, namely:

1. the user U submits a service request to the agent A;
2. A requires the collaboration of M to obtain the service asked by U;
3. M provides the collaboration;
4. A provides U with the response to the service request;
5. U provides A with a feedback.
6. A updates his internal trust model.

3   Trust-Based Clustering

This section describes the structure of the agents (trust model and agent behaviour) and the mechanism that are exploited to detect agent clusters.

3.1   The Agent Structure: Trust Model and Interactions between Agents

Given two agents A and M, the trust value that A assigns to M represents the subjective measure of the trustworthiness that A has in another agent. It comes from the past direct experience of A with M.

At the beginning, since the agent has not had any interaction with the other agents and thus he has no knowledge about the environment, the trust measure assigned to any other agent can not be evaluated (we can say his trust value is null).

Now we describe how the trust measure of an agent w.r.t. any other agent is computed and updated.

Consider the i-th step carried out by the agent A (recall the agent interactions described in Section 2). Here, the agent A is required to provide a service of type T, where the type T is one of the available service types. As a consequence, A thinks of using the collaboration of another agent M.

After the request is forwarded from A to M and M provides the response, A receives from U a feedback about the quality of the response. Each feedback is a real number belonging to $[0, 1]$. A feedback equal to 0 (1, resp.) means minimum (maximum, resp.) quality of the response. Let $FB_1, \ldots, FB_s$ be the feedbacks that evaluate the quality of the responses given by M to the request services $S_1, \ldots, S_s$ (recall, of type T).

At the i-th step, the trust measure $TRUST^i(A, M, T)$ assigned by A to M about services of type T is computed as:

$$TRUST^i(A, M, T) = \alpha \cdot TRUST^{i-1}(A, M, T) + (1 - \alpha) \cdot \frac{\sum_{1 \leq j \leq s} FB_j}{s}$$

where $TRUST^{i-1}(A, M, T)$ is the previous value of trust assigned by A to M for service type T and α is a real value belonging to $[0, 1]$ representing the importance given to the past evaluations of the reliability with respect to the current evaluation (in other words, α measures the importance given to the *memory* with respect to the current time).

Observe that, in case this is the first interaction of A with M, the value of α is set to zero since no past evaluation can be used when updating the trust.

3.2   Detection of the Clusters

At the end of each step i, each agent sends to a central *trust agency* the value of the trust value assigned to the other agents. The trust agency, for each agent and for any service type T, computes a global measure of the agent trust in the agent community, called *global reputation* for service type T. Let $A_1, \ldots A_p$ be all the agents that have sent trust values at the step i to the security agency about M in information of type T. The *global reputation* $GR^i(M, T)$ at the i-th step of the agent M in service of type T is defined as:

$$GR^i(M, T) = \frac{\sum_{1 \leq j \leq p} TRUST^i(A_j, M, T)}{p}$$

i.e., the arithmetic average of all the received trust values.

We define $GR(M, T)$ the value of the global reputation of the agent M at the final step.

The values $GR(M, T)$ are decreasingly ordered by the security agency. Then, a cluster C_k of agents is detected if $GR(MIN_{k+1}, T) - GR(MAX_k, T) > \tau$, where MIN_k (resp. MAX_k) is the agent of the cluster k having the minimum (resp. maximum) global reputation and τ is a suitable threshold that it is assumed as the minimum distance between two contiguous values of GR measure that must hold in order to consider those values as belonging to different clusters. It is important to highlight that the activity of the agency is only limited to compute and order the GR values, and thus detecting the clusters. All these activities have a computational cost $O(n)$, where n is the number of the users. Instead, the core activity of the multi-agent system, which is represented by the computation of the trust values, is distributed over the n agents associated with the users.

4   Evaluation

In this section, we describe some experiments we have performed in order to evaluate the effectiveness of our approach in detecting expertise-based clusters in a multi-agent system. The experiments have been carried out by using a prototypical simulator of the scenario described in Section 2. We have built this simulator as an extension of the well-known **A**gent **R**eputation and **T**rust (ART) platform [1] that allows the simulation of a competition among agents equipped with a trust model.

Our prototype simulates the behaviour of the competitive agents. In particular, each simulated agent has a specific *expertise* which is a real value ranging in $[0, 1]$ assigned by the simulator. For simplicity, in this simulation we have assumed that only one service type exists. The error in responding to a service request generated by the simulator for an agent having an expertise e is described by a normal distribution with mean 0 and standard deviation equal to $1 - e$. In other words, the higher e, the more precise the responses of the simulated agent. The agent's expertise does not change throughout the game and the agents know their levels of expertise. Moreover, the simulator does not inform agents about the other agents' expertise levels. The true values of the control requests presented by the agents to other agents are generated uniformly at random.

In a first experiment, in order to evaluate the capability of our approach to detect clusters of agent having different expertise, we consider 68 agents modelled as described above, and four cases, called A, B, C, and D, each of them corresponding to a different clustering in the space of the expertise values $[0..1]$ (see Figure 2). More in particular:

- Case A represents the situation in which the agents belong to only two clusters. The first cluster $C1$ is composed by agents having an expertise randomly generated in the interval $[0..0.05]$, while the second one, called $C2$ contains agents randomly generated in the interval $[0.95..1]$. This is thus the case in which there exists only agents having a very small expertise, or agents having a very high expertise.
- Case B represents the situation in which a third cluster $C3$ has been added with respect to the case A, containing agents randomly generated in the interval $[0.475..0.525]$. This cluster thus represents agents having medium-level expertise, and it can be consider intermediate between $C1$ and $C2$.

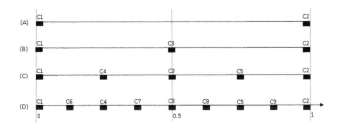

Fig. 2. Different cases of clustering the expertise space for the Experiment 1

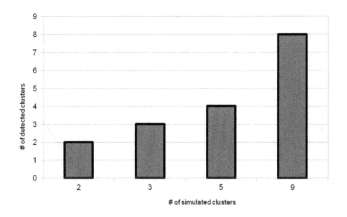

Fig. 3. Number of detected clusters vs number of simulated clusters

- Case C represents a situation in which two other clusters $C4$ and $C5$ has been added with respect to case B, where $C4$ contains agents having expertise randomly generated in $[0.225..0.275]$, while $C5$ contains agents with expertise randomly generated in $[0.725..0.775]$. In other words, $C4$ is a cluster intermediate between $C1$ and $C3$, while $C5$ is intermediate between $C3$ and $C2$.
- Case D adds 4 more clusters $C6$, $C7$, $C8$ and $C9$ with respect to the case C, that are intermediate between $C1$ and $C4$, $C4$ and $C3$, $C3$ and $C5$, $C5$ and $C2$, respectively.

In each case, the agents are equally distributed over the available clusters.

We perform a 100 steps-simulation for each of the above cases, and for each case we use the approach described in Section 3.2 for detecting clusters based on the GR trust measure. In our experiment, we consider τ equal to the dimension of the interval of each simulated cluster (i.e., 0.05).

Figure 3 shows that our approach detects the right number of clusters in case A and B, while it makes an error in case C (where the number of simulated clusters

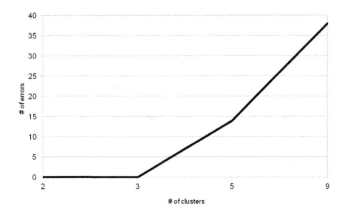

Fig. 4. Number of clustering errors vs number of clusters

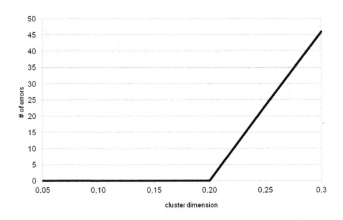

Fig. 5. Number of clustering errors vs cluster dimension

is 5 while the number of detected clusters results equal to 4) and in case D (where the number of simulated clusters is 9 while the number of detected clusters results equal to 8). Moreover, Figure 4 shows that the number of clustering error (i.e., the number of agents that belongs to an original cluster i and are erroneously assigned to a cluster $j \neq i$) is 0 both in case A and B, while it is greater than 0 in cases C and D, increasing with the number of simulated clusters. This result shows that our approach is particularly effective when the number of clusters is small (e.g., 2 or 3 clusters) and it is acceptable when the number of clusters is limited enough (e.g., 5 clusters), while it shows limitations when the number of clusters becomes medium-high.

As a second experiment, we analyze the situation of the case B (three clusters), for different cases of cluster dimensions. In particular, we consider as dimension of the clusters the values: 0.05 - 0.1 - 0.15 - 0.20 - 0.25 - 0.30. Figure 5 shows that the clustering error becomes greater than 0 for a dimension larger than 0.20, and that the error increases with the cluster dimension.

5 Conclusion

The agents of a multi-agent system usually provide services with different levels of QoS, due to the different expertise that each agent has in a given service type. For the user, it would be very useful to know a clustering of the agents based on their expertise, since this would allow him to select those agents corresponding to a given expertise and consequently to a given level of QoS. For this purpose, since the agent expertise is not known, it is necessary to have a measure to detect the existing clusters. In this paper, we have proposed to use a trust measure, computed in a distributed manner by the whole agent community, in order to form clusters of agents having similar values. Since the trust strictly depends on the agent expertise, we have assumed that the detected clusters reflect the actual clusters of expertise. Some preliminary experiments we have performed demonstrate that this assumption is valid if the number of clusters is small (2 or 3 clusters), and is already acceptable for a limited number of clusters (e.g., 5 clusters). Instead, the results get worse when the number of clusters increases, generating a significant confusion between some clusters. Moreover, the results show also that the approach is effective if the size of the cluster is sufficiently small in the expertise space. These results, although preliminary, confirms the conclusions drawn in our previous work, where we demonstrate the possibility to use trust for detecting misbehaving agents, so partitioning the agents in only two classes. Moreover, our approach generalizes those conclusions, showing that the trust can be effective also for clustering with a limited number of clusters and a limited cluster size. Our ongoing research is now devoted to explore more sophisticated trust techniques capable of detecting expertise clusters also in presence of a minor cluster resolution, i.e., when the number of clusters is high and the cluster size is large.

Acknowledgement. This work has been partially supported by TENACE PRIN Project (n. 20103P34XC) funded by the Italian Ministry of Education, University and Research.

References

1. ART-Testbed (2011), http://megatron.iiia.csic.es/art-testbed/
2. Buccafurri, F., Comi, A., Lax, G., Rosaci, D.: A Trust-Based Approach for Detecting Compromised Nodes in SCADA Systems. In: Klusch, M., Thimm, M., Paprzycki, M. (eds.) MATES 2013. LNCS, vol. 8076, pp. 222–235. Springer, Heidelberg (2013)
3. Buchegger, S., Boudec, J.L.: Performance analysis of the CONFIDANT protocol. In: Proceedings of the 3rd ACM International Symposium on Mobile Ad hoc Networking and Computing, pp. 226–236. ACM Press, Lausanne (2002)

4. Buttyan, L., Hubaux, J.: Stimulating cooperation in self-organizing mobile ad hoc networks. Mobile Networks Applications 8, 579–592 (2003)
5. Breban, S., Vassileva, J.: A coalition formation mechanism based on inter-agent trust relationships. In: Proceedings of the 1st Conference on Autonomous Agents and Multi-Agent Systems, Bologna, Italy, pp. 306–308
6. Garruzzo, S., Rosaci, D.: The roles of reliability and reputation in competitive multi agent systems. In: Meersman, R., Dillon, T.S., Herrero, P. (eds.) OTM 2010. LNCS, vol. 6426, pp. 326–339. Springer, Heidelberg (2010)
7. Gomez, M., Sabater-Mir, J., Carbo, J., Muller, G.: Improving the ART-Testbed, thoughts and reflections. In: Proceedings of the Workshop on Competitive Agents in the Agent Reputation and Trust Testbed at CAEPIA 2007, Salamanca, Spain, pp. 1–15 (2007)
8. Jöosang, A., Ismail, R., Boyd, C.: A Survey of Trust and Reputation Systems for Online Service Provision. Decision Support System 43(2), 618–644 (2005)
9. Khosravifar, B., Gomrokchi, M., Bentahar, J., Thiran, P.: Maintenance-based Trust for Multi-Agent Systems. In: Proc. of the 8th Int. Conf. on Autonomous Agents and Multiagent Systems, pp. 1017–1024. Int. Foundation for Autonomous Agents and Multiagent Systems (2009)
10. Marti, S., Giuli, T.J., Lai, K., Baker, M.: Mitigating routing misbehavior in mobile ad hoc networks. In: Proc. of the 6th Annual International Conference on Mobile Computing and Networking, pp. 255–265. ACM Press, Boston (2000)
11. Malik, Z., Bouguettaya, A.: Rateweb: Reputation assessment for trust establishment among web services. VLDB Journal 18(4), 885–911 (2009)
12. Moya, J.M., Araujo, A., Bankovic, Z., de Goyenech, J., Vallejo, J.C., Malagon, P., Villanueva, D., Fraga, D., Romero, E., Blesa, J.: Improving Security for SCADA Sensor Networks with Reputation Systems and Self-Organizing Maps. Sensors 9, 9380–9397 (2009)
13. Na, S.J., Choi, K.H., Shin, D.R.: Reputation-based Service Discovery in Multi-Agents Systems. In: Proc. of the IEEE Int. Work. on Semantic Computing and Applications, pp. 326–339. Springer (2010)
14. Nepal, S., Malik, Z., Bouguettaya, A.: Reputation Management for Composite Services in Service-Oriented Systems. Int. J. Web Service Res. 8(2), 29–52 (2011)
15. Sarvapali, D.H., Ramchurn, S.D., Jennings, N.R.: Trust in Multi-Agent Systems. The Knowledge Engineering Review 19, 1–25 (2004)
16. Rosaci, D., Garruzzo, S.: Agent clustering based on semantic negotiation. ACM Transactions on Autonomous and Adaptive Systems 3(2) (2008)
17. Rosaci, D.: Trust measures for competitive agents. Knowledge-Based Systems 28, 38–46 (2012)
18. Sabater, J., Sierra, C.: Review on Computational Trust and Reputation Models. Artificial Intelligence Review 24, 33–60 (2005)
19. Becker Villamil, M., Raupp Musse, S., Luna de Oliveira, L.P.: A model for generating and animating groups of virtual agents. In: Rist, T., Aylett, R.S., Ballin, D., Rickel, J. (eds.) IVA 2003. LNCS (LNAI), vol. 2792, pp. 164–169. Springer, Heidelberg (2003)
20. Wang, H., Shi, Y., Zhou, X., Zhou, Q., Shao, S., Bouguettaya, A.: Web Service Classification Using Support Vector Machine. In: IEEE International Conference on Tools with Artificial Intelligence, pp. 3–6

Indeterministic Belief Structures*

Barbara Dunin-Kęplicz[1] and Andrzej Szałas[1,2]

[1] Institute of Informatics, University of Warsaw, Poland
{keplicz,andrzej.szalas}@mimuw.edu.pl
[2] Department of Computer and Information Science, Linköping University, Sweden

Abstract. The current paper falls into a bigger research programme concerning construction of modern belief structures applicable in multiagent systems. In previous papers we approached individual and group beliefs via querying paraconsistent belief bases. This framework, covering deterministic belief structures, turned out to be tractable under some natural restrictions on implementation. Moreover, we have indicated a four-valued query language 4QL as an implementation tool guaranteeing tractability and capturing all PTIME-constructible belief structures.

In this paper we generalize our approach to the nondeterministic case. This is achieved by adjusting the key abstractions of epistemic profiles and belief structures to this new situation. Importantly, tractability of the approach is still maintained.

1 Epistemic Profiles and Belief Structures

Uncertainty and ignorance in the information shared by agents' are inherent properties of autonomous, decentralized, situated systems. Such systems include multiagent systems operating in dynamic and unpredictable environments where information is typically delivered by a variety of information sources. Unfortunately, this information in often of law quality and imprecise due to:

- limited accuracy of sensors,
- time restrictions,
- unfortunate combinations of environmental conditions,
- limited reliability of physical devices.

This unfortunate combination of features leads to *inconsistencies* that may appear on many different levels: individual, between agents, between agents and groups and/or between different groups. In many real-world scenarios, inconsistencies together with missing information invoke nondeterminism in agent's activities, even though agent's decision processes are typically assumed to be deterministic. Inevitably, incomplete and/or inconsistent information has to be interpreted, sometimes leading to alternative sets of beliefs which, in turn, may result in different decisions.

In logic-based approaches to multiagent systems beliefs have typically been modeled using different combinations of multi-modal logics [8,9], non-monotonic logics [14],

---

* Supported by the Polish National Science Centre grants 2011/01/B/ST6/02769 and 2012/05/B/ST6/03094.

G. Jezic et al. (eds.), *Agent and Multi-Agent Systems: Technologies and Applications*,
Advances in Intelligent Systems and Computing 296,
DOI: 10.1007/978-3-319-07650-8_7, © Springer International Publishing Switzerland 2014

probabilistic reasoning [16] or fuzzy reasoning [19], just to mention some of them. However, most of the proposed solutions turn out to be unsatisfactory due to either the lack of tools for inconsistency/ignorance handling or to high complexity. In our approach [6,7] we have already suggested a shift in perspective. Rather than reasoning in modal logics of high complexity we have indicated a tractable approach based on querying paraconsistent belief bases. To achieve the required expressiveness and modeling convenience, we adapted two additional truth values: *inconsistent* and *unknown*, representing inconsistent and missing information. Importantly, in the proposed framework, incomplete and/or inconsistent beliefs may be resolved non-monotonically.

We build our approach on *epistemic profiles and belief structures*, as introduced in [6,7] and further applied, e.g., in [4,5]. In short, an epistemic profile encapsulates agents' reasoning capabilities, including individual methods of both disambiguation of inconsistencies and completing missing information. Resulting belief structures consist of *constituents* and *consequents*: an agent starts with *constituents*, which are further transformed into *consequents* via the agent's *individual epistemic profile*. While constituents contain sets of beliefs acquired by perception, expert-supplied knowledge, communication with other agents and other ways, consequents contain final, "mature" beliefs.

During belief formation in multiagent systems, initial and intermediate beliefs are confronted with other beliefs originating from a variety of sources. In effect, the transformed beliefs can deviate from the initial ones. Strangely enough, in extreme cases final beliefs may even contradict the initial ones. For example, when beliefs about safety on a road are to be formed, constituents may consist of beliefs about weather conditions, temperature, air pressure and moisture levels. When, for example, the weather is sunny, temperature and pressure are high but information about the moisture level is missing, the associated epistemic profile may transform these beliefs into a consequent stating that the situation on a road is safe.

In [6,7] belief structures are deterministic in the sense that, given a set of constituents, consequents are uniquely determined. Though this is a typical situation in real-world applications, deterministic belief structures do not allow to model situations resulting from nondeterministic actions with effects that cannot be foreseen a priori [3]. Therefore, the consequents cannot be uniquely determined, too. In our example, even though there are good reasons to believe that moisture level is low, it might be undetermined whether a given car will skid or not. It is reasonable to consider two consequents: one, where the car skids and another, where the car does not. This motivated us to introduce *indeterministic* belief structures, i.e., structures allowing for more than one consequent. We will show both how to query such structures and how to combine them into more complex structures. We will also define belief operators, show some of their properties and demonstrate their use.

Analogically to [6,7], as an implementation tool we suggest 4QL, a four-valued rule-based query language designed in [10,15,12].[1] Our approach is strongly influenced by ideas underlying 4QL, which allows for negation in premises and conclusions of rules. It provides simple, yet powerful constructs (modules and external literals) for expressing non-monotonic rules reflecting, among others, lightweight forms of default reasoning,

---

[1] See http://4ql.org, where an open source experimental interpreter of 4QL is available.

auto-epistemic reasoning, defeasible reasoning, and the local closed world assumption. Importantly, 4QL enjoys tractable query computation and captures all tractable queries (see [11] for details). Therefore, 4QL is a natural implementation tool opening the space for a diversity of applications by providing firm foundations for paraconsistent knowledge bases.

The rest of this paper is structured as follows. First, in Section 2 we introduce belief bases and discuss the underlying querying machinery. In Section 3 we we discuss our new understanding of beliefs. Next, in Section 4, we motivate and introduce indeterministic belief structures. Section 5 illustrates our solutions by an example. In Section 6 we discuss implementation issues and complexity of the approach. Finally, Section 7 concludes the paper.

2 Belief Bases

In order to construct paraconsistent belief bases, we deal with the classical first-order language over a given vocabulary without function symbols. We assume that *Const* is a fixed set of constants, *Var* is a fixed set of variables and *Rel* is a fixed set of relation symbols.

Definition 2.1. *A literal is an expression of the form $R(\bar{\tau})$ or $\neg R(\bar{\tau})$, with τ being a sequence of arguments, $\bar{\tau} \in (Const \cup Var)^k$, where k is the arity of R. Ground literals over Const, denoted by $\mathcal{G}(Const)$, are literals without variables, with all constants in Const. If $\ell = \neg R(\bar{\tau})$ then $\neg \ell \stackrel{\text{def}}{=} R(\bar{\tau})$.*   ◁

Though we use the classical first-order syntax, the presented semantics substantially differs from the classical one. Namely,

- truth values t, i, u, f (true, inconsistent, unknown, false) are explicitly present;
- the semantics is based on sets of ground literals rather than on relational structures.

This allows one to deal with the lack of information as well as inconsistencies. As 4QL is based on the same principles, it can immediately be used as the implementation tool.

Table 1. Truth tables for $\wedge$, $\vee$, $\rightarrow$ and $\neg$ (see [18,10,12])

$\wedge$	f	u	i	t		$\vee$	f	u	i	t		$\rightarrow$	f	u	i	t		$\neg$	
f	f	f	f	f		f	f	u	i	t		f	t	t	t	t		f	t
u	f	u	u	u		u	u	u	i	t		u	t	t	t	t		u	u
i	f	u	i	i		i	i	i	i	t		i	f	f	t	f		i	i
t	f	u	i	t		t	t	t	t	t		t	f	f	t	t		t	f

The semantics of propositional connectives is summarized in Table 1. Observe that definitions of $\wedge$ and $\vee$ reflect minimum and maximum w.r.t. the ordering:

$$f < u < i < t, \tag{1}$$

as advocated, e.g., in [2,10,15,18]. Such a truth ordering appears to be natural and reflecting intuitions of the classical two-valued logic. For example, a conjunction is true if all its operands are true, etc.

Let $v : Var \longrightarrow Const$ be a *valuation of variables*. For a literal ℓ, by $\ell(v)$ we understand the ground literal obtained from ℓ by substituting each variable x occurring in ℓ by constant $v(x)$.

Definition 2.2. The *truth value* of a literal ℓ w.r.t. a set of ground literals L and valuation v, denoted by $\ell(L, v)$, is defined as follows:

$$\ell(L, v) \overset{\text{def}}{=} \begin{cases} \mathbf{t} \text{ if } \ell(v) \in L \text{ and } (\neg \ell(v)) \notin L; \\ \mathbf{i} \text{ if } \ell(v) \in L \text{ and } (\neg \ell(v)) \in L; \\ \mathbf{u} \text{ if } \ell(v) \notin L \text{ and } (\neg \ell(v)) \notin L; \\ \mathbf{f} \text{ if } \ell(v) \notin L \text{ and } (\neg \ell(v)) \in L. \end{cases}$$

$\lhd$

For a formula $\alpha(x)$ with a free variable x and $c \in Const$, by $\alpha(x)_c^x$ we understand the formula obtained from α by substituting all free occurrences of x by c. Definition 2.2 is extended to all formulas in Table 2, where α and β denote first-order formulas, v is a valuation of variables, L is a set of ground literals, and the semantics of propositional connectives appearing at righthand sides of equivalences is given in Table 1.

Table 2. Semantics of first-order formulas

– if α is a literal then $\alpha(L, v)$ is defined in Definition 2.2;
– $(\neg \alpha)(L, v) \overset{\text{def}}{=} \neg(\alpha(L, v))$;
– $(\alpha \circ \beta)(L, v) \overset{\text{def}}{=} \alpha(L, v) \circ \beta(L, v)$, where $\circ \in \{\vee, \wedge, \rightarrow\}$;
– $(\forall x \alpha(x))(L, v) \overset{\text{def}}{=} \min_{a \in Const} \{(\alpha_a^x)(L, v)\}$, where min is the minimum w.r.t. ordering (1);
– $(\exists x \alpha(x))(L, v) \overset{\text{def}}{=} \max_{a \in Const} \{(\alpha_a^x)(L, v)\}$, where max is the maximum w.r.t. ordering (1).

If S is a set then by $\text{FIN}(S)$ we understand the set of all finite subsets of S. We further assume that *Const* is always finite and by $\mathbb{C} \overset{\text{def}}{=} \text{FIN}(\mathcal{G}(Const))$ we denote the set of all finite sets of ground literals over the set of constants *Const*.

Definition 2.3. *By a* belief base *over a set of constants Const we understand any finite set Δ of finite sets of ground literals over Const, i.e. any finite set $\Delta \subseteq \mathbb{C}$.*

$\lhd$

3   Understanding Beliefs

By *information ordering* we understand the ordering on truth values shown in Figure 1. This ordering reflects the process of gathering and fusing information. Starting from the lack of information, in the course of belief acquisition, evidences supporting or denying hypotheses are collected. This finally permits one to decide about the truth value of the hypotheses.

Fig. 1. Information ordering on truth values

Definition 3.1. *Let Δ be a belief base, $v : Var \longrightarrow Const$ be a valuation of variables and α be a formula. We define the* belief operator *by:*

$$\mathrm{Bel}_{\Delta,v}(\alpha) \stackrel{\mathrm{def}}{\equiv} \mathrm{LUB}\{\alpha(D,v) \mid D \in \Delta\},$$

where LUB *denotes the least upper bound wrt the ordering shown in Figure 1.* ◁

Note that whenever α in Definition 3.1 is closed (i.e., does not contain free variables), the valuation v becomes redundant. From now on, for the sake of simplicity we shall often restrict considerations to closed formulas and will use notation $\mathrm{Bel}_\Delta()$ rather than $\mathrm{Bel}_{\Delta,v}()$. For clarity let us indicate that:

$$\mathrm{Bel}_\Delta(t) = t \text{ when } t \in \{\mathsf{t}, \mathsf{i}, \mathsf{f}, \mathsf{u}\}. \tag{2}$$

Example 3.2. Consider $\Delta = \{\{p(a), \neg q(a)\}; \{q(a), \neg p(a)\}\}$. According to Definition 3.1 we have $\mathrm{Bel}_\Delta(p(a)) = \mathsf{i}, \mathrm{Bel}_\Delta(q(a)) = \mathsf{i}$ and $\mathrm{Bel}_\Delta(p(a) \vee q(a)) = \mathsf{t}$, so,

$$\mathrm{Bel}_\Delta(p(a)) \vee \mathrm{Bel}_\Delta(q(a)) \not\equiv \mathrm{Bel}_\Delta(p(a) \vee q(a)). \tag{3}$$

◁

Remark 3.3. Note that sets $D \in \Delta$ appearing in Definition 3.1 can be considered as four-valued worlds. Comparing to Kripke-like semantics for beliefs (see, e.g., [8,13]), at this point the main difference are:

- we do not require fixed, rigid structure connecting worlds via accessibility relations;
- we use four rather than two truth values. ◁

4 Indeterministic Belief Structures

In our framework we intend to implement a shift from idealized reasoning in multi-modal systems of high complexity to querying paraconsistent belief-bases representing possibly incomplete and imprecise information. To ensure realism in modeling beliefs, a key abstraction of epistemic profile has been introduced in [6,7]. It permits to isolate and separately characterize agent's individual reasoning capabilities. This way heterogeneity of agents informational stance in ensured by, firstly, adjusting reasoning capabilities to application in question, and then, individualizing them. Importantly, a holistic aspect of traditional group beliefs that seem to be unnecessary in real-life applications may be abandoned in favor of capturing selected/needed aspects of information solely. This new quality built in epistemic profiles allows one to construct modern belief structures.

Let us now define the concepts of indeterministic belief structures and epistemic profiles. Further on we fix a finite set of constants *Const*. Recall that $\mathbb{C} = \text{FIN}(\mathcal{G}(Const))$ is the set of all finite sets of ground literals over *Const*.

Definition 4.1

- By a *constituent* we understand any set $C \in \mathbb{C}$;
- by an *indeterministic epistemic profile* we understand any function $\mathcal{E}$ of the sort $\text{FIN}(\mathbb{C}) \longrightarrow \text{FIN}(\mathbb{C})$;
- by an *indeterministic belief structure over an indeterministic epistemic profile* $\mathcal{E}$ we mean $\mathcal{B}^{\mathcal{E}} = \langle \mathcal{C}, \mathcal{F} \rangle$, where:
  - $\mathcal{C} \subseteq \mathbb{C}$ is a nonempty set of constituents;
  - $\mathcal{F} \stackrel{\text{def}}{=} \mathcal{E}(\mathcal{C})$ is the set of *consequents* of $\mathcal{B}^{\mathcal{E}}$. ◁

Note that epistemic profiles of [6,7] are functions of the sort $\text{FIN}(\mathbb{C}) \longrightarrow \mathbb{C}$. That is, they basically are deterministic epistemic profiles with $\mathcal{F}$ consisting of one consequent.

Remark 4.2. When comparing indeterministic belief structures to Kripke structures:

- rather than accessibility relations we have much more flexible epistemic profiles;
- accessibility relations connect a world to a set of worlds while epistemic profiles transform sets of worlds into sets of worlds;
- Kripke structures are static while epistemic profiles transform arbitrary sets of worlds (constituents) into suitable sets of worlds (consequents), so are better adjusted to the dynamics of the environment. ◁

We are now ready to define the semantics of formulas in belief structures, where we assume that the language is an extension of the classical language by adding the operator $\text{Bel}_{\mathcal{B}^{\mathcal{E}}}()$ indexed by a belief structure. Note that rather than indexing $\text{Bel}()$ by belief structures one can index them, as frequently done, by agents and refer to belief structures via agents. For simplicity we omit this intermediate step.

A formula is $\text{Bel}()$-*free* if it does not contain belief operators.

Definition 4.3. Let $\mathcal{B}_1^{\mathcal{E}} = \langle \mathcal{C}_1, \mathcal{F}_1 \rangle$ and $\mathcal{B}_2^{\mathcal{P}} = \langle \mathcal{C}_2, \mathcal{F}_2 \rangle$ *be indeterministic belief structures and* v *be a valuation of variables in Const. The semantics of formulas is defined by:*

$$\alpha(\mathcal{B}_1^{\mathcal{E}}, v) \stackrel{\text{def}}{=} \begin{cases} \text{Bel}_{\mathcal{C}_1, v}(\alpha) & \text{when } \alpha \text{ is Bel}()\text{-free;} \\ \text{Bel}_{\mathcal{F}_2, v}(\beta) & \text{when } \alpha \text{ is of the form } \text{Bel}_{\mathcal{B}_2^{\mathcal{P}}}(\beta) \text{ and } \beta \text{ is Bel}()\text{-free,} \end{cases}$$

where $\text{Bel}_{\mathcal{C}_1, v}(\alpha)$ *and* $\text{Bel}_{\mathcal{F}_2, v}(\beta)$ *are defined in Definition 3.1.*[2] ◁

The above definition can be extended for all formulas by defining the semantics of connectives and quantifiers as in Section 2 and nested $\text{Bel}()$ operators starting from the innermost ones. The intuition behind the above definition is that we treat both constituents and consequents as basis for beliefs. Note that in [6,7] constituents have been

---

[2] Note that, in the simplest case, $\mathcal{B}_1^{\mathcal{E}}$ and $\mathcal{B}_2^{\mathcal{E}}$ can be identical.

treated as "direct perception" of the world, so have been referenced to using $\bigcup_{C \in \mathcal{C}} C$.

Both the previous and the current approach are compatible when only literals are considered. They differ, however, on more complex formulas. We have decided to modify the definition to make the approach uniform wrt. complex formulas, too.

Example 4.4. Consider a belief structure $\mathcal{B}^{\mathcal{E}} = \langle \mathcal{C}, \mathcal{F} \rangle$, where:

- constituents are $\mathcal{C} = \{\{p(a), \neg q(a)\}; \{q(a), \neg p(a)\}\}$;
- consequents are $\mathcal{F} = \{\{p(a)\}; \{q(a), \neg p(a)\}\}$.

Let $v(x) = a$. Then, according to Definition 4.3,

$$
\begin{aligned}
&\left(p(x) \wedge \mathrm{Bel}_{\mathcal{B}^{\mathcal{E}}}\left(p(x) \vee q(x)\right)\right)\left(\mathcal{B}^{\mathcal{E}}, v\right) \equiv \\
&\quad \left(p(x)\right)\left(\mathcal{B}^{\mathcal{E}}, v\right) \wedge \left(\mathrm{Bel}_{\mathcal{B}^{\mathcal{E}}}\left(p(x) \vee q(x)\right)\right)\left(\mathcal{B}^{\mathcal{E}}, v\right) \equiv \\
&\quad \mathrm{Bel}_{\mathcal{C},v}\left(p(x)\right) \wedge \mathrm{Bel}_{\mathcal{F},v}\left(p(x) \vee q(x)\right) \equiv \\
&\quad \underbrace{\mathrm{Bel}_{\mathcal{C}}\left(p(a)\right)}_{\mathbf{i}} \wedge \underbrace{\mathrm{Bel}_{\mathcal{F}}\left(p(a) \vee q(a)\right)}_{\mathbf{t}} \equiv \mathbf{i}.
\end{aligned}
$$
◁

Note that typical requirements as to belief operators are satisfied as shown by the following proposition.

Proposition 4.5. *For any formula α and belief structure $\mathcal{B}^{\mathcal{E}}$:*

$$
\begin{aligned}
&\left(\neg \mathrm{Bel}_{\mathcal{B}^{\mathcal{E}}}\left(\mathfrak{f}\right)\right)\left(\mathcal{B}^{\mathcal{E}}\right) = \mathbf{t} && \text{(consistency of beliefs)} \\
&\left(\mathrm{Bel}_{\mathcal{B}^{\mathcal{E}}}\left(\alpha\right) \to \mathrm{Bel}_{\mathcal{B}^{\mathcal{E}}}\left(\mathrm{Bel}_{\mathcal{B}^{\mathcal{E}}}\left(\alpha\right)\right)\right)\left(\mathcal{B}^{\mathcal{E}}\right) = \mathbf{t} && \text{(positive introspection)} \\
&\left(\neg \mathrm{Bel}_{\mathcal{B}^{\mathcal{E}}}\left(\alpha\right) \to \mathrm{Bel}_{\mathcal{B}^{\mathcal{E}}}\left(\neg \mathrm{Bel}_{\mathcal{B}^{\mathcal{E}}}\left(\alpha\right)\right)\right)\left(\mathcal{B}^{\mathcal{E}}\right) = \mathbf{t} && \text{(negative introspection). ◁}
\end{aligned}
$$

Remark 4.6. A belief structure may use other belief structures. The usefulness of such constructions has been demonstrated, e.g., in [4,5,6,7]. One important application depends on constructing belief structures for groups of agents. In such cases it is natural to assume that consequents of group members are constitutents of a group. In the case of indeterministic belief structures such a construction has to be adjusted. Namely, we deal with possibly many consequents, so rather than referring to input consequents directly, one has to reference them via a suitable belief operator. If a given belief structure uses consequents of another belief structure $\mathcal{B}^{\mathcal{E}}$ then checking the status of a belief, expressed as a formula α, is done by using $\mathrm{Bel}_{\mathcal{B}^{\mathcal{E}}}\left(\alpha\right)$.   ◁

5   An Example

Consider a patrol robot which detects a potential intruder (moving object) and informs of security issues in the environment. Such a robot is typically equipped with a camera and other detectors/tracking devices (see, e.g., [17]). Let us also assume that ultrasound sensors are located in certain places of the environment. Such a situation can be modeled by a belief structure, $\mathcal{B}^{\mathcal{E}} = \langle \mathcal{C}, \mathcal{F} \rangle$, where:

- $\mathcal{C} \subseteq \mathbb{C}$ consists of two constituents: *Camera* and *Detectors*;
- $\mathcal{F} = \mathcal{E}(\mathcal{C})$ is a set of consequents reflecting the current robot's target places.

More precisely,

- *Camera* contains literals $intruder(P)$ (there is a visible intruder in place P);
- *Detectors* contains literals $detected(P)$ (a moving object is detected in place P);
- $\mathcal{E}$ returns as many consequents as there are different places submitted by *Camera* and *Detectors*; each consequent represents a possible target place provided by literal $target(P)$.

If there is only one consequent,[3] say $\{target(p_1)\}$, then $\text{Bel}_{\mathcal{B}^{\mathcal{E}}}\left(target(p_1)\right)$ is true. That is, the patrol robot believes that its target place currently is p_1.

If there are more consequents, making the case indeterministic, then the situation changes. Assume for simplicity that there are two consequents:

$$\{target(p_1), \neg target(p_2)\} \text{ and } \{target(p_2)\}, \text{ where } p_1 \neq p_2.$$

Then we have that $\text{Bel}_{\mathcal{B}^{\mathcal{E}}}\left(target(p_1)\right) = \text{t}$ and $\text{Bel}_{\mathcal{B}^{\mathcal{E}}}\left(target(p_2)\right) = \text{i}$, meaning that the patrol robot believes that its target place is p_1 and its belief as to target p_2 is inconsistent, perhaps due to a faulty detection. On the basis of this information the robot can make a choice between moving to place p_1 and p_2, not necessarily choosing p_1 as p_2 may be considered more important.

If the consequents are: $\{target(p_1), \neg target(p_2)\}$ and $\{\neg target(p_1), target(p_2)\}$ then both $\text{Bel}_{\mathcal{B}^{\mathcal{E}}}\left(target(p_1)\right)$ and $\text{Bel}_{\mathcal{B}^{\mathcal{E}}}\left(target(p_2)\right)$ are inconsistent. On the other hand, $\text{Bel}_{\mathcal{B}^{\mathcal{E}}}\left(target(p_1) \vee target(p_2)\right)$ is true. Note that information whether the disjunction $target(p_1) \vee target(p_2)$ holds is important, as it restricts possible robot's choices to places p_1, p_2 only.

Observe that nondeterminism is needed not only when a potential intruder is detected. For example, when a robot is patrolling certain paths, it is better when it makes a more or less random walk rather than always repeating the same routine. Rigid patrolling patterns simplify potential activities of intruders.

6   Implementation and Complexity

As an implementation tool we indicate 4QL [10,12,15]. The semantics of 4QL is defined by *well-supported models* , i.e., models consisting of (positive or negative) ground literals, where each literal is a conclusion of a derivation starting from facts. 4QL programs are structured in the form of modules consisting of rules and facts. For any module, the corresponding well-supported model is uniquely determined. That is,

> each module can be treated as a finite set of ground literals and this set can be computed in deterministic polynomial time [10,12].

All constituents and consequents are finite sets of ground literals. Therefore there is a natural correspondence between 4QL modules and (indeterministic) belief structures.

The following theorem follows from analogous results for 4QL [10,11,12].

Theorem 6.1. *If an indeterministic belief structure $\mathcal{B}^{\mathcal{E}}$ is implemented using 4QL then it enjoys* PTIME *data complexity.*

If an indeterministic belief structure is PTIME *constructible then it can be implemented in 4QL (assuming a linear order on domain elements).*   ◁

---

[3] Meaning that the case is deterministic.

7   Conclusions

In the current paper we have generalized our previous formalism [6,7] to allow one to model nondeterminism without compromising computational feasibility. The main motivations stem from multiagent systems operating in dynamic and unpredictable environments, where nondeterminism is an inherent property. Indeterministic epistemic profiles are well adjusted to these difficult characteristics as they transform arbitrary sets of worlds (constituents) into suitable sets of worlds (consequents).

Let us emphasize that nondeterminism is often a desired property in modeling autonomous behavior. In particular,

- it can be applied in modeling nondeterminism of robots or mobile agents when their behavior should be patternless;
- it can be needed to model semantical structures associated with various knowledge representation formalisms, including non-monotonic reasoning like default reasoning having many extensions, all or some of which may be used as conseqents;
- it is useful in modeling unpredictable effects of actions, like potentially caused by an inaccurate model of the environment;
- it is well-suited in modeling actions with non-typical effects.

Indeterministic belief structures share commonly agreed properties like consistency of beliefs, positive and negative introspection and are not distributive over disjunction. In many applications, especially those involving nondeterminism, the latter is a desirable property, not present in the deterministic case. Therefore, belief structures introduced in the current paper substantially strengthen modeling capabilities of the belief structures–based formalism.

References

1. Barbucha, D., Le, M., Howlett, R., Jain, L. (eds.): Proc. 7th KES Conference on Agent and Multi-Agent Systems - Technologies and Applications. Frontiers in Artificial Intelligence and Applications, vol. 252. IOS Press (2013)
2. de Amo, S., Pais, M.: A paraconsistent logic approach for querying inconsistent databases. International Journal of Approximate Reasoning 46, 366–386 (2007)
3. Dunin-Keplicz, B., Radzikowska, A.M.: Nondeterministic actions with typical effects: Reasoning about scenarios. In: Meyer, J.-J.C., Schobbens, P.-Y. (eds.) ModelAge-WS 1997. LNCS (LNAI), vol. 1760, pp. 143–156. Springer, Heidelberg (2000)
4. Dunin-Keplicz, B., Strachocka, A.: Perceiving rules under incomplete and inconsistent information. In: Leite, J., Son, T.C., Torroni, P., van der Torre, L., Woltran, S. (eds.) CLIMA XIV 2013. LNCS, vol. 8143, pp. 256–272. Springer, Heidelberg (2013)
5. Dunin-Keplicz, B., Strachocka, A., Szałas, A., Verbrugge, R.: Perceiving speech acts under incomplete and inconsistent information. In: Barbucha, et al. (eds.) [1], pp. 255–264
6. Dunin-Keplicz, B., Szałas, A.: Epistemic profiles and belief structures. In: Jezic, G., Kusek, M., Nguyen, N.-T., Howlett, R.J., Jain, L.C. (eds.) KES-AMSTA 2012. LNCS, vol. 7327, pp. 360–369. Springer, Heidelberg (2012)
7. Dunin-Keplicz, B., Szałas, A.: Taming complex beliefs. In: Nguyen, N.T. (ed.) Transactions on Computational Collective Intelligence XI. LNCS, vol. 8065, pp. 1–21. Springer, Heidelberg (2013)

8. Dunin-Kęplicz, B., Verbrugge, R.: Teamwork in Multi-Agent Systems. A Formal Approach. John Wiley & Sons, Ltd. (2010)
9. Fagin, R., Halpern, J., Moses, Y., Vardi, M.: Reasoning About Knowledge. The MIT Press (2003)
10. Małuszyński, J., Szałas, A.: Living with inconsistency and taming nonmonotonicity. In: de Moor, O., Gottlob, G., Furche, T., Sellers, A. (eds.) Datalog 2010. LNCS, vol. 6702, pp. 384–398. Springer, Heidelberg (2011)
11. Małuszyński, J., Szałas, A.: Logical foundations and complexity of 4QL, a query language with unrestricted negation. Journal of Applied Non-Classical Logics 21(2), 211–232 (2011)
12. Małuszyński, J., Szałas, A.: Partiality and inconsistency in agents' belief bases. In: Barbucha, et al. (eds.) [1], pp. 3–17
13. Meyer, J.-J.C., van der Hoek, W.: Epistemic Logic for Computer Science and Artificial Intelligence. Cambridge University Press (1995)
14. Mueller, E.: Commonsense Reasoning. Morgan Kaufmann (2006)
15. Szałas, A.: How an agent might think. Logic Journal of the IGPL 21(3), 515–535 (2013)
16. Thrun, S., Burgard, W., Fox, D.: Probabilistic Robotics (Intelligent Robotics and Autonomous Agents). The MIT Press (2005)
17. Tseng, C., Lin, C., Shih, B., Chen, C.: Sip-enabled surveillance patrol robot. Robotics and Computer-Integrated Manufacturing 29(2), 394–399 (2013)
18. Vitória, A., Małuszyński, J., Szałas, A.: Modeling and reasoning with paraconsistent rough sets. Fundamenta Informaticae 97(4), 405–438 (2009)
19. Zadeh, L.: Fuzzy sets. Information and Control 8, 333–353 (1965)

Implementation of Agent-Based Games Recommendation System on Mobile Platforms

Pavle Skočir, Iva Bojić, and Gordan Ježić

University of Zagreb
Faculty of Electrical Engineering and Computing
Unska 3, HR-10000 Zagreb, Croatia
{pavle.skocir,iva.bojic,gordan.jezic}@fer.hr

Abstract. Because of Google Play and App Store, today numerous different games are offered to every smartphone user. This diversity in supply is undoubtedly a good thing, but it also virtually disables users to find games they would like to play. However, it was shown that users tend to spend more money on purchasing recommended games when these recommendations are done using personal recommenders. In this paper we present an agent-based recommender for mobile platforms in which recommendations are made taking into account user game experience. Inputs for our recommender, which are collected both inside and outside the games, are stored in a semantic database. Based on collected information, user and game profiles are made that are then used in our recommendation algorithm. The focus of the paper is on how to do the implementation of the proposed system in real-world environments and obtain all the necessary data and how to make recommendations based on generated user and game profiles.

Keywords: Google Play, App Store, mobile games, user experience, user profiles, semantic database, MARS recommender system.

1 Introduction

Nowadays users are offered large amounts of content in form of applications for smartphones, videos on services like YouTube, or movies on IPTV platforms. In order to get users more familiar with the available content and to help them find what they might be interested in, service providers are developing personalized services that recommend specific content for each user. The available content recommendations can be made by using different methods. Generally, each recommendation system takes user interests as an input and generates personalized lists of recommended items for each user as an output [1].

Games as a content type require a higher level of user involvement compared to other content types. While playing games, users interact more with their devices than when, for example, listening to music. In the latter the only interactions that users have with their media players are for stopping, pausing, rewinding the content, while in the former various features about user interaction with games can be monitored. Such features include time needed for passing a certain level, collected points or touch gestures on smartphone screens.

G. Jezic et al. (eds.), *Agent and Multi-Agent Systems: Technologies and Applications,*   67
Advances in Intelligent Systems and Computing 296,
DOI: 10.1007/978-3-319-07650-8_8, © Springer International Publishing Switzerland 2014

In our previous work [2][3], we proposed an agent-based model for monitoring user experience through several parameters: user motivation, her/his progress through the game and how successful she/he was in order to find an appropriate genre and difficulty of the game which should be recommended to the user. Together with the appropriate genre and difficulty, our recommender also takes into account the download history and similarities between different games to provide users with new recommendations. This model has been implemented for a custom game [3]. In this paper, we focus on implementing our model in real-world environments, i.e. monitoring user interactions while playing existing games on smartphones and obtaining information about those games that were then used as inputs for our recommendation system. To the best of our knowledge, the approach of collecting information about users by monitoring user interactions while playing games is not so abundantly used like obtaining this information through ratings or reviews of purchased products [4].

The rest of the paper is organized as follows. Section 2 introduces the current achievements regarding recommendation systems, with focus on game recommenders. In Section 3 we present the model of our game provisioning system, while Section 4 describes the process of collecting inputs for our model from existing mobile platforms and within the existing games. Section 5 presents adaptation of our recommendation algorithm to enable working with the new set of data, different than originally proposed one. Section 6 concludes the paper.

2   Related Work

Even before App Store[1] and Google Play[2], it was clear that one-size-fits-all approach would not hold in case of mobile games and that recommendation systems would be needed in order to adapt gameplay [5]. General purpose recommendation systems can be classified in three main categories: content-based, collaborative and hybrid approach [6]. In content-based recommenders, users are recommended items similar to the ones they preferred in the past, while when using collaborative recommendations, users will be recommended items that other people with similar tastes and preferences to theirs liked in the past. Finally, a hybrid approach combines two aforementioned approaches.

Usually, there are three phases in every recommendation process: collecting the inputs, filtering the data and giving the outputs. Recommenders used for recommending games can gather inputs outside the games or inside the games [5]. When information about the games is collected outside, these services used for collecting are usually not strictly game-related and are used for plenty other purposes. On the other hand, when collecting information about users only inside the games, each game needs to be modified to collect the data. In our work we combine both approaches in order to achieve better results. Namely, we collect information about every specific game from App Store or Google Play, and combine them with information about user interaction inside the games.

---

[1] Initial release of App Store was on July 9, 2008.

[2] Initial release of Google Play (i.e. Android Market) was on October 23, 2008.

2.1   Collecting Inputs Outside the Games

Outside the games, information about users can be collected explicitly or implicitly. The explicit form of information collection relays on users giving their feedback, while the implicit way of collecting the data is mostly done without user knowledge. Nowadays, the most popular way to explicitly collect user opinions about the games they downloaded and tried is to collect their reviews and game ratings through App Store or Google Play. Both of these stores provide ranking for the top games in different game genres including both free and paid games, and for the top grossing games. Additionally, App Store provides information about best new games, current and previous editors' choices.

The easiest way to collect information about the users implicitly is to create their profiles that are then used to recommend personalized services. Roh and Jin developed the *Wallet App* for smartphones to gather personal information about the users [7]. In their system, each user profile consisted of static information such as user age, gender and address, as well as dynamic ones such as frequently used stores and favorite purchased items. Their goal was to develop a personalized advertisement recommendation system which would encourage users to spend more money on different apps. Namely, in 2009 Jannach and Hegelich [8] already showed that when using personalized recommendations instead of non-personalized ones a significant increase in viewed and sold items could be achieved. According to their results, effectiveness of online sale could be increased by up to 3.6 % by providing personalized item recommendations.

Which App? [9] is a general purpose recommender system that implements a hybrid approach (i.e. combines both techniques based on collaborative filtering and content-based filtering) and monitors user activities. The data that is being collected is: which applications are used, how many times the application has been used, during how much time it has been used, if it was used combined with other applications, how many times it has been updated, etc. Similar to work done by Roh and Jin, information collected through this app is also being used for building user profiles. However, unlike the *Wallet App*, this application collects information that can model user routines of using her/his smartphone better. The disadvantage is of course that this application has to run continuously in the background which increases battery consumption.

Another interesting way of collecting user opinion about the games they played implicitly was proposed in [10]. Sorensen based her approach on information collected from Twitter. Twitter is a service for a micro blogging where users can "follow" other users they are interested in. Before her research, much research that aimed at describing recommendations based on the relationships between users who were already "connected" on Twitter was done. However, her work did not leverage information about prior relations between different users, but was based on an overlap between words used by different users within the same game domain. In this way by extracting user tweets about different games, she managed to give valuable recommendations about unknown games to Twitter users within the same domain of interests without a prior knowledge about their relationships.

2.2   Collecting Inputs Inside the Games

Research conducted by Shin et al. [11] showed that fewer than a half of installed applications on mobile phones are actually being used, while Bohmer et al. showed that the average session with an application lasts less than a minute, even though users spend almost an hour a day using their smartphones [12]. It is thus very important to track user interactions in games in order to conclude whether users like them or not. Inputs inside the games can be also collected explicitly or implicitly. The easiest way to explicitly collect information about the user inside the games is to allow her/him to choose a difficulty level at the beginning of the game. This choice of a difficulty level is usually set on a single linear scale from easy to hard and it reflects user opinion of her/his abilities to successfully finish the game.

It is far more complex to implicitly collect information inside the games than outside them and to the best of our knowledge it was only done by Yang et al. in [13]. Yang et al. developed a mobile game recommender that was based on discovering of mobile games with similar gameplays taking into account player touch gestures while playing games. Gameplay is the specific way in which players interact with a specific game and the evaluation results from their work showed that touch gestures could serve as robust signatures of gameplay and that their recommender could give accurate recommendations of similar games simply based on users touch gestures. Consequently, common rank lists of the games or social recommendation approaches for discovering new games could be substituted or at least expanded with their system that allows players to search for games with a particular gameplay (e.g. slow-paced games for elderly people).

3   Agent-Based Model of Recommendation System

The model of our recommendation system called MARS - A Multi-Agent Recommendation System for Games on Mobile Phones is shown in Figure 3. In our model we used software agents because of their ability to automate business interactions and negotiate on behalf of their users. The model presented in this paper is a modification of our previous MARS model described in [3] and is based on three main processes: playing games, analyzing game consumption data and recommending games. While users are playing a certain game, User Agents monitor their interaction with the game. User interaction with the game determines the following parameters: user progress, motivation and success. These parameters model user experience and detailed description about how we chose and computed them can be found in [2,3]. After User Agents collect data about user interaction, they send the data to MARS Agent for analysis. Additionally, MARS Agent collects information about games (e.g. game genre, user ratings) from two most popular online stores: Google Play and App Store. The results of the analysis are used to recommend new games to User Agents who can then show these recommendations to users. If users like the recommended game, they can download it from Google Play or App Store and play it on their smartphones.

Data about games and user experience, inputs for MARS recommendation algorithm, is stored into a semantic database to enable logging of relationships among data which can be used later on for advanced reasoning. Based on these inputs, the recommendation algorithm calculates the list of games to be recommended specifically for each user. In order to collect the inputs, User Agents capture and send information about different events during playing of the game to MARS Agent: time moments when the whole game was started/finished and when different levels were started/finished. Additionally, MARS Agent collects information about games from different content providers (Google Play and App Store) and then calculates game recommendation lists for each user based on genres and difficulties of available games, the user motivation to play the game, user success in the game and progress that she/he is making in the game.

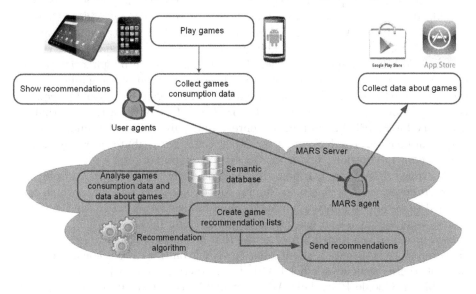

Fig. 1. The model of MARS recommendation system

4   Collecting Inputs for MARS Recommender

In order to collect data describing user experience, we had to monitor gameplay and based on user interactions within the game conclude how much she/he is motivated to play the game, how successful she/he is and whether she/he is making any progress. Moreover, we had to collect information about games from content providers. In this section we will describe how we collect information needed for our recommendation algorithm on two most popular mobile platforms - Android and iOS and its content stores Google Play and App Store. We want to recommend games regardless of the used platform, i.e. we want to use collected information from both platforms in order to produce game recommendations. To the best of our knowledge, this is a novel approach because currently game recommendations are performed only within the same platform.

4.1   Collecting Inputs Outside the Games

Outside the games, we collect information from Google Play and App Store. The recommendation algorithm that was previously proposed in [2] was based on information about genres and difficulties of the games. Both Google Play and App Store provide information about game genres, but it is not possible to collect information about game difficulties. Therefore, we needed to adapt our recommendation algorithm to give recommendations using the available information about games: e.g. game author, device platform, game genre, game price and rating. In this section we listed all available information about games from Google Play and App Store, and in Section 5 we discuss changes in recommendation algorithm that had to be done.

All games from App Store are publicly available on iTunes, a media library developed by Apple that lists different kinds of content[3] (e.g. videos, music, games, magazines). When using iTunes library, MARS Agent can find all available games for iOS platform. To find information in such a way that can be easily stored in MARS semantic database, iTunes lookup service by game *id* can be used. When opening game description in iTunes library, *id* of the game is written in the URL. For example, for Candy Crush Saga game, the lookup request is: http://itunes.apple.com/lookup?id=553834731. One of the attributes that is necessary for MARS functioning is genres, and it lists all genres under which game can be categorized. Some of the other available attributes are: screenshot URLs, game description, minimum iOS version supported, etc.

On Google Play users can find all games officially available for Android devices. However, there is no service similar to lookup service on iTunes which can be used to obtain all information about games in such a way that can be easily stored. Google Play is optimized to provide an insight into applications of a registered user, and recommends new games that are compatible with user devices. However, since information about games available on this store is considered important, there exists an unofficial API[4] for extracting information about games available on Google Play. This API is available as a Java library and can be used in every Java application. Some of the available attributes are game genres, average user rating, user rating count, file size, etc.

Since we want to analyze game data, it was necessary to store it into MARS semantic database. Some of the attributes we consider important and necessary for our system functioning are: author ID, author name, price, currency, genre, name, average rating, rating count, game ID, description, and downloads count - necessary for calculating game similarity; supported devices for determining compatibility with users' device; and screenshots for showing how the game looks like. Since we use a semantic database, we also stored data that is specific for a certain platform, e.g. minimum iOS version supported fetched from App Store or a minimum Android version supported fetched from Google Play.

---

[3] List of the games available on App Store is available at the following link:
https://itunes.apple.com/us/genre/ios-games/id6014?mt=8

[4] Unofficial API for downloading information about application on Google Play is available at: https://code.google.com/p/android-market-api/

4.2   Collecting Inputs Inside the Games

Inside the games we collect information about user experience that is in correlation with user involvement in games: time spent in playing each level of the game, the result of playing a certain level (weather it was successfully completed or not), and total time spent in playing the game. Monitoring of the aforementioned parameters can be easily implemented into a new developed game, like the one mentioned in [3]. However, implementing monitoring into already existing games proved to be quite a challenging task.

The easiest way would be to monitor game parameters from another application. However, on iOS it is impossible to monitor an application from external one since each application has its *sandbox*[5] in which it stores its local settings and data that are not accessible from other applications. On the other hand, this approach is possible to be implemented on Android OS by registering events necessary for user monitoring into a log that is then analyzed by User Agent. However, the problem with this approach is that logging has to be implemented into all games. This is possible for open source games, but the downside of this approach is that monitoring could not be generalized for every game because every game is unique in a sense that different methods are called to create new levels, or that user level results cannot be captured uniformly for all games.

Since we concluded that monitoring of existing games would be very demanding and almost impossible to achieve after the game is developed, we wanted to explore if monitoring could be implemented during the development of the game. Games can be developed by using Android or iOS API, but such games are usually very limited, with limited graphic options and playability. To enhance user experience by providing more powerful visual features and making games more interesting to users, game engines and frameworks can be used [14]. They simplify game development process because they contain various functions which can be used for scene construction, animations, etc. One of the most popular game engines is Unity3D[6] that is available for both Android and iOS platforms. This engine is commercial and its usage is charged, but there also frameworks like Corona SDK[7] and libGDX[8] that are free.

Our idea was to modify game frameworks by adding User Agent that would send event information (e.g. time at which some level was started/finished) to MARS Agent. We used specialized frameworks for each platform: Andengine[9] and Cocos2D[10]. However, this approach also had aggravating factors. For example, the same method can be called when starting a new level or when only

---

[5] Sandbox is an environment in which every application has limited access to data and settings within OS, network resources, etc.

[6] Unity3D game engine is available at: `http://unity3d.com`

[7] Corona SDK framework is available at:
   `http://coronalabs.com/products/corona-sdk`

[8] libGDX framework is available at: `http://libgdx.badlogicgames.com`

[9] Andengine for Android available at: `http://andengine.org`

[10] Cocos2D for iOS available at: `http://www.cocos2d-iphone.org`

changing the scenario within the same level. Thus, we concluded that we cannot implement User Agent functionalities into the existing game frameworks.

This problem could be solved if there was a way to officially define and standardize that a certain method has to be called when a new level is started or finished. However, in order to achieve that, the consent of framework developers and users would be needed. By trying out different methods, we concluded that currently it is impossible to implement User Agent functionalities in existing game engines or frameworks to enable efficient gameplay monitoring. Nevertheless, it is possible to implement our model in the existing games by integrating User Agent functionalities within the game code. In our case we choose Logic game for Android and Climbers game for iOS. In these games, User Agent is implemented as a separate class that consists of five main methods: *levelStarted*, *levelCompleted*, *levelFailed*, *gameStarted* and *gameFinished*. Each of these methods calls the *sendRequest* method that sends a message containing a timestamp of the occurred event to MARS Agent. Additionally, in the message that is sent when level is finished (called by *levelCompleted* and *levelFailed* methods), information whether the level was successfully completed or not is attached.

5   Adaptation of MARS Recommendation Algorithm

MARS Agent collects information about games and events during gameplay for each user and calculates time spent within each level and within the whole game. This information is stored in semantic database on server and used to form user and game profiles. User profiles store information about levels of motivation, progress and success for each user and each game. We use three levels for each parameter: not motivated, averagely motivated and highly motivated; not making progress, making progress and no more room for progress; and successful below the average, within the average or above the average. For instance, user x who plays game y can be seen as averagely motivated, making progress and successful within the average. These parameters are calculated based on data sent by User Agent which are compared to historic data about that game for that user, and historic data about that game for other users. Parameters included in game profiles are filtered from information obtained from Google Play and App Store: genre, game author, description, price and rating. Since the game difficulty parameter, previously proposed in [2], could not be obtained, recommendation algorithm had to be adapted and user and game profiles enhanced in order to be able to function on a wider number of games on mobile platforms.

User profiles store information about user levels of motivation, progress and success for each game she/he played and they are grouped in such a way that for each game genre, user experiences are aggregated so that they present user average motivation, progress and success levels for that genre. User profiles that have the same user experience values are grouped together. On the other hand, games are grouped based on similarities of information stored in their profiles: genre, game author, description, price and rating. For each game, similarity factors with other games are calculated. If two games belong to the same genre, have the same author or rating, and have similar price and words in their description, they are going to have a larger similarity factor than games of different genres/authors/ratings and totally different set of words in descriptions and prices.

Our adapted version of the algorithm uses a hybrid approach to recommend games because it includes both user profiles grouping (i.e. collaborative approach) and game profiles grouping (i.e. content-based approach). Besides user and game profile groups, an input for the algorithm is also user motivation for the game she/he is currently playing. The algorithm then makes a list of games, which the user did not play before, that were played the most by users belonging to the same user profile group. Moreover, the algorithm also makes a second list of games. If user is considered averagely of highly motivated for playing the current game, a game from the same game profile group is put into the list. On the other hand, if she/he is not considered motivated for playing this game, random games from different game profile groups are put into the list.

The algorithm then combines two aforementioned lists into one list from which it randomly chooses k games. Our recommendation list uses information collected from both platforms which is, to the best of our knowledge, a novel approach because currently game recommendations are performed only within a same platform. The produced game list is then filtered so that it contains only games available on user smartphone. Those games are then sent to the User Agent and displayed in the advertisement area of the currently played game.

6 Conclusion

In this paper we presented steps we had to take to enable implementation of Multi-Agent Recommendation System for Games on Mobile Phones (MARS) on most popular mobile platforms - iOS and Android. To integrate MARS system with aforementioned platforms, two types of agents needed to be implemented - User Agent that monitors user experience and MARS Agent that collects data about games, analyses user experience data obtained from User Agent, and creates game recommendation lists.

The most challenging aspect of MARS implementation process was to enable gameplay monitoring. We explored different ways of external application monitoring and came to conclusion that it was not possible to enable logging of interesting events within games without going into the game code. Furthermore, we also tried to implement gameplay monitoring in different frameworks used for game development and concluded that this approach was also not possible without standardization of methods which are to be used for particular events.

Therefore, we implemented MARS gameplay monitoring into only two open-source games: Logic game for Android and Climbers game for iOS. Since one of the parameters important for previous version of MARS algorithm was not possible to be obtained - difficulty of the game, we had to improve the existing algorithm. Another contribution of this paper is the definition of user and game profiles that enable calculating similarities between users and games and which are used as inputs for improved recommendation algorithm. For future work, we plan to implement the adapted version of MARS algorithm with newly defined profiles. After the implementation, the influence of our approach on user satisfaction and downloading recommended games will be tested.

Acknowledgements. We would like to acknowledge support of our students: Marin Glibic, Mladen Kubat, Marko Milos, Marko Pavelic and Stjepko Zrncic.

References

1. Pripuzic, K., Zarko, I., Podobnik, V., Lovrek, I., Cavka, M., Petkovic, I., Stulic, P., Gojceta, M.: Building an IPTV VoD Recommender System: An Experience Report. In: Proceedings of the 12th International Conference on Telecommunications, pp. 155–162 (2013)
2. Skocir, P., Marusic, L., Marusic, M., Petric, A.: The MARS – A Multi-Agent Recommendation System for Games on Mobile Phones. In: Jezic, G., Kusek, M., Nguyen, N.-T., Howlett, R.J., Jain, L.C. (eds.) KES-AMSTA 2012. LNCS, vol. 7327, pp. 104–113. Springer, Heidelberg (2012)
3. Skocir, P., Jezic, G.: A Multi-Agent System for Games Trading on B2B Market Based on Users Skills and Preferences. In: Advanced Methods and Technologies for Agent and Multi-Agent Systems, pp. 303–312 (2013)
4. Oh, J., Kim, D., Lee, U., Lee, J.G., Song, J.: Facilitating Developer-user Interactions with Mobile App Review Digests. In: Proceedings of Extended Abstracts on Human Factors in Computing Systems, pp. 1809–1814 (2013)
5. Medler, B.: Using Recommendation Systems to Adapt Gameplay. In: Discoveries in Gaming and Computer-Mediated Simulations: New Interdisciplinary Applications, pp. 64–77 (2011)
6. Adomavicius, G., Tuzhilin, A.: Toward the Next Generation of Recommender Systems: A Survey of the State-of-the-art and Possible Extensions. IEEE Transactions on Knowledge and Data Engineering 17, 734–749 (2005)
7. Roh, J.H., Jin, S.: Personalized Advertisement Recommendation System Based on User Profile in the Smart Phone. In: Proceedings of the 14th International Conference on Advanced Communication Technology, pp. 1300–1303 (2012)
8. Jannach, D., Hegelich, K.: A Case Study on the Effectiveness of Recommendations in the Mobile Internet. In: Proceedings of the Third ACM Conference on Recommender Systems, pp. 205–208 (2009)
9. Costa-Montenegro, E., Barragans-Martinez, A., Rey-Lopez, M., Mikic-Fonte, F., Peleteiro-Ramallo, A.: Which App? A recommender System of Applications in Markets by Monitoring Users' Interaction. In: Proceedings of the International Conference on Consumer Electronics, pp. 353–354 (2011)
10. Sorensen, D.R.: Recommending Games on Twitter. Master's thesis, The Royal School of Library and Information Science, Copenhagen, Denmark (2013)
11. Shin, C., Hong, J.H., Dey, A.K.: Understanding and Prediction of Mobile Application Usage for Smart Phones. In: Proceedings of the ACM Conference on Ubiquitous Computing, pp. 173–182 (2012)
12. Böhmer, M., Hecht, B., Schöning, J., Krüger, A., Bauer, G.: Falling Asleep with Angry Birds, Facebook and Kindle: A Large Scale Study on Mobile Application Usage. In: Proceedings of the 13th International Conference on Human Computer Interaction with Mobile Devices and Services, pp. 47–56 (2011)
13. Yang, H.T., Chen, D.Y., Hong, Y.X., Chen, K.T.: Mobile Game Recommendation Using Touch Gestures. In: Proceedings of NetGames, pp. 1–6 (2013)
14. Gregory, J.: Game Engine Architecture. Taylor & Francis (2009)

Combination of Textual and Structural Context for Retrieving Multimedia Elements

Sana Fakhfakh, Mohamed Tmar, and Walid Mahdi

Laboratory MIRACL, Institute of Computer Science and Multimedia of Sfax,
Sfax University, Tunisia
sanafakhfakh@yahoo.fr, {mohamed.tmar,walid.mahdi}@isimsf.rnu.tn
http://www.miracl.rnu.tn

Abstract. We study in this paper the combination of textual and structural context for multimedia retrieval in XML (Extensible Markup Language) document. We propose a geometric method that use implicitly of textual and structural context of XML elements and we are particularly interested to improve the effectiveness of various structural factors for multimedia retrieval. Experimental evaluation is carried out using the INEX Ad Hoc Task 2007 and the ImageCLEF Wikipedia Retrieval Task 2010.

Keywords: Geometric distance, textual context, multimedia retrieval, XML element, structure.

1 Introduction

The goal of an Multimedia Information Retrieval (MIR) is to retrieve relevant information to a users query. Following the transition of using plain-text documents to using semi-structured documents (HTML or XML format), together with multimedia contents (photos, music, videos, etc) are justified by quick change of scopes of application in the internet. However the user often fails, to find the best information in the context of his multimedia need. XML (Extensible Markup Language) document includes textual element and multimedia element such as image, audio and video. These elements are organized according to structure which includes information notably although there is not only one manner to organize contents. However, the choice of structure depends greatly on context of use of the textual contents. In the literature, there are two main approaches in Multimedia Information Retrieval (MIR): content-based approaches and context-based approaches. Content-based MIR approaches use specific features of low level according to type of media. We can cite for example image retrieval that exploits visual features (color, texture, forms, etc). These methods have proven effective with media "image" in well defined fields such as medical field this is due to requirement for thorough knowledge of distinctive media. This type of research can be applied to only one type of media in system due to lack of semantic representation in media content. Context-based MIR approaches do not depend on type of media in question. Indeed, these methods rely on information

G. Jezic et al. (eds.), *Agent and Multi-Agent Systems: Technologies and Applications,*
Advances in Intelligent Systems and Computing 296,
DOI: 10.1007/978-3-319-07650-8_9, © Springer International Publishing Switzerland 2014

surrounding the multimedia element representing its semantic description. Multimedia retrieval based on textual context is most used, although the structural context remains an obvious source which plays a part paramount in understanding of structured documents. In this paper, we are interested in Context-based MIR techniques, and more precisely in MIR based on textual and structural context in XML documents. Image context is composed all textual information surrounding the image. For retrieve image presentated in Figure 1, we can use text surrounding image such as document title, image name, image caption, etc

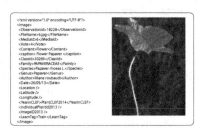

Fig. 1. Example of a multimedia element context

The textual context remains insufficient in most of time. In this context, [2] say: "Ignore the document structure is to ignore its semantics". There are other sources of evidence that were used as visual descriptors, information from link around the image, structure of XML document. Indeed, We focus on XML documents don't have a homogeneous structure. What makes the structure as new source of evidence. The idea is to calculate the relevancy score of media element based on information from the textual and structural context to answer a specific information needs of user, expressed as query composed of set of keywords. And seeking the most appropriate manner to combine two sources of evidence: text and structure. Our main inspiration is to use the structure to involve each textual information depending on its position in XML document, that is textual information that gives the best possible description of multimedia element. The rest of paper is organised as follows. Section 2 presents an overview of existing work in multimedia retrieval. In Section 3, we describe our structural indexing system combining conceptual information for semi-structured documents dedicated to approximate retrieval data. Section 4 and 5 presents the details of our proposed method and the results of her applying on two data sets "INEX 2007" and "ImageCLEF 2010". The last section provides our conclusions and future works.

2   Related Work

The advent of structured documents has caused new problems in information retrieval world, and more specifically in multimedia elements retrieval. These problems are strongly related to nature of these documents that provide the structure as a new source of evidence. Thus, nowadays, XML documents include multimedia elements of different types (audio, video and image)implicitly embedded in the textual elements. These multimedia elements (such as physical objects) do not contain enough information to be able to answer a given query. Therefore, the calculation of relevance score of multimedia element must be linked to textual and structural information provided by other nodes XML. Several works deal XML document as a flat source of information and ignore the structure of XML documents. Indeed, XML document is used to describe a set of data by a structure that provides a semantic lexicon. Thus, it facilitates the presentation of information in terms of interpretation and exploitation. Replying to this need, new works appear in the field of multimedia retrieval that takes in account the structure as source of relevant information. Existing work in structured retrieval of multimedia elements is decomposed in two classes. The first class includes some works which proceed to adopt some traditional technical of retrieval information as language model. In this context, the team $CWI/UTwente$ performs a step of filtering results to keep the fragments containing at least one multimedia element [3]. The second class includes the specific work to be structured multimedia retrieval. This class uses the structure as a source of evidence in the process of selection of multimedia elements. As first step, [1] proposed a method which combines structure of XML document (XPath) with the use of links (XLink). This method consist to divide XML document into regions. Each region represent a area of ancestors of the multimedia element. His score is calculated in function of the scores of each region. This method exploits vertical structure only. In a second time, [4] have used the addition of horizontal structure to the notion of hierarchy. [4] use a method called "CBA" (Children, Brothers, Ancestors), which takes into consideration the information carried by the children , brothers and fathers nodes for calculate the relevance of multimedia elements. The authors propose an alternative method "OntologyLike" which is based on the identification of XML document to ontology. To calculate the similarity between nodes the authors use similarity measures that are mainly based on the number of edges to calculate the distance between nodes. There are other approaches to multimedia retrieval are based on exploitation of links in XML document [5]. This work was improved by proposing a hybrid approach that combines structure with using of links who is consider as semantic links [5]. This method above to divide the document into regions according the hierarchical structure and the location of image in document. This factor plays a role in the weighting of links for compute the score of image. In this paper, we propose a new metric for multimedia retrieval in XML documents which involves the use of geometric distances to calculate the relevance of each node from the multimedia node. This method consists of placing the nodes of XML document in Euclidean space and define each node by a vector of coordinates to calculate then the distance between each pair of nodes. This distance will play a beneficial role in to calculate the score of multimedia element.

3   Indexing System

We propose a indexing system $MXS - index$ composed by two parts: part of textual indexing and part of structural indexing. Our indexing methodology as schematized in Figure 2. The first part consists of four main steps: Pretreatment, term extraction and term weighing using NLP (Natural Language Processing) techniques to extract the candidate XML nodes of the resulting indexing. The first step is to split text into a set of sentences, prune the stop words for each XML node of the corpus and radicalize terms using the algorithm PORTER [6]. The second step is term extraction and the last step is calculating term importance. That is a fundamental step in information retrieval process and it is determined through term frequency (tf) and inverse document frequency(idf). In Second part, we built structural index using information extract from XML tree and geometric metric. Each XML node will presented by characteristic vector. We start by extract geometric proprieties. And we compute coordinates of each XML nodes. This part is accompanied by generating XML data model which processes ancestor, descendant and proximity relationships (Figure 2). The step of selection of descriptors of each node consists in associating each XML node own these textural and structural descriptors to better combine.

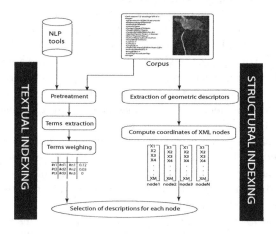

Fig. 2. Architecture of our indexing model $MXS - index$

4   Proposed Approach

We focus on techniques for multimedia retrieval based on textual and structural context in XML documents. XML documents cannot be effectively exploited by classical techniques of IR, which regard document as a bog of words. Therefore, the calculation of relevance score of multimedia element must be linked to textual and structural information provided by other nodes XML [9]. Thus, it facilitates the presentation of information in terms of interpretation and exploitation. Replying to this need, we propose a new method in the field of multimedia retrieval

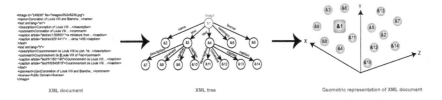

XML document   XML tree   Geometric representation of XML document

Fig. 3. The steps of passing an XML document to geometric representation

that takes into account the structure as a source of evidence and its impact on search performance. We present a new source of evidence dedicated to multimedia retrieval based on the intuition that each textual node contains information that describes semantically a multimedia element. And the participation of each text node in the score of a multimedia element varies with its position in there XML document. To compute the geometric distance, we initially place the nodes of each XML document in an Euclidean space to calculate the coordinates of each node by algorithm 1. Then, we compute the score of a multimedia element depending on the distance between each textual node. Figure 3 shows the steps of passing an XML document to a geometric representation of the XML elements in a Euclidean space. The first step consist to present a XML document as XML tree to take into account XML document properties. An XML tree is described by a set of relationships between nodes. Formally an XML tree is a pair $A = (E, R)$ where E is a set of XML elements and $R \subset E^2$, $((p, q) \in R$ if p is the parent of q) is a set of relations satisfying:

$$\exists! r \in E, \forall q \in E - \{r\}, (r, q) \in R \tag{1}$$

With r is the root of the tree.

$$\forall p \in E - \{r\}, \exists! q \in E, (p, q) \in R \tag{2}$$

Each node has a parent except the root r. In second step, we will spend to presentation of XML tree in a geometric representation. This step is mainly based on equalities extraction in XML tree according to our proposed hypotheses. The XML tree representation allowed us to unveil certain relationships of neighboring, brotherhood and offspring. Indeed, the distance d which separate two or more brothers with their common ancestors iteratively is the same. And brothers of the same hierarchical level are equidistant. These distances are defined according to the relationship of contiguity and semantic similarity between nodes. These distances are not quantized but will be extracted in function of the position of each textual node in XML tree. All these properties result in: For all $q_i = (x_{i1}, x_{i2} \cdots x_{im})$ and $q_j = (x_{j1}, x_{j2} \cdots x_{jm})$ where Q is a set of vectors in $\mathbb{R}^m$.

- In the same hierarchy, if there are more than two brothers then their adjacent nodes are equidistant:

property 1

$$\forall q_i, q_j, q_k \in Q, \; if \; A_1(q_i) = A_1(q_j) = A_1(q_k)$$
$$d(q_i, q_j) = d(q_i, q_k)$$

– The distance between any node and its descendants is the same:
property 2

$$\forall q_i, q_j, q \in Q, n \in \mathbb{N}, A_n(q_i) = A_n(q_j) = q$$
$$d(q_i, q) = d(q_j, q)$$

With $\forall n \in \mathbb{N}^*$, we define function A_n by: $\forall q \in E$,

$$A_n(q) = \begin{cases} \{q\} \; \text{if } n = 0 \\ A_{n-1}(p) \; \text{if } \exists \, p \in \; E, \; (p,q) \in \; R \text{ and } n > 0 \\ \varnothing \; \text{else} \end{cases}$$

From these relationships, we can generate system of equations taking into account for kinship relationships nodes based on hierarchy and adjacency. These relationships are decried by equalities in this order (these equations are only examples)(Figure 3):

$$d(n_1, n_2) = d(n_1, n_3)$$
$$d(n_1, n_2) = d(n_1, n_4)$$
$$d(n_1, n_7) = d(n_1, n_8)$$
$$d(n_1, n_7) = d(n_1, n_9)$$

These distances are defined according to the relationship of contiguity and semantic similarity between nodes. They are not quantized but will be extracted in function of the position of each textual node in the XML tree. The resulting system is nonlinear, its resolution requires the use of an approximate resolution iteratively method where we used iterative solution method (see Algorithm 1). The process begins by assigning to each XML node a random vector it. Tries to improve the coordinate values of each node according to an error value (the sum of the squared deviations). At each iteration, the coordinates are improved together with the minimization of this error. The algorithm stops when the error reaches its minimum value (no improvement is possible). Let Q the set of vectors obtained at a given iteration during the running of the algorithm, the error is defined by:

$$error(Q) = \sum_{\substack{q_i, q_j, q_k \in Q, \\ A_1(q_i) = A_1(q_j) = A_1(q_k)}} (d(q_i, q_j) - d(q_i, q_k))^2$$

$$+ \sum_{\substack{q_i, q_j, q \in Q, n \in \mathbb{N}, \\ A_n(q_i) = A_n(q_j) = q}} (d(q_i, q) - d(q_j, q))^2$$

Algorithm 1. Resolution algorithm approximate nonlinear system of equations

Require: $(Q = (q_1, q_2 \cdots q_{|Q|}), R)$:an XML tree as $q_i = (q_{i1}, q_{i2} \cdots q_{im})$ $\forall i \in [1, |Q|]$ m:dimension

$\quad$**for** $(i, j) \in [1, |Q|]^2$ **do**
$\quad\quad q_{ij} \leftarrow$ random value
$\quad$**end for**
$\quad Q_1 \leftarrow (q_1, q_2 \cdots q_{|Q|})$
$\quad$**repeat**
$\quad\quad P \leftarrow Q_1$
$\quad\quad$**for** $(i, j) \in [1, |Q|]^2$ **do**
$\quad\quad\quad Q_2 \leftarrow (q_1, q_2 \cdots q_{i-1}, q_i + d_j(1), q_{i+1} \cdots q_{|Q|})$
$\quad\quad\quad Q_3 \leftarrow (q_1, q_2 \cdots q_{i-1}, q_i + d_j(\varepsilon), q_{i+1} \cdots q_{|Q|})$
$\quad\quad\quad Q_4 \leftarrow (q_1, q_2 \cdots q_{i-1}, q_i + d_j(1 - \varepsilon), q_{i+1} \cdots q_{|Q|})$
$\quad\quad\quad t \leftarrow 0$
$\quad\quad\quad$**while** $error(Q_1) > error(Q_2) > error(Q_3) > error(Q_4)$ **do**
$\quad\quad\quad\quad Q_4 = (q_1, q_2 \cdots q_{i-1}, q_i + 2^t d_j(1), q_{i+1} \cdots q_{|Q|})$
$\quad\quad\quad\quad t = t+1$
$\quad\quad\quad$**end while**
$\quad\quad\quad t \leftarrow 0$
$\quad\quad\quad$**while** $error(Q_1) < error(Q_2) < error(Q_3) < error(Q_4)$ **do**
$\quad\quad\quad\quad Q_1 = (q_1, q_2 \cdots q_{i-1}, q_i - 2^t d_j(1), q_{i+1} \cdots q_{|Q|})$
$\quad\quad\quad\quad t = t+1$
$\quad\quad\quad$**end while**
$\quad\quad\quad$**while** $|error(Q_1) - error(Q_2)| > \varepsilon$ **do**
$\quad\quad\quad\quad Q_5 \leftarrow \dfrac{Q_1 + Q_2}{2}$
$\quad\quad\quad\quad$let $Q_5 = (P_1, P_2 \cdots P_{|Q|})$
$\quad\quad\quad\quad$**if** $error(p_1, p_2 \cdots p_{i-1}, p_i - d_j(\varepsilon), p_{i+1} \cdots p_{|Q|}) > error(p_1, p_2 \cdots p_{i-1}, p_i + d_j(\varepsilon), p_{i+1} \cdots p_{|Q|})$
$\quad\quad\quad\quad$**then**
$\quad\quad\quad\quad\quad Q_1 \leftarrow Q_5$
$\quad\quad\quad\quad$**else**
$\quad\quad\quad\quad\quad Q_2 \leftarrow Q_5$
$\quad\quad\quad\quad$**end if**
$\quad\quad\quad$**end while**
$\quad\quad$**end for**
$\quad$**until** $P = Q_1$

Where m is the dimension of the Euclidean space and $\forall v \in \mathbb{R}$, $D_j(v) = (d_1, d_2 \cdots d_m)$ is such as:

$$d_k = \begin{cases} 0 \ if \ k \neq j \\ v \ otherwise \end{cases}$$

4.1 Multimedia Element Representation by Textual and Structural Context

A multimedia element (eg *image*) does not contain textual content. Its score is based on textual nodes in its neighborhood. The transition from the XML tree structure representation of elements in an Euclidean space, where we exploit the dissimilarity distances separating a multimedia node and other textual nodes, is performed by extracting the equations satisfying the properties defined earlier and the application of algorithm 1. To calculate the distance between a node n and multimedia element H, we will try to use several geometric distances such as Manhattan distance, Euclidean distance and Minkowski distance between their respective feature vectors q_n and q_H described by the following equations:

$$dist_{Manhattan}(n, H) = \sum_{i=1}^{m} | q_n - q_H | \qquad (3)$$

$$dist_{Euclidean}(n, H) = \sqrt{\sum_{i=1}^{m}(q_n - q_H)^2} \tag{4}$$

$$dist_{Minkowski}(n, H) = \sqrt[p]{\sum_{i=1}^{m}| q_n - q_H |^p} \tag{5}$$

With m is the dimension of the Euclidean space and $p=1$. q_n is defined by: $q_n=(xn_{i1}, xn_{i2} ... xn_{im})$ with xn are the vector characteristics of node n. And q_H is defined by: $q_H =(xH_{i1}, xH_{i2} ... xH_{im}$ with xH represent the coordinates compose the vector characteristics of a node H. We calculate the score for each textual node depending on the frequency of each term (tf) and the number of elements in the corpus according to the number of elements containing the term (idf). A textual node is presented by: $n = (n_1, n_2 \cdots n_{|v|})$ where n_i is the weight of the term t_i, v is the set of indexing terms:

With
$$n_i = tf(t_i, n) \times idf(t_i) \tag{6}$$

$$idf(t_i) = log(\frac{N}{N_i}) \tag{7}$$

Where N is the total number of XML elements in the corpus, N_i is the number of elements that contain the term t_i and $tf(t_i, n)$ is the frequency of the term t_i in node n. The score of textual node depends on the weight of each indexing term. A query is made by the list $v = (v_1, v_2 \cdots v_{|v|})$ where $v_i \in \{0, 1\}$ (0:not exist, 1:exist) according membership t_i at the query. The score of textual node n for the query q is defined by:

$$rsv(q, n) = q \times n^T = \sum_{i=1}^{|V|} q_i \times n_i \tag{8}$$

Where μ is the set of textual elements. The score of multimedia node H is defined by:

$$rsv(q, H) = \sum_{n \in \mu} \frac{rsv(q, n)}{dist(n, H)} \tag{9}$$

With $dist(n, H)$ is the distance (Manhattan distance or Euclidean distance or Minkowski distance) between the feature vectors corresponding to the nodes n and H. This equation leads to assign the importance of contribution of all nodes in computing the score of multimedia element that shows its beneficial impact in multimedia retrieval.

5 Evaluation and Results

We evaluate our system into two databases extracted from two collections : INEX 2007 (Initiative for the Evaluation of XML Retrieval) Ad Hoc task [7] and ImageCLEF 2010 Wikipedia image retrieval task [8]. These databases are composed by XML documents extracted from Wikipedia (Table 1). The aim of the experiments in this section is to show the effectiveness of XML structure in multimedia retrieval. For this purpose, we evaluated separately the use of textual context only (**TC**), as well as the combination of the two (**TC and TS**). For INEX 2007 and ImageCLEF 2010 test set, we respectively obtain the following MAP values: 0.2376 and 0.1674 using textual context only. We compare between the use of Manhattan distance, Euclidean distance and Minkowski distance. We observed that the difference of results between the three distances is very signicant in the INEX 2007 test set and ImageCLEF 2010 test set. The Euclidean distance gets a most suitable representation of multimedia element which is none other than the dissimilarity distance between XML nodes. Indeed, the evaluation results show that this distance provides a MAP which is equal to 0.2572 as MAP with using "ImageCLEF 2010" collection. The result has been improved significantly with the "INEX 2007" collection to 0.3102 as MAP relative to Manhattan distance (0.2376 for "INEX 2007" collection and 0.1754 for "ImageCLEF 2010" collection) and Minkowski distance (0.2876 for "INEX 2007" collection for and 0.2245 for "ImageCLEF 2010" collection). This increase is due to nature of "INEX 2007" collection who includes XML documents with

Table 1. INEX 2007 and ImageCLEF 2010 collections

Company	INEX 2007	ImageCLEF 2010
Task	Collection XML Ad Hoc	Wikipedia Retrieval
Number of XML document	659,388	237,434
Number of image	246,730	237,434
Topics	19	70

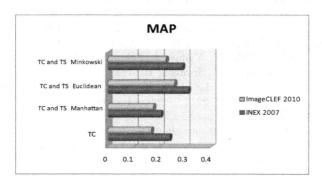

Fig. 4. Results of the impact our approach on INEX 2007 and ImageCLEF 2010 based in *MAP*(Mean Average Precision)

heterogeneous structure. So in "INEX 2007" collection we find documents with high depth. This factor highlights structural information and amplifies effect textual information based on computed distances . For against, our system is more stable with "ImageCLEF 2010" collection, this is due to rapid convergence of results. With our measure, we have shown that combined use of textual and structural context can properly determine the relevance of multimedia element, and the structure plays a primordial role in multimedia retrieval (Figure4).

6   Conclusions

This approach allowed us to calculate the score of element multimedia according the textual context provided by nodes in proximity and structural context from distance between nodes and multimedia element. This method was evaluated with using of two collections "INEX 2007" and "ImageCLEF 2010". In this work, we studied the impact of textual and structural context on multimedia element retrieval, where the user need can be a multimedia element (text). We plan to investigate the impact of a mixture of text and multimedia element(text+image) with to using visual descriptors.

References

1. Kong, Z., Lalmas, M.: XML Multimedia Retrieval. In: Consens, M.P., Navarro, G. (eds.) SPIRE 2005. LNCS, vol. 3772, pp. 218–223. Springer, Heidelberg (2005)
2. Schlieder, T., Holger, M.: Querying and Ranking XML Documents. Journal of the American Society for Information Science and Technolog 53, 489–503 (2002)
3. Westerveld, T., Rode, H., van Os, R., Hiemstra, D., Ramírez, G., Mihajlović, V., de Vries, A.P.: Evaluating Structured Information Retrieval and Multimedia Retrieval Using PF/Tijah. In: Fuhr, N., Lalmas, M., Trotman, A. (eds.) INEX 2006. LNCS, vol. 4518, pp. 104–114. Springer, Heidelberg (2007)
4. Torjmen, M., Pinel-Sauvagnat, K., Boughanem, M.: Using textual and structural context for searching Multimedia Elements. IJBIDM 5, 323–352 (2010)
5. Aouadi, H., Torjmen, M., Ben Jemaa, M.: Combination of document structure and links for multimedia object retrieval. J. Information Science 38, 442–458 (2012)
6. Porter, M.: An Algorithm for Suffix Stripping, pp. 130–137 (1980)
7. Fuhr, N., Kamps, J., Lalmas, M., Malik, S., Trotman, A.: Overview of the INEX 2007 Ad Hoc Track. In: Fuhr, N., Kamps, J., Lalmas, M., Trotman, A. (eds.) INEX 2007. LNCS, vol. 4862, pp. 1–23. Springer, Heidelberg (2008)
8. Popescu, A., Theodora, T., Kludas, J.: Overview of the Wikipedia Retrieval Task at Image CLEF 2010. In: CLEF 2010 LABs and Workshops, CLEF 2010 (2010)
9. Hliaoutakis, A., Varelas, G., Voutsakis, E., Euripides, G.M.: Information Retrieval by Semantic Similarity. International Journal on Semantic Web and Information Systems (IJSWIS), 55–73 (2006), Special Issue of Multimedia Semantics

Multi-agent Non-linear Temporal Logic with Embodied Agent Describing Uncertainty

Vladimir Rybakov[1,2]

[1] School of Computing, Mathematics and DT, Manchester Metropolitan University,
John Dalton Building, Chester Street, Manchester M1 5GD, U.K
[2] Siberian Federal University, 79 Svobodny Prospect, Krasnoyarsk, 660041, Russia
V.Rybakov@mmu.ac.uk

Abstract. We study multi-agent non-linear temporal Logic $\mathbf{T_{Kn}^{Em,Int}}$ with embodied agent. Our approach models interaction of the agents and various aspects for computation of uncertainty in multi-agent environment. We construct algorithms for verification satisfiability and truth statements in the logic $\mathbf{T_{Kn}^{Em,Int}}$. Found computational algorithms are based at refutability of rules in reduced form at special finite frames of effectively bounded size. We show that our chosen framework is rather flexible and it allows to express various approaches to uncertainty and formalizing meaning of the embodied agent.

Keywords: multi-agent logic, interacting agents, non-linear temporal logic, embodied agent.

1 Introduction

This paper primarily deals with models for computational logic of multi-agent systems. In general, multi-agent systems (MAS) are collections of problem-solving entities that work together upon their environment for achieving both their individual goals and their joint goals. Development and modeling MAS may integrate many technologies and concepts from artificial intelligence (AI), CS, IT, Mathematics and other areas of computing as well as other disciplines. It is widely accepted nowadays that computational logic provides a well-defined, general, and rigorous framework for studying the syntax, semantics and procedures for the various tasks in individual agents, as well as the interaction between, and integration amongst, agents in multi-agent systems. Background of such a logic is usually multi-modal (or temporal) logic using for modeling agent knowledge modal operations K_i. In particular, this approach was usefully implemented in analysis of common and distributed agent's knowledge. A collection of summarized to 1996 research outputs may be found in e.g. Fagin et al [10]. Tools of this technique take issue in multi-agent epistemic logic. They help to describe the properties (specifications) with explicit, mathematically preciseness, which simplifies identification.

These techniques use logical languages for reasoning about agent's knowledge and properties (e.g. various technique of mathematical (symbolic) logic is widely

G. Jezic et al. (eds.), *Agent and Multi-Agent Systems: Technologies and Applications*,
Advances in Intelligent Systems and Computing 296,
DOI: 10.1007/978-3-319-07650-8_10, © Springer International Publishing Switzerland 2014

used (cf. [12, 13]); in particular, multi-agent modal logics were implemented. Logical language is turned out to be indeed useful for these aims, cf. for a summary, – Wooldridge, 2000, [33].

Initiation of usage of logical language in knowledge representation may also be referred to e.g. Brachman and Schmolze (1985, [7]), Moses and Shoham (1993, [14]), Nebel (1990, [16]), Quantz and Schmits (1994, [21]), Rychtycki (1996, [32]).

Though technique and research outputs in MAS are various, diverse and work well in many contemporary areas, it seems, most popular area is applications in IT, – cf. Nguyen et al [18–20], Arisha et al [1], Avouris [2], Hendler [11]. Nonetheless, pure theoretical research for logic of MAS is also very popular. In particular, it was connected with attempts to clearly formalize what is a shared knowledge and what is a common knowledge It seems, first ideas concerning these problems appeared in Barwise (1988, [8]), Niegerand and Tuttle (1993, [17]), Dvorek and Moses (1990, [9]). Since a time, an approach to common knowledge logics in multi-modal framework was summarized in the book Fagin R., Halpern J., Moses Y., Vardi M. (1995, [10]).

In modeling of multi-agents reasoning an important question is how to represent interaction of agents, exchange of information (cf. e.g., Sakama et al [22]). Study of multi-modal agents logics and temporal agents-logics, representing these features, were undertaken in a series of works of the author. A kernel part in these works was representation the case when the logics describe interacting agents. In Rybakov, 2009, [28] some technique to handle interactions was found, and, as a consequence, it was proved that the multi-agent Linear Temporal Logic (with UNTIL and NEXT and with interacting agents, or dually, common knowledge) is decidable (in particular, – that the satisfiability problem for this logic is also decidable) and some algorithms solving the problem were found (cf. also Rybakov [27]). Besides, research of just multi-agent logics (as modal and temporal) with aim to find solution of satisfiability problem was earlier undertaken in Rybakov [29–31], Babenyshev and Rybakov [3–6]. Recently solution for satisfiability problem in non-linear temporal logic with only interacting agents was found in McLean and Rybakov [15].

The current paper considers multi-agent non-linear temporal logic $\mathbf{T}_{\mathbf{Kn}}^{\mathbf{Em,Int}}$ with embodied agent. Here we model interaction of the agents and various aspects for computation of uncertainty in multi-agent environment. Paper suggests an algorithm for verification satisfiability and truth statements in the logic $\mathbf{T}_{\mathbf{Kn}}^{\mathbf{Em,Int}}$. We show that our chosen framework is rather flexible and it allows to express various approaches to uncertainty and formalizing meaning of the embodied agent.

2   Background, Basic Notation, Definitions, Preliminaries

The viewpoint on essence of an embodied agent might be diverse, depending on chosen model and intended implementations. Though rather common view is that an embodied agent is an interface agent: an intelligent agent that interacts with the environment through a physical body within that environment.

We would like to model this understanding by semantics based at logical Kripke-Hintikka like models representing branching time including (standard but interacting) agents (it is also a point of novelty here). The second aim is to represent in this framework the conception of uncertainty via agent's interaction and embodied agent.

The basic background idea of our representation is: we have a web network with local cluster of web-network connections available for a local admin (embodied agent, interface agent), yet we have a whole network of connections, represented by web links (we interpret links forward as the time). To follow this line we start from description of a symbolic model for such representation.

The models for our semantics are based upon standard models for branching time with new subsidiary operations. In more details, Kripke/Hintikka-like frame $\mathbf{F}$ in our approach is a model $\mathbf{F} := \langle W, R, R_e, R_1, \ldots, R_n \rangle$, where W is a (base) set of states (worlds, which model web sites). Properties and essence of the operations in this frame are described below.

Essence of the Binary Operations in Frames

The relation R is a binary relation on W (Time-relation, it models, for example, web connections, or runs of computations. Then aRb means that there is a web-connection from state a to state b (e.g. by clicking link buttons, some amount of steps in a computational procedure, etc.)). We view at R as time; it is assumed to be reflexive and transitive (which corresponds well with (i) standard understanding of time in a run of a computation, and (ii) models transitions in runs of computations, (iii) passing via web connections, etc). Formally we may fix this by laws laid upon the frame: $\forall a \in W,\ aRa;\quad \forall a, b, c \in W,\ aRb\ \&\ bRc \Rightarrow aRc$. The states from $\mathbf{F}$ – symbols from W – form with respect to R clusters. A cluster $C(a)$ generated by $a \in W$ is the set $\{b \mid b \in W, aRb\ \&\ bRa\ \}$.

The choice of R to be non-linear corresponds well to usual structures of web networks, where links (connections) depend both on time and possible choice of one of available links (illustrated as branching time) and present physical structure of network (hardware) (so transaction overall look as non-linear routs).

The relation R_e is a binary relation on W for an embodied agent: interface relation. We assume R_e to be the equivalence relations on any $C(a)$, where

$$\forall b, c \in C(a)(bR_ec);\ \text{ in particular }\ \forall b, c \in C(a)\forall i[(bR_ic) \Rightarrow (bR_ec)].$$

The background for this definition is that R_e is the relation for the embodied agent: interface relation. That is the interface agent may achieve via web links any web site in the zone of its responsibility (within $C(a)$). And the links within this zone are reversible, it may do any backtrack. This, it seems, corresponds very well with standard understanding of local admin in a network. Next, any relation R_i (agent i accessibility relation) is reflexive, transitive and symmetric relation (i.e. $aRb \Rightarrow bRa$) on $C(a)$ for any $a \in W$. It corresponds one-to one with definitions of R_i in standard multi-agent model with autonomous agents.

The interpretation of agent's accessibility relation R_i, – via eg. internet connections, – is as follows: being logged at a web-site a, i-agent may access by R_i some other web sites from the cluster $C(a)$ (in accordance with possession of access rules/passwords) - and switch between sites in its disposal freely, back and forth. Yet i cannot jump to another sites outside $C(a)$ without permitting (convoy) from administrator. Say, also we may interpret relations R_i as computational runs: there are several computational threads imitated as relations R_i – any thread is a computational agent, while the relation R_e holds a cluster of local computations around an time tick.

To express and elicit information which might be collected and computed via this framework we will use language with syntax based at a hybrid of a non-linear temporal logic and some multi-agent logic with new subsidiary logical operations. Language of our logic consists of standard language of Boolean logic extended with temporal and agent knowledge operations. So, it contains potentially infinite set of propositional letters P. Its logical operations include standard Boolean logical operations and usual unary agent knowledge operations K_i, $1 \leq i \leq m$ and the unary operation K_e - for knowledge of the embodied agent. It, as well as in [15], contains the operation for knowledge via agent's interaction **KnI**. This operation is a dual counterpart of the common knowledge operation introduced, e.g. in Fagin et al [10]) for common knowledge in multi-modal logic. The language also contains the unary logical operation U with meaning 'uncertain'.

Later on we will introduce some more logical operations definable in the chosen language. To express dynamics of current processes we also directly use unary temporary operations $\mathbf{P}^+$ (with meaning 'possible in future' by a sequence of computational steps) and $\mathbf{P}^-$ (with meaning *possible, so to say in past, – by a sequence of backtracks*). The formation rules for formulas are standard: any propositional letter is a formula,

(i) if α and β are formulas, then $\alpha \wedge \beta$ is a formula;
(ii) if α and β are formulas, then $\alpha \vee \beta$ is a formula;
(iii) if α and β are formulas, then $\alpha \to \beta$ is a formula;
(iv) if α is a formula, then $\mathbf{P}^+\alpha$ is a formula;
(v) if α is a formula, then $\mathbf{P}^-\alpha$ is a formula;
(vi) if α is a formula, then for any i (and for $i = e$) $K_i\alpha$ is a formula;
(vii) if α is a formula, then $\mathbf{KnI}\alpha$ is a formula;
(vii) if α is a formula, then $\mathrm{U}\alpha$ is a formula.

Accepted meaning for these operations is as follows. $K_i\varphi$ means: the agent i knows φ in the current state; $\mathbf{P}^+\varphi$ says that there is a state (web site) b accessible from the current state a by a sequence of links, were the statement (formula) φ is true at b. So to say, there is a state, accessible in future, where φ is true. $\mathbf{P}^-\varphi$ means that there is a state b accessible from the current state a by a sequence of backtracks, were the statement (formula) φ is true at b.

In own turns, $\mathbf{KnI}\varphi$ means: in the current state, the statement φ *may be known by interaction between agents*. $\mathrm{U}\varphi$ has meaning the statement φ is uncertain (has uncertain truth value).

Computational Rules

Computational rules for truth of compound formulas (statements) are as follows. Assume we have given a frame $\mathbf{F} := \langle W, R, R_e, R_1, \ldots, R_n \rangle$, a set of propositional letters P and a valuation V of P in $\mathbf{F}$ which is a mapping of P into the set of all subsets of the set W. Thus, $\forall p \in P, V(p) \subseteq W$. If, for an element $a \in W$, $a \in V(p)$ we say that the statement p is true in the state a. In the notation below $(\mathbf{F}, a) \Vdash_V \varphi$ is meant to say the formula φ in true at the state a in the model $\mathbf{F}$ w.r.t. the valuation V. The rules for computation of truth values of compound formulas are given below:

$$\forall p \in P, \ \forall a \in W \ \ (\mathbf{F}, a) \Vdash_V p \iff a \in V(p);$$

$$(\mathbf{F}, a) \Vdash_V \varphi \wedge \psi \iff [(\mathbf{F}, a) \Vdash_V \varphi \text{ and } [(\mathbf{F}, a) \Vdash_V \psi];$$

$$(\mathbf{F}, a) \Vdash_V \varphi \vee \psi \iff [(\mathbf{F}, a) \Vdash_V \varphi \text{ or } [(\mathbf{F}, a) \Vdash_V \psi];$$

$$(\mathbf{F}, a) \Vdash_V \varphi \rightarrow \psi \iff [not[(\mathbf{F}, a) \Vdash_V \varphi] \text{ or } [(\mathbf{F}, a) \Vdash_V \psi];$$

$$(\mathbf{F}, a) \Vdash_V \neg\varphi \iff \text{ not } [(\mathbf{F}_C, a) \Vdash_V \varphi];$$

$$(\mathbf{F}, a) \Vdash_V K_i\varphi \iff (and \ for \ i = e \ )\forall b \in W[(aR_ib) \implies (\mathbf{F}, b) \Vdash_V \varphi];$$

$$(\mathbf{F}, a) \Vdash_V \mathbf{P}^+\varphi \iff \exists b \in W \ [(aRb) \text{ and } (\mathbf{F}, b) \Vdash_V \varphi];$$

$$(\mathbf{F}, a) \Vdash_V \mathbf{P}^-\varphi \iff \exists b \in W \ [(bRa) \text{ and } (\mathbf{F}, b) \Vdash_V \varphi];$$

$$(\mathbf{F}, a) \Vdash_V \mathbf{KnI}\varphi \Leftrightarrow \exists a_{i1}, a_{i2}, \ldots, a_{ik} \in W$$

$$[aR_{i1}a_{i1}R_{i2}a_{i2} \ldots R_{ik}a_{ik}]\&(\mathbf{F}, a_{ik}) \Vdash_V \varphi];$$

$$(\mathbf{F}, a) \Vdash_V \mathbf{U}\varphi \iff [(\mathbf{F}, a) \Vdash_V \mathbf{KnI}\varphi \text{ and } (\mathbf{F}, a) \Vdash_V \mathbf{KnI}\neg\varphi];$$

So, as in [15], we assume that a statement φ has the uncertain truth value in the current world (state) if agents may, passing to each other information, conclude that φ might be true in some state of the current environment, but that φ can also be false in some state.

Approach Using Embodied Agent and Other Variations of Uncertainty

Another understanding of uncertainty might be given via embodied agent:

$$(\mathbf{F}, a) \Vdash_V \mathbf{U}\varphi \iff [(\mathbf{F}, a) \Vdash_V K_e\varphi \text{ and } (\mathbf{F}, a) \Vdash_V K_e\neg\varphi].$$

That is we think that φ is uncertain if the embodied agent may discover that it is somewhere true and somewhere false. It is stronger version of uncertainty comparing with the one suggested above since knowledge on the embodied agent may be bigger than the one for all agents obtained via their interaction. Yet more stronger version of uncertainty may be expressed via possibility to discover contradictory information in both future and past:

$$(\mathbf{F}, a) \Vdash_V \mathbf{U}\varphi \iff [(\mathbf{F}, a) \Vdash_V \mathbf{P}^{Sign_a}\varphi \text{ and } (\mathbf{F}, a) \Vdash_V \mathbf{P}^{Sign_b}\neg\varphi,]$$

where $Sign_a, Sign_b \in \{+, -\}$. That is in this view, φ is uncertain if regardless where - in future or past - this statement might be true and might be false. This is rather strongest version of uncertainty within our accepted model. A variation which is weaker is:

$$(\mathbf{F}, a) \Vdash_V U\varphi \iff [(\mathbf{F}, a) \Vdash_V \mathbf{P}^{Sign_a}(\mathbf{KnI}\varphi \wedge \mathbf{KnI}\neg\varphi)].$$

This is a weaker but more subtle approach - the statement φ is uncertain if somewhere in past or future there is a state where agents via their interaction may discover that it is true and that it is false.

So, the approach we suggest is rather flexible and may express very various views on uncertainty. It is important to say, that definitions for our computation of uncertainty work similarly for all pointed approaches and we may accept any we wish for final postulating our logic.

Now on we would like to point another possible definitions for knowledge of *embodied agent*. We may use:

$$(\mathbf{F}, a) \Vdash_V K_e\varphi \iff [(\mathbf{F}, a) \Vdash_V \mathbf{P}^+\varphi \vee \mathbf{P}^-\neg\varphi].$$

This is rather drastically differs from the one offered earlier, and it interprets knowledge of the embodied agent not as purely knowledge, but as to point that embodied agent definitely may always discover that φ is true in future or otherwise in past. Again, we may accept for our approach this definition as well. Now we need to recall some definitions necessary for the sequel.

Given a model $\mathcal{M} := \langle \mathbf{F}, V \rangle$ based at a frame $\mathbf{F}$ with a base set W and a valuation V, and a formula φ, (i) φ is *satisfiable* in $\mathcal{M}$ (denotation – $\mathcal{M} \Vdash_{Sat}\varphi$) if there is a state b of $\mathcal{M}$ ($b \in W$) where φ is true: $(\mathbf{F}, b) \Vdash_V \varphi$. (ii) φ is *valid* in $\mathcal{M}$ (denotation – $\mathcal{M} \Vdash \varphi$) if, for any b of W, the formula φ is true at b ($(\mathbf{F}, b) \Vdash_V \varphi$) w.r.t. V.

For a frame $\mathbf{F}$ and a formula φ, φ is satisfiable in $\mathbf{F}$ (denotation $\mathbf{F} \Vdash_{Sat}\varphi$) if there is a valuation V in the frame $\mathbf{F}$ such that $\langle \mathbf{F}, V \rangle \Vdash_{Sat}\varphi$. φ is valid in $\mathbf{F}$ (notation $\mathbf{F} \Vdash \varphi$) if $not(\mathbf{F} \Vdash_{Sat}\neg\varphi)$.

Definition 1. *The logic* $\mathbf{T}_{\mathbf{Kn}}^{\mathbf{Em,Int}}$ *is the set of all formulas which are valid in all frames* $\mathbf{F}$ *(i.e. valid at all frames w.r.t. all valuations). A formula* φ *is said to be a theorem of* $\mathbf{T}_{\mathbf{Kn}}^{\mathbf{Em,Int}}$ *if* $\varphi \in \mathbf{T}_{\mathbf{Kn}}^{\mathbf{Em,Int}}$.

We say a formula φ is *satisfiable* iff there is a valuation V in a Kripke frame $\mathbf{F}$ which makes φ satisfiable: $\langle \mathbf{F}, V \rangle \Vdash_{Sat}\varphi$. Clearly, a formula φ is satisfiable iff $\neg\varphi$ is not a theorem of $\mathbf{T}_{\mathbf{Kn}}^{\mathbf{Em,Int}}$: $\neg\varphi \notin \mathbf{T}_{\mathbf{Kn}}^{\mathbf{Em,Int}}$, and vice versa, φ is a theorem of $\mathbf{T}_{\mathbf{Kn}}^{\mathbf{Em,Int}}$ ($\varphi \in \mathbf{T}_{\mathbf{Kn}}^{\mathbf{Em,Int}}$) if $\neg\varphi$ is not satisfiable.

The prime aim of our paper is to find algorithm which may compute satisfiability in this logic and to compute if a statement if logically true - is a theorem. That is a very popular goal in Logic in Computer Science and AI.

3 Computation of Satisfiability and Truth

In this section we will use the approach borrowed from our work [15], which will be very convenient to implement for our case (actually it is just extension to implement embodied agent and new conceptions for uncertainty). The main step we need is transformation of formulas to the ones with no nested modalities at all i.e. - temporal, agents knowledge and other operations, and yet the formula in question to be just a disjunction of conjuncts with only letters, applications of modal-like operations to the letters, or yet their negations. For this, we initially convert formulas to rules and then use ready technique. The representation of formulas in such form is necessary to find algorithms (to avoid infinite loops or chains).

To recall notation and definitions, a rule $\mathbf{r}$ is an expression in the form $\mathbf{r} := \frac{\varphi_1(x_1,\ldots,x_n),\ldots,\varphi_l(x_1,\ldots,x_n)}{\psi(x_1,\ldots,x_n)}$. Here the expressions $\varphi_1(x_1,\ldots,x_n),\ldots,\varphi_l(x_1,\ldots,x_n)$ and $\psi(x_1,\ldots,x_n)$ are formulas constructed out of letters $x_1,\ldots,x_n$. The letters $x_1,\ldots,x_n$ are the variables of $\mathbf{r}$, we use the notation $x_i \in Var(\mathbf{r})$. A meaning of a rule $\mathbf{r}$ is that the statement (formula) $\psi(x_1,\ldots,x_n)$ follows from statements (formulas) $\varphi_1(x_1,\ldots,x_n), \ldots, \varphi_l(x_1,\ldots,x_n)$. Recall definition from [15]:

Definition 2. *A rule $\mathbf{r}$ is said to be* valid *in a Kripke model $\langle \mathbf{F}, V \rangle$ (notation $\mathbf{F} \Vdash_V r$) if $[\forall a \ ((\mathbf{F},a) \Vdash_V \bigwedge_{1 \leq i \leq l} \varphi_i)] \Rightarrow \forall a \ ((\mathbf{F},a) \Vdash_V \psi)$. Otherwise we say $\mathbf{r}$ is refuted in $\mathbf{F}$, or refuted in $\mathbf{F}$ by V, and write $\mathbf{F} \nVdash_V \mathbf{r}$. A rule $\mathbf{r}$ is valid in a frame $\mathbf{F}$ (notation $\mathbf{F} \Vdash \mathbf{r}$) if, for any valuation V, $\mathbf{F} \Vdash_V \mathbf{r}$*

For any formula φ we can convert it into the rule $x \to x/\varphi$. Clearly,

Lemma 1. *A formula φ is a theorem of $\mathbf{T}_{\mathbf{Kn}}^{\mathbf{Em,Int}}$ iff the rule $(x \to x/\varphi)$ is valid in any frame $\mathbf{F}$.*

A rule $\mathbf{r}$ is said to be in *reduced normal form* if $\mathbf{r} = \varepsilon/x_1$ where

$$\varepsilon := \bigvee_{1 \leq j \leq l} (\bigwedge_{1 \leq i \leq n} [x_i^{t(j,i,0)} \wedge (\mathbf{P}^+ x_i)^{t(j,i,1)} \wedge (\mathbf{P}^- x_i)^{t(j,i,2)} \wedge (\neg \mathbf{K}_e \neg x_i)^{e_{j,i}} \wedge$$

$$\bigwedge_{1 \leq q \leq n} (\neg \mathbf{K}_q \neg x_i)^{t(j,i,q,1)} \wedge \mathbf{KnI} x_i^{t(j,i,3)} \wedge (\mathbf{U} x_i)^{t(j,i,4)}]),$$

all x_s are certain letters (variables), $t(j,i,z), t(j,i,k,z), e_{j,i} \in \{0,1\}$ and, for any formula α above, $\alpha^0 := \alpha$, $\alpha^1 := \neg \alpha$.

Definition 3. *Given a rule $\mathbf{r_{nf}}$ in reduced normal form, $\mathbf{r_{nf}}$ is said to be a normal reduced form for a rule $\mathbf{r}$ iff, for any frame $\mathbf{F}$, $\mathbf{F} \Vdash \mathbf{r} \Leftrightarrow \mathbf{F} \Vdash \mathbf{r_{nf}}$.*

Theorem 1. *There exists an algorithm running in (single) exponential time, which, for any given rule $\mathbf{r}$, constructs its normal reduced form $\mathbf{r_{nf}}$.*

For readers interested in proof of this statement, cf. Theorem 1 and its proof in [15]. As we know, the decidability of our logic (in particular decidability of the satisfiability problem) will follow (by this theorem) if we find an algorithm recognizing rules in reduced normal form which are valid in all frames $\mathbf{F}$. Very important starting point to implement this technique is to efficiently bound the size of clusters under consideration in order to efficiently define the interaction of agents. As in [15] we will use the same step as been earlier implemented in Lemma 8 in Rybakov [28] for simply linear temporal multi-agent logic.

Lemma 2. *A rule* $\mathbf{r_{nf}}$ *in reduced normal form is refuted in a frame* $\mathbf{F}$ *if and only if* $\mathbf{r_{nf}}$ *can be refuted in a frame with time clusters of size square exponential from* $\mathbf{r_{nf}}$.

If this is curried out, the rest is a standard work using filtration technique and other instruments of non-classical mathematical logic. As result we obtain

Lemma 3. *A rule* $\mathbf{r_{nf}}$ *in reduced normal form is refuted in a frame* $\mathbf{F}$ *iff* $\mathbf{r_{nf}}$ *can be refuted in a finite frame* $\mathbf{F}_1$ *by a valuation* V, *where the size of the frame* $\mathbf{F}_1$ *has effective upper bound computable from the size of* $\mathbf{r_{nf}}$.

Based at Theorem 1, Lemma 1 and Lemma 3 we obtain our main technical result: an algorithm for computation of satisfiability and decidability of our logic.

Theorem 2. *The logic* $\mathbf{T_{Kn}^{Em,Int}}$ *is decidable; the satisfiability problem for logic* $\mathbf{T_{Kn}^{Em,Int}}$ *is decidable.*

Conclusion, Future Work

We investigate multi-agent non-linear temporal Logic $\mathbf{T_{Kn}^{Em,Int}}$ with embedded agent. In suggested framework we model interaction of the agents and various aspects for definition of uncertainty in multi-agent environment. The aim of the paper is to construct algorithms for verification satisfiability and truth statements for $\mathbf{T_{Kn}^{Em,Int}}$. We find computational algorithms based at refutability of rules in reduced from at special finite frames of effectively bounded size. It is shown that our chosen framework is rather flexible and allows to handle various approaches to uncertainty and definitions of the embedded agent.

Future subsequent research may concern various aspects in suggested approach. In particular, pointed technique may be extended to handle more subtle aspects of agents interaction and duties of the embodied agent. E.g. interaction of agents is represented now as just passing information, without considering intermediate conflicts, voting etc. Functions of the embodied agent are also shown as pure universal modality or modal-like operation of kind *possible*. Pure logical problems, as axiomatizability, complexity issues are open up to now. Yet it is interesting to extend our approach to components of fuzzy logic - with numeric values for agents knowledge and believes.

References

1. Arisha, K., Ozcan, F., Ross, R., Subrahmanian, V.S., Eiter, T., Kraus, S.: Impact: A platform for collaborating agents. IEEE Intelligent Systems 14(2), 64–72 (1999)
2. Avouris, N.M.: Co-operation knowledge-based systems for environmental decision-support. Knowledge-Based Systems 8(1), 39–53 (1995)
3. Babenyshev, S., Rybakov, V.: Logic of Plausibility for Discovery in Multi-agent Environment Deciding Algorithms. In: Lovrek, I., Howlett, R.J., Jain, L.C. (eds.) KES 2008, Part III. LNCS (LNAI), vol. 5179, pp. 210–217. Springer, Heidelberg (2008)
4. Babenyshev, S., Rybakov, V.: Decidability of Hybrid Logic with Local Common Knowledge Based on Linear Temporal Logic LTL. In: Beckmann, A., Dimitracopoulos, C., Löwe, B. (eds.) CiE 2008. LNCS, vol. 5028, pp. 32–41. Springer, Heidelberg (2008)
5. Babenyshev, S., Rybakov, V.: Logic of Discovery and Knowledge: Decision Algorithm. In: Lovrek, I., Howlett, R.J., Jain, L.C. (eds.) KES 2008, Part II. LNCS (LNAI), vol. 5178, pp. 711–718. Springer, Heidelberg (2008)
6. Babenyshev, S., Rybakov, V.: Describing Evolutions of Multi-Agent Systems. In: Velásquez, J.D., Ríos, S.A., Howlett, R.J., Jain, L.C. (eds.) KES 2009, Part I. LNCS, vol. 5711, pp. 38–45. Springer, Heidelberg (2009)
7. Brachman, R.J., Schmolze, J.G.: An overview on the KL-ONE knowledge representation system. Cognitive Science 9(2), 179–226 (1985)
8. Barwise, J.: Three Views of Common Knowledge. In: Vardi (ed.) Proc. Second Confeuence on Theoretical Aspects of Reasoning about Knowledge, pp. 365–379. Morgan Kaufmann, San Francisco (1988)
9. Dwork, C., Moses, Y.: Knowledge and Common Knowledge in a Byzantine Environment: Crash Failures. Information and Computation 68(2), 156–183 (1990)
10. Fagin, R., Halpern, J., Moses, Y., Vardi, M.: Reasoning About Knowledge, p. 410. The MNT Press, Cambridge (1995)
11. Hendler, J.: Agents and the semantic web. IEEE Intelligent Systems 16(2), 30–37 (2001)
12. Kifer, M., Lozinski, L.: A Logic for Reasoning with Inconsistency. J. Automated Deduction 9, 171–115 (1992)
13. Kraus, S., Lehmann, D.L.: Knowledge, Belief, and Time. Theoretical Computer Science 98, 143–174 (1988)
14. Moses, Y., Shoham, Y.: Belief and Defeasible Knowledge. Artificial Intelligence 64(2), 299–322 (1993)
15. McLean, D., Rybakov, V.: Multi-Agent Temporary Logic $TS4_{K_n}^U$ Based at Non-linear Time and Imitating Uncertainty via Agents Interaction. In: Rutkowski, L., Korytkowski, M., Scherer, R., Tadeusiewicz, R., Zadeh, L.A., Zurada, J.M. (eds.) ICAISC 2013, Part II. LNCS, vol. 7895, pp. 375–384. Springer, Heidelberg (2013)
16. Nebel, B.: Reasoning and Revision in Hybrid Representation Systems. LNCS, vol. 422. Springer, Heidelberg (1990)
17. Neiger, G., Tuttle, M.R.: Common knowledge and consistent simultaneous coordination. Distributed Computing 5(3), 334–352 (1993)
18. Nguyen, N.T., Jo, G.-S., Howlett, R.J., Jain, L.C. (eds.): KES-AMSTA 2008. LNCS (LNAI), vol. 4953. Springer, Heidelberg (2008)
19. Nguyen, N.T., Huang, D.S.: Knowledge Management for Autonomous Systems and Computational Intelligence. Journal of Universal Computer Science 15(4) (2009)

20. Nguyen, N.T., Katarzyniak, R.: Actions and Social Interactions in Multi-agent Systems. Special issue for International Journal of Knowledge and Information Systems 18(2) (2009)

21. Quantz, J., Schmitz, B.: Knowledge-based disambiguation of machine translation. Minds and Machines 9, 99–97 (1996)

22. Sakama, C., Son, T.C.: Interacting Answer Sets. In: Dix, J., Fisher, M., Novák, P. (eds.) CLIMA X. LNCS, vol. 6214, pp. 122–140. Springer, Heidelberg (2010)

23. Rybakov, V.V.: A Criterion for Admissibility of Rules in the Modal System $S4$ and the Intuitionistic Logic. Algebra and Logic 23(5), 369–384 (1984) (Engl. Translation)

24. Rybakov, V.V.: Admissible Logical Inference Rules. Studies in Logic and the Foundations of Mathematics, vol. 136. Elsevier Sci. Publ., North-Holland (1997) ISBN: 0444895051

25. Rybakov, V.V.: Logical Consecutions in Discrete Linear Temporal Logic. Journal of Symbolic Logic (ASL, USA) 70(4), 1137–1149 (2005)

26. Rybakov, V.: Until-Since Temporal Logic Based on Parallel Time with Common Past. Deciding Algorithms. In: Artemov, S., Nerode, A. (eds.) LFCS 2007. LNCS, vol. 4514, pp. 486–497. Springer, Heidelberg (2007)

27. Rybakov, V.: Logic of knowledge and discovery via interacting agents - Decision algorithm for true and satisfiable statements. Inf. Sci (Elsevier, North-Hollnd – New York) 179(11), 1608–1614 (2009)

28. Rybakov, V.: Linear Temporal Logic LTK_K extended by Multi-Agent Logic K_n with Interacting Agents. J. Log. Comput. 19(6), 989–1017 (2009)

29. Rybakov, V.V.: Representation of Knowledge and Uncertainty in Temporal Logic LTL with Since on Frames Z of Integer Numbers. In: König, A., Dengel, A., Hinkelmann, K., Kise, K., Howlett, R.J., Jain, L.C. (eds.) KES 2011, Part I. LNCS, vol. 6881, pp. 306–315. Springer, Heidelberg (2011)

30. Rybakov, V.V.: Agents' Logics with Common Knowledge and Uncertainty: Unification Problem, Algorithm for Construction Solutions. In: König, A., Dengel, A., Hinkelmann, K., Kise, K., Howlett, R.J., Jain, L.C. (eds.) KES 2011, Part I. LNCS, vol. 6881, pp. 171–179. Springer, Heidelberg (2011)

31. Rybakov, V.: Multi-Agent Logic based on Temporary Logic $TS4_{K_n}$ serving Web Search. In: Grana, M., et al. (eds.) Advances in Knowledge-Based and Intelligent Information and Engineering Systems, KES 2012. Frontiers in Artificial Intelligence and Applications, vol. 243, pp. 108–117 (2012)

32. Rychtyckyi, N.: DLMS: An evaluation of KL-ONE in the automobile industry. In: Aiello, L.C., Doyle, J., Shapiro, S. (eds.) Proc. of the 5-th Int. Conf. on Principles of Knowledge Representation and Reasoning (KR 1996), pp. 588–596. Morgan Kaufmann, San Francisco, Cambridge, Mass (1996)

33. Wooldridge, M.: Reasoning about rational agents. MIT Press (2000)

Modelling Travel Routes in Transport Systems by Means of Timed and Hybrid Petri Nets

František Čapkovič

Institute of Informatics, Slovak Academy of Sciences
Dúbravská cesta 9, 845 07 Bratislava, Slovak Republic
Frantisek.Capkovic@savba.sk
http://www.ui.sav.sk/home/capkovic/capkhome.htm

Abstract. Timed Petri nets (TPN) and first order hybrid Petri nets (FOHPN) are tested here in order to model transport systems and to find the suitable travel routes in different non-standard situations during the increased traffic density (i.e. at the bounded traffic or congestion). This work extends our previous work where the flexible routes in transport systems were found by means of the place/transition Petri nets (P/T PN). While at usage of the TPN only the time parameters are assigned to the P/T PN model transitions, the FOHPN model is different and it yields the continuous flows of vehicles.

Keywords: Agent, cooperation, first order hybrid Petri nets, hybrid Petri nets, place/transition Petri nets, timed Petri nets, transport systems.

1 Introduction and Preliminaries

Place/transition Petri nets (P/T PN) [11, 12] are the effective tool for modelling discrete event systems (DES). However, their extended (modified) version - timed Petri nets (TPN) - are suitable for DES behaviour in time. Hybrid Petri nets (HPN) [6, 8] are suitable for modelling hybrid systems (HS) containing both the continuous part and the discrete one. Simplified version of HPN - the first order HPN (FOHPN) [2–5, 7, 13] - are especially suitable for modelling HS because of the existence of the handy simulation tool HYPENS for Matlab [13, 14]. P/T PN are (as to their structure) bipartite directed graphs with two kinds of nodes - places $p_i \in P$, $i = 1, \ldots, n$, and transitions $t_j \in T$, $j = 1, \ldots, m$, and two kinds of edges - the set $F \subseteq P \times T$ of edges from places to transitions and the set $G \subseteq T \times P$ of edges from transitions to places. But moreover, P/T PN have also dynamics - the evolution of marking of their places given as $\mathbf{x}_{k+1} = \mathbf{x}_k + \mathbf{B}.\mathbf{u}_k$, where $\mathbf{x}_k = (\sigma_{p_1}, \ldots, \sigma_{p_n})^T$ with $\sigma_{p_i} \in \{0, 1, \ldots, c\}$ (here c, being the capacity of the places, may be either infinite or finite) is the marking vector expressing the state of the marking of the particular places, $\mathbf{u}_k = (\gamma_{t_1}, \ldots, \gamma_{t_m})^T$ with $\gamma_{t_j} \in \{0, 1\}$ is the vector of the states of transitions (disabled or enabled). The matrix $\mathbf{B} = \mathbf{G}^T - \mathbf{F}$ expresses the structure. $\mathbf{F}$ (**Pre**) and $\mathbf{G}^T$ (**Post**) are the incidence matrices of the arcs corresponding, respectively, to the sets F and G.

G. Jezic et al. (eds.), *Agent and Multi-Agent Systems: Technologies and Applications*, 97
Advances in Intelligent Systems and Computing 296,
DOI: 10.1007/978-3-319-07650-8_11, © Springer International Publishing Switzerland 2014

P/T PN do not depend on time. Their transitions, places, arcs and tokens do not contain any time specifications. In TPN [15] time specifications are defined. In general, the time specification can be assigned to TPN places, transitions, directed arcs, even to tokens. However, in this paper the time specifications will be assigned exclusively to the TPN transitions (more precisely to the P/T PN transitions). Namely, in the deterministic case they will represent time delays of the transitions while in the non-deterministic cases they will express a kind of probability distribution of timing (e.g. exponential, discrete uniformed, Poisson's, Rayleigh's, Weitbull's). HPN in general are [6] an extension of Petri nets (PN). HPN model the hybrid systems where discrete and continuous variables coexist. HPN have two groups of places and transitions - discrete and continuous. Consequently, three kinds of directed arcs exist in HPN: (i) the arcs between discrete places and discrete transitions; (ii) the arcs between continuous places and continuous transitions; (iii) the arcs between discrete places and continuous transitions as well as the arcs between the continuous places and discrete transitions. While HPN discrete places and transitions handle discrete tokens, the HPN continuous places and transitions handle continuous variables (different kinds of flows). FOHPN are a simplified kind of HPN. FOHPN are defined in details in [2–4, 7, 13]. The comprehensive definition will not be introduced here, of course. But to give the basic idea about FOHPN it is necessary to introduce that the set of places $P = P_d \cup P_c$, where P_d is a set of discrete places (figured by circles) and P_c is a set of continuous places (figured by double concentric circles). Analogically, the set of transitions $T = T_d \cup T_c$, where T_d is a set of discrete transitions (figured by rectangles) and T_c is a set of continuous transitions (figured by double rectangles). T_d contains a subset of immediate (no-timed) transitions like those used in P/T PN and/or a subset of timed transitions (deterministic and/or non-deterministic like in TPN). Consequently, the FOHPN marking consists of two parts: (i) discrete - integer number of tokens in discrete places; (ii) continuous - an amount of a fluid in continuous places. The instantaneous firing speed (IFS) [2–4], determining an amount of fluid per a time units in a time instance τ, is assigned to each of the continuous transition T_j. For all time instances τ holds $V_j^{min} \leq v_j(\tau) \leq V_j^{max}$, where min and max denote the minimal and maximal values of the speed $v_j(\tau)$. IFS is piecewise constant. The rules for enabling the continuous transitions are as follows. An empty continuous place P_i is filled through its enabled input transition. So, the fluid can flow to its output transition. The continuous transition T_j is enabled in the time τ [2–4, 13] if and only if its input discrete places $p_k \in P_d$ have marking $m_k(\tau)$ at least equal to the element $Pre_{dc}(p_k, T_j)$ of the incidence matrix $\mathbf{Pre}_{dc}$ of arcs from discrete places to continuous transitions and all of its input continuous places $P_i \in P_c$ satisfy the condition that their markings $M_i(\tau) \geq 0$ - i.e. the places P_i are filled. If all of the input continuous places of the transition T_j have non-zero marking then T_j is strongly enabled, otherwise T_j is weakly enabled. The continuous transition T_j is disabled if some of its input places is not filled. Namely, T_j cannot take more fluid from any empty input continuous place than the amount entering the place from other transitions. This corresponds to the principle of mass conservation.

Segments of a transport system (e.g. whole city or a part of country) with an increased traffic density are understood to be agents. The agents are mutually connected by the roads. The cooperation among the adjacent agents is realized by means of passing the vehicles in/out each other. Here, we will be interested only in the interior of one of such agents (segments) and test its internal dynamic behaviour. Firstly, the segment will be modelled by the TPN and tested by simulation in Matlab by means of the HYPENS tool. Secondly, the modelling will be performed by FOHPN and tested by means of the HYPENS tool too.

2 Case Study

Let us start from the P/T PN model of the segment of a transport system given in Fig. 1 studied in [1]. The P/T PN places (denoted by circles) represent the intersections of roads while the arcs between the places containing the P/T PN transitions (denoted by rectangles) represent the roads between the intersections. In order to find the route from the crossroad modelled by the place p_1 at respecting the bounded throughput of the particular roads at the increased traffic density (i.e. when in some roads either the traffic is bounded or a congestion occurs) to the crossroad modelled by the place p_6, usage of TPN and FOHPN will be tried. Consider that the distance of each road section is given by $d_{12} = 40$m, $d_{24} = 30$m, $d_{13} = 30$m, $d_{34} = 20$m, $d_{45} = 50$m, $d_{65} = 30$m, $d_{76} = 30$m, $d_{73} = 40$m, $d_{37} = 40$m, $d_{87} = 40$m, $d_{18} = 30$m, where d_{ij} is the distance from intersection point i to j. Assume that the average vehicle speed of each road section is estimated by an Adaptive Gray Threshold Traffic Parameters Measurement (AGTTPM) [9, 10] system as $v_{12} = 10$m/s, $v_{24} = 5$m/s, $v_{13} = 5$m/s, $v_{34} = 5$m/s, $v_{45} = 10$m/s, $v_{65} = 10$m/s, $v_{76} = 3.33$m/s, $v_{73} = 10$m/s, $v_{37} = 5$m/s, $v_{87} = 10$m/s, $v_{18} = 10$m/s (where m/s stands for meters per second), the estimated travel time τ_{ij} from the place p_i to p_j can be computed as the ratio of the distance to the average vehicle speed and assigned to the corresponding transition as $(\tau_{12} = 4\text{s}) \rightarrow t_1$, $(\tau_{24} = 6\text{s}) \rightarrow t_2$, $(\tau_{13} = 6\text{s}) \rightarrow t_3$, $(\tau_{34} = 4\text{s}) \rightarrow t_4$, $(\tau_{45} = 5\text{s}) \rightarrow t_5$, $(\tau_{56} = 3\text{s}) \rightarrow t_6$, $(\tau_{76} = 9\text{s}) \rightarrow t_7$, $(\tau_{73} = 4\text{s}) \rightarrow t_8$, $(\tau_{37} = 8\text{s}) \rightarrow t_9$,

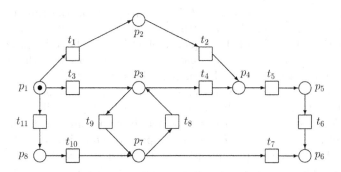

Fig. 1. The P/T PN-based model of the segment of the transport system

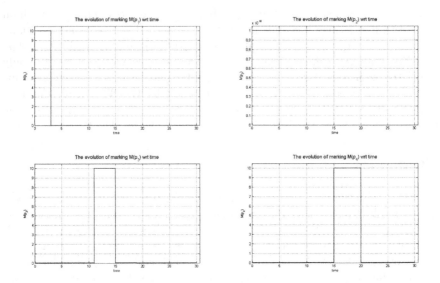

Fig. 2. TPN marking of the places p_1-p_4 with respect to time in the deterministic case. The particular courses give us information about the occupation of the corresponding crossroads. The crossroad modelled by p_2 (its marking is 1.10^{-10} is passed not a bit.

$(\tau_{87} = 4\text{s}) \rightarrow t_{10}$, $(\tau_{18} = 3\text{s}) \rightarrow t_{11}$ (where s stands for seconds). These estimated travel times can be incorporated into the transitions of the TPN model, namely, in the deterministic case as their time delays, while in the non-deterministic case as a kind of probability distributions of their timing. The graphical results obtained by means of the HYPENS tool (it is able to model not only FOHPN but also TPN) are given in Fig. 2, Fig. 3. While in Fig. 2 the marking of TPN places p_1-p_4 are displayed, in Fig. 3 the marking of TPN places p_5-p_8 are shown. The results correspond to the input parameters $\mathbf{m}_0 \equiv \mathbf{x}_0^T = (10, 0, 0, 0, 0, 0, 0, 0)$ and the vector of the TPN transitions weights (it is the internal HYPENS parameter) $\alpha = (5, 3, 3, 5, 4, 6, 1, 5, 2, 5, 6)$ depending on the time delays (the maximal priority is assigned to the transition with the shortest time delay). It can be seen that the order of the sequence of activating the TPN places in time is: $p_1 \rightarrow p_8 \rightarrow p_7 \rightarrow p_4 \rightarrow p_5 \rightarrow p_6$. Just this sequence represents the route upon which the passing from p_1 to p_6 happens in the shortest time.

Now, let us use the FOHPN model given in Fig. 4. The sense of the discrete blocks can be explained on the block $\{p_1, p_2, p_3, p_4, t_1, t_2, T_i\}$ displayed in Fig. 5 as follows. The discrete place p_1 has to be active (i.e. to have the token) in order that the continuous transition T_i might be open. When p_2 is active, T_1 is closed. The active place p_3 makes possible to open the closed T_i, while the active place p_4 makes possible to close the opened T_i. The transitions t_1, t_2 may be either deterministic (with deterministic time delays only) or non-deterministic (with a kind of the probability distribution of its timing). The flow through the T_i is

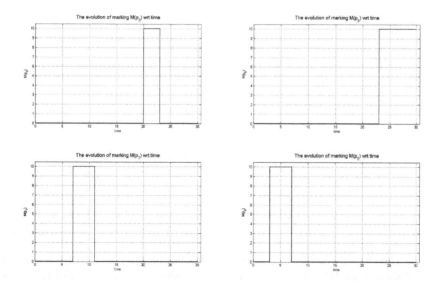

Fig. 3. TPN marking of the places $p_5 - p_8$ with respect to time in the deterministic case show the occupation of the corresponding crossroads. All of them are passed during the corresponding time interval.

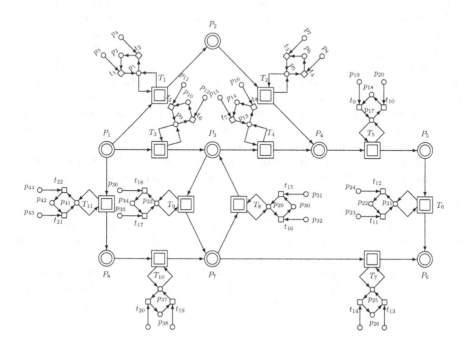

Fig. 4. The FOHPN model of the situation in the transport segment

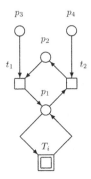

Fig. 5. The P/T PN-based discrete block in FOHPN

liable to the rules (mentioned above) explained in details in [2–4, 13] concerning the evolution of FOHPN continuous marking. Using the model in HYPENS tool we can obtain the graphical simulation results in deterministic case as the courses given in Fig. 6, Fig. 7, where the following input parameters were used: the initial continuous marking $M_0 = (100, 0, 0, 0, 0, 0, 0, 0)$, the initial discrete marking $m_0 = (1\,0\,1\,1\,1\,0\,1\,1\,1\,0\,1\,1\,1\,0\,1\,1\,1\,0\,1\,1\,1\,0\,1\,1\,1\,0\,1\,1\,1\,0\,1\,1\,1\,0\,1\,1\,1\,0\,1\,1\,1\,0\,1\,1)$, the limits of IFS for continuous transitions $V^{min} = (0\,0\,0\,0\,0\,0\,0\,0\,0\,0\,0)$, $V^{max} = (11, 6, 6, 6, 11, 11, 3.7, 11, 6, 11, 11)$. Simultaneously, the parameters of the discrete uniform probability distribution - $f_x = 1/(b-a)$ when $x \in (a, b)$ and $f_x = 0$ otherwise - for the discrete transitions timing are $\delta = 15 * ([2\ 6], [2\ 6], [4\ 8], [4\ 8], [4\ 8], [4\ 8], [2\ 6], [2\ 6], [3\ 7], [3\ 7], [1\ 5], [1\ 5], [7\ 11], [7\ 11], [2\ 6], [2\ 6], [6\ 10], [6\ 10], [2\ 6], [2\ 6], [1\ 5], [1\ 5])$, where the pairs $[a_i\ b_i]$, $i = 1, \ldots, 22$, create the parameters for particular discrete transitions $t_1 - t_{22}$. The weights of the discrete transition were not predefined while the weights of the continuous transitions are: $(5, 3, 3, 5, 4, 6, 1, 5, 2, 5, 6)$. It can be seen from the results that at passing the routes the crossroads modelled by the continuous places P_5 and P_8 are attended less than the other ones. More or less it is confirmed also by the graphical simulation results obtained at using the exponential probability distribution of timing the discrete transitions: $f_x = \lambda.e^{-\lambda.x}$ for $x \geq 0$ and $f_x = 0$ otherwise. These results are given in Fig. 8, Fig. 9. In this non-deterministic case the parameters were the same, except the parameters characterizing the probability distribution. Here, the parameters of the exponential probability distribution are: $\lambda = (4, 40, 6, 60, 6, 60, 4, 40, 5, 50, 3, 30, 9, 90, 4, 40, 8, 80, 4, 40, 3, 30)$. It means that the transport is largely realized throughout the other places (out of the P_5 and P_8). The results obtained by using the exponential probability distribution are more smoothed than those gained at the discrete uniform one.

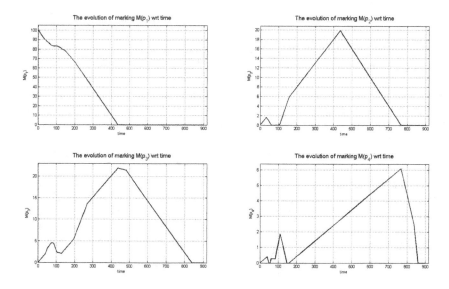

Fig. 6. FOHPN continuous marking of $P_1 - P_4$ with respect to time in non-deterministic case with the discrete uniform probability distribution of timing the discrete transitions

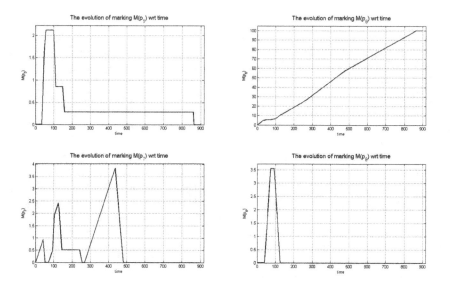

Fig. 7. FOHPN continuous marking of $P_5 - P_8$ with respect to time in non-deterministic case with the discrete uniform probability distribution of timing the discrete transitions

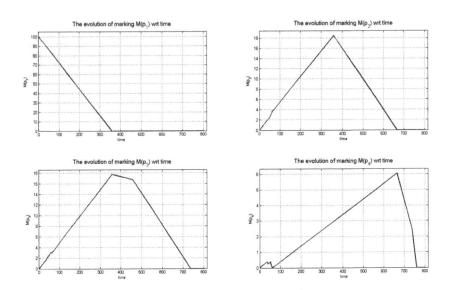

Fig. 8. FOHPN continuous marking of $P_1 - P_4$ with respect to time in non-deterministic case with the exponential probability distribution of timing the discrete transitions

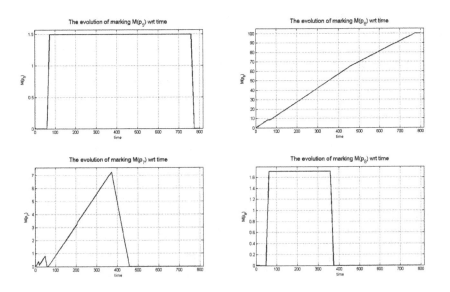

Fig. 9. FOHPN continuous marking of $P_5 - P_8$ with respect to time in non-deterministic case with the exponential probability distribution of timing the discrete transitions

3   Conclusion

The main idea of this paper is to point out the further possibilities of modelling the transport systems throughput at the increased traffic density. The information from the AGTTPM system makes possible to find the suitable routes. This work extends our previous work [1] where the P/T PN-based approach was presented. While there the time specifications were missing, here, at the application of TPN and FOHPN, the time specifications can be applied in a wide range. Consequently, the simulation of the time behaviour of the transport segment (agent) of a global transport system is possible. Such a procedure is very useful, because it yields the flows of the vehicles in time. Then, the cooperation among the adjacent segments (agents) can be realized by means of passing the vehicles in/out each other - i.e. by means of the mutual exchange of the vehicles. Of course, here (on the prescribed limited space) only one segment is dealt with. The TPN model presented here was built directly from the P/T PN model presented in [1]. Namely, the time specifications - i.e. either simple time delays in deterministic case of timing or a kind of the probability distributions of timing - were assigned to the P/T PN transitions. In such a way the TPN model of the segment arose. Using the model for simulation in Matlab by means of the HYPENS tool the graphical simulation results can be obtained in the form of stepped time functions. To illustrate the soundness of such an approach the graphical results for the deterministic case (where exclusively the deterministic time delays were used) were introduced and verbally described in order to document the abilities of the approach. The approach yields the most suitable passing the roads with respect to prescribed conditions. Then, the FOHPN model was proposed and used for the simulation in Matlab by means of the HYPENS tool. In such a way the continual courses of the vehicle flows passing the intersections of the roads can be found in the form of the continuous piecewise-linear real time functions. To illustrate the soundness of such and approach two graphical simulation results were presented. They correspond to non-deterministic case of timing the discrete transitions of the FOHPN. While in the former illustration the discrete uniform probability distribution was used at timing the discrete transitions of FOHPN model, in the latter illustration the exponential probability distribution at timing the discrete transitions was used. The results were corroborated by parameters used at the simulations and verbally interpreted.

Acknowledgement. The theoretical research was partially supported by the Slovak Grant Agency for Science (VEGA) under the current grant # 2/0039/13. The theoretical results were applied at solving the practical project, videlicet: This contribution is the result of the project implementation: Technology research for the management of business processes in heterogeneous distributed systems in real time with the support of multimodal communication, code ITMS: 26240220064, supported by Operational Programme Research & Development funded by the ERDF. The author thanks both institutions for the support of his research.

References

1. Čapkovič, F.: Travel Routes Flexibility in Transport Systems. In: Barbucha, D., Le, M.T., Howlett, R.J., Lakhmi, C.J. (eds.) Advanced Methods and Technologies for Agent and Multi-Agent Systems, pp. 30–39. IOS Press, Amsterdam-Berlin-Tokyo-Washington, DC3 (2013)
2. Balduzzi, F., Giua, A., Menga, G.: First-Order Hybrid Petri Nets: A Model for Optimization and Control. IEEE Trans. on Robotics and Automation 16, 382–399 (2000)
3. Balduzzi, F., Giua, A., Seatzu, C.: Modelling and Simulation of Manufacturing Systems Using First-Order Hybrid Petri Nets. International Journal of Production Research 39, 255–282 (2001)
4. Balduzzi, F., Di Febbraro, A., Giua, A., Seatzu, C.: Decidability Results in First-Order Hybrid Petri Nets. Discrete Event Dynamic Systems 11, 41–58 (2001)
5. Dotoli, M., Fanti, M., Giua, A., Seatzu, C.: First-Order Hybrid Petri Nets. An Application to Distributed Manufacturing Systems. Nonlinear Analysis: Hybrid Systems 2, 408–430 (2008)
6. David, R., Alla, H.: On Hybrid Petri Nets. Discrete Event Dynamic Systems: Theory and Applications 11, 9–40 (2001)
7. Dotoli, M., Fanti, M., Iacobellis, G., Mangini, A.M.: A First-Order Hybrid Petri Net Model for Supply Chain Management. IEEE Transactions on Automation Science and Engineering 6, 744–758 (2009)
8. Ghomri, L., Alla, H.: Modeling and Analysis of Hybrid Dynamic Systems Using Hybrid Petri Nets. In: Kordic, V. (ed.) Petri Net Theory and Applications, ch. 6, pp. 113–130. I-Tech Education and Publishing, Vienna (2008)
9. Li, L., Ru, Y., Hadjicostis, C.N.: Least-cost transition firing sequence estimation in labeled Petri nets. In: Proceedings of the 45th IEEE International Conference on Decision and Control, San Diego, CA, USA, December 13-15, pp. 416–421 (2006)
10. Liu, Y., Dai, Y., Dai, Z.: Real-time adaptive gray threshold measurement in extracting traffic parameters. In: Proceedings of the 3rd International Symposium on Computational Intelligence and Industrial Applications, ISCIIA 2008, Dali, Yunnan Province, China, November 21-26, pp. 255–265 (2008)
11. Murata, T.: Petri Nets: Properties, Analysis and Applications. Proceedings of the IEEE 77, 541–580 (1989)
12. Peterson, J.L.: Petri Nets Theory and the Modelling of Systems. Prentice-Hall Inc., Englewood Cliffs (1981)
13. Sessego, F., Giua, A., Seatzu, C.: HYPENS: A Matlab Tool for Timed Discrete, Continuous and Hybrid Petri Nets. In: van Hee, K.M., Valk, R. (eds.) PETRI NETS 2008. LNCS, vol. 5062, pp. 419–428. Springer, Heidelberg (2008)
14. Sessego, F., Giua, A., Seatzu, C.: HYPENS Manual (2008),
   http://www.diee.unica.it/auto-matica/hypens/Manual_HYPENS.pdf
15. Popova-Zeugmann, L.: Time Petri Nets: Theory, Tools and Applications, Part 1, Part 2 (2008),
   http://www2.informatik.hu-berlin.de/~popova/1-part-short.pdf ,
   http://www2.informatik.hu-berlin.de/~popova/2-part-short.pdf

Distributed Regret Matching Algorithm for a Dynamic Route Guidance

Tai-Yu Ma

CEPS/INSTEAD, 3, Avenue de la Fonte L-4364 Esch-sur-Alzette, Luxembourg
`tai-yu.ma@ceps.lu`

Abstract. This paper proposes a distributed self-learning algorithm based on the regret matching process in games for a dynamic route guidance. We incorporate a user's past routing experiences and en-route traffic information into their optimal route guidance learning. The numerical study illustrates that the proposed self-guidance method can effectively reduce the travel times and delays of guided users in congested situation.

Keywords: game, multi-agent, distributed learning, route guidance, Nash Equilibrium.

1 Introduction

A route guidance system aims at providing road users with optimal route recommendations in order to assist them in reaching their destinations with the shortest travel time. The system can predict the travel times in a near future and recommends the shortest routes to travelers. In the past, different route guidance systems have been proposed. The problem can be formulated as a fixed point problem, where users are assigned to routes with the constrains that no guided users can travel a better route by unilaterally changing their current route choices [1-2]. As path flows depend on complex interactions between the travel demand and the network capacity supply, traditional derivative-based methods cannot be used to solve such a dynamic route guidance problem.

The widely used solution algorithm is based on an iterative route flow adjustment process, which consists in shifting flows on a day-to-day basis, towards faster routes [1-2]. In opposition with the route-based flow adjustment method, a few distributed methods, based on control theory, have been suggested [3-6]. These methods aim at optimizing the flow assignment at each network node, so as to minimize a global cost function, i.e. to minimize the total system travel times/delays or the gap between the nominal least travel times and users' experienced travel times. Although the above algorithmic studies provide some effective route guidance strategies, they only focus on the "supply" aspect and neglect the "behavior" of the driver (i.e. the reaction to received en-route traffic information). To address this issue, Peeta and his colleagues [7-8] conducted a series of empirical studies to model the driver's behavior in terms of route choice, and then suggested hybrid models for a consistent route guidance.

G. Jezic et al. (eds.), *Agent and Multi-Agent Systems: Technologies and Applications*,
Advances in Intelligent Systems and Computing 296,
DOI: 10.1007/978-3-319-07650-8_12, © Springer International Publishing Switzerland 2014

Recently, the agent-base approach has received an increasing level of interest as far as transportation system modeling and simulations are concerned. One important research issue concerns the non-cooperative Nash equilibrium learning process, a common problem arising in artificial intelligence, economics and automatic control theory. With the recent advances in game theory, some game theoretical learning processes have been proposed for a Nash equilibrium learning in multi-agent systems [9-10] and applied in congestion games [11-13] and system/user optimal routing problem solving [14-15].

In this study, we propose a game-theoretical based self-learning algorithm to solve a dynamic route guidance problem. The proposed method is a distributed self-learning algorithm based on the driver's (agent's) routing experiences in the past. Previously experienced routing performances are assumed to be stored in each guided vehicle, and updated in a consistent way. The objective is to propose a self-guidance algorithm based on the regret matching process in games [10] for a dynamic route guidance. We illustrate that the distributed multiagent regret matching process can achieve a coarse Nash equilibrium for the dynamic route guidance problem and reduce effectively travel delay of guided users on a given road network.

2   Dynamic Route Guidance Problem Formulation

Let's consider a road network represented by a directed graph $G(N, E)$ where N is a set of nodes and E is a set of links. A finite set of guided drivers (users) aim at reaching their respective destinations with the shortest travel times. Following Bottom [1], the dynamic route guidance problem can be formulated as a fixed point problem designed to assign to each user a set of partial paths until they reach their final destination, taking into account the fact that no any guided user can find a better route by unilaterally changing their route choice (Nash equilibrium). The problem can be mathematically formulated as an infinite dimensional variational inequality problem for a dynamic traffic assignment at each node level, as shown in [2][16]:

For any destination $d \in D$ and node $s \in N$, find user equilibrium flows $\mathbf{f}^*$ such that

$$\begin{cases} C_r^*(t) = \min_{r \in R_{sd}}(C_r(t)) & \Rightarrow \quad f_r^*(t) > 0 \\ C_r^*(t) > \min_{r \in R_{sd}}(C_r(t)) & \Rightarrow \quad f_r^*(t) = 0 \end{cases} \tag{1}$$

where $f_r(t)$ is the traffic flow on a path r at time t. R_{sd} is a set of routes connecting s and d. $C_r(t)$ is the travel time on path r when entering the path at time t. $\mathbf{f}^*$ is the path flow vector on minimum travel time paths, constrained by the flow conservation and the non-negativity of the path flow, so that:

$$\sum_{r \in R_{sd}} f_r(t) = q_{sd}(t), \forall s \in N, \forall d \in D \tag{2}$$

$$f_r(t) \geq 0, \forall t \in [t_0, T] \tag{3}$$

where $q_{sd}(t)$ is travel demand from s to d at time t. T is the period studied.

The above user equilibrium states that all used paths are the shortest paths, and no user can reduce their travel times by unilaterally changing their route choice. The problem is difficult to solve, due to the fact that the travel time on a given path depends on the interactions of the users' route choices and the traffic flow dynamics. The above Nash equilibrium problem can be approximated by minimizing a gap function $G(C(\mathbf{f}))$ which measures the distance to a user equilibrium state as [17]:

$$\underset{f \in \Phi}{\text{Min }} G(C(\mathbf{f})) = \frac{\sum_{h \in H, k \in K, r \in R_{hk}, m \in \Omega_{hkr}} [C_m(t) - C_{hk}^*]}{\sum_{h \in H, k \in K} q_{hk} C_{hk}^*} \tag{4}$$

where m corresponds to a specific user. Ω_{hkr} is a set of users on path r with an OD pair k and a departure time interval h. $C_m(t)$ is user m's experienced travel time with departure time t. q_{hk} is the travel demand for an OD pair k and a departure time interval h. C_{hk}^* is the minimum travel time with respect to h and k. Φ is a feasible flow set satisfying eq. 2-3.

3   Distributed Learning Algorithm in Games to Solve a Dynamic Route Guidance Problem

We consider a repeated N-player game setting where users learn (adjust) their routes on a day-to-day (pre-trip) basis and within-day (en-route) basis. The route choice decision is made at a local (node/link) level, i.e. each user chooses their next outgoing links following a strategy (i.e. a probability distribution over the set of outgoing links). Based on the assumed information acquisition mechanism described below, each user learns their best route based on their regrets of not having taken a specific decision in the past.

3.1   Basic Assumptions

We assume the following system characteristics for the dynamic route guidance problem.

- Each guided vehicle is assumed to be equipped with a GPS device capable of storing in a digital map the experimented travel times on all routes previously travelled and is able to receive traffic congestion information (information concerning delays on specific network links). Such in-vehicle information system is similar to that described in [18].
- A road information service provider sends updated incident / road congestion information (estimation) every 5 minutes, which means that each vehicle properly equipped can receive real-time information about traffic congestion.

- Although various factors may influence driver's behavior (route change) [7], we assume that each driver completely follows the route guidance advised by their vehicle.

Note that the system assumption by which each vehicle can receive traffic congestion information is very moderate, as is already made available in most current road traffic information services.

3.2   Information Processing Mechanism and Learning Algorithms in Games

Consider a road network represented by a directed graph $G(N, E)$ in which each guided vehicle updates the expected travel times on all links of the network, based on past travel experiences. The following definitions are used for our routing problem formulation.

Definition 1: an *agent* is a vehicle equipped so as to store, compute and learn optimal route guidance based on past routing experiences. A *user* is a driver who can comply with or ignore the advice given by the agent.

Definition 2: a *partial route* $r_i(s)$ (or $r_i(e)$) is a route starting at a non-origin-destination node $s \in N \setminus \{O, D\}$ (or link $e \in E$) ending at user's destination d_i.

Definition 3: a *payoff function* of partial route is the user's experienced travel time on the partial route.

The payoff function of partial route $r_i(s)$ from node s to user i destination is the travel times experienced by the agent, defined as

$$U_i^w(r_i(s)) = \sum_{e \in r_i(s)} C_i^w(e) \tag{5}$$

where $C_i^w(e)$ is agent i's experienced travel time on link e on day (iteration) w. Based on the agent experienced travel times, the following information is computed:

- Empirical average payoff function $\overline{U}_i^w(e)$ of link: the function represents agent i's average experienced travel times by travelling through link $e \in E$ to his destination up to day w-1, defined as

$$\overline{U}_i^w(e) = \begin{cases} [\sum_{\tau=0}^{w-1} I_{\{e \in r_i(o_i)\}} U_i^\tau(r_i(e))] / n_e^w, & \text{if } e \in r_i(o_i) \\ \overline{U}_i^{w-1}(e), & \text{otherwise} \end{cases} \tag{6}$$

where $I_{\{\bullet\}}$ is the indicator function. n_e^w is defined as $\sum_{\tau=0}^{w-1} I_{\{e \in r_i(o_i)\}}$. $r_i(o_i)$ is agent i's used route from his origin node o_i to his destination d_i.

- Empirical average payoff function $\overline{U}_i^{\,w}(s)$ of node: the function represents agent i's average experienced travel times from $s \in N$ to his destination, up to day w-1, defined as

$$\overline{U}_i^{\,w}(s) = \begin{cases} [\sum_{\tau=0}^{w-1} I_{\{s \in r_i(o_i)\}} U_i^{\tau}(r_i(s))]/n_s^w, & \text{if } s \in r_i(o_i) \\ \overline{U}_i^{\,w-1}(s), & \text{otherwise} \end{cases} \qquad (7)$$

where $n_s^w = \sum_{\tau=0}^{w-1} I_{\{s \in r_i(o_i)\}}$ is the number of days (iterations) during which node s is part of agent i's route $r_i(o_i)$ from origin node o_i.

- Potential function $V_i^{\,w}(u)$: each node $u \in N$ is associated with a potential function $V_i^{\,w}(u)$, defined as *minimum average travel time* from node u to agent i's destination up to day w-1:

$$V_i^{\,w}(u) = \min_{e \in Y(u)} \overline{U}_i^{\,w}(e), \forall i \in I \qquad (8)$$

where $Y(u)$ is the set of downstream (outgoing) links of node u. $\overline{U}_i^{\,w}(e)$ is the average travel time going through link e until the agent's destination, up to day w-1.

We set the initial value of $V_i(u)$ as the shortest travel time from $u \in N$ to user i's destination, based on a free flow travel time. Note that the above payoff function and potential function are updated at the end of each day for each agent.

Based on the above information processing and storage mechanism, each agent gains some empirical information on the network performance, based on their own experiences up to day w-1. Given an agent currently located at node u, we assume that they can receive real-time traffic incident/congestion information, and update their empirical average payoff function estimates for each outgoing link based on the external information as follows:

$$\tilde{U}_i^{\,w}(e(u,u')) = \overline{C}_i^{\,w}(e(u,u')) + \xi_i^{\,w}(e(u,u')) + V_i^{\,w}(u') \qquad (9)$$

where $\overline{C}_i^{\,w}(e(u,u'))$ is the average travel time of link $e(u,u')$ for agent i up to day w-1. $\xi_i^{\,w}(e(u,u'))$ is the external information about any incident / delay on the link, estimated by the traffic information provider as the average queuing delay over a most recent discretized time interval. $V_i^{\,w}(u')$ is the potential function from u' to agent i's destination. The initial average travel time of a link is set as its free flow travel time.

Let us consider now an agent's learning process in repeated N-player games. The learning process is based on the regret matching process for non-cooperative Nash equilibrium computed in a multi-agent system [10]. Let's consider a finite set of agents $I = \{1,2,...,n\}$ who progressively select their actions (next outgoing links) towards their respective destinations. At any time t, an agent's strategy consists in choosing a next outgoing link with the minimum estimated travel times to destination. When arriving at any node s, the action set is the set of outgoing links of the node, denoted as $A_i(s)$. A payoff function for agent i is defined as $U_i : A(s) \rightarrow \Re$, where $A(s) = \times_{i=1,\ldots,n} A_i(s)$ is the Cartesian product of all agents chosed actions. An action profile of all agents is defined as $a(s) = (a_i(s), a_{-i}(s))$, where $a_{-i}(s) = \{a_1(s),\ldots, a_{i-1}(s), a_{i+1}(s),\ldots, a_n(s)\}$ is the action profile of other agents except agent i, i.e. profile of actions towards each agent's respective destination $d_{j \neq i}$. Note that the payoff function $U_i(a_i(s))$ represents agent i's experienced travel time from node s to destination. A pure Nash equilibrium is achieved if, for all agents $i \in I$, the following condition holds:

$$a_i^*(s) \in \arg \min_{a_i \in A_i(s)} U_i(a_i(s), a_{-i}^*(s)), \tag{10}$$

for any $s \in N$. The above condition states that at Nash equilibrium, each agent chooses their optimal action $a_i^*(s)$, which gives the best payoff among the array of possible actions. Note that in an agent local routing context, the game to Nash equilibrium consists in selecting a series of optimal outgoing links so that the equilibrium can be achieved.

In a repeated game, each player (agent) tracks the payoffs received in the past and computes the empirical average payoff for each action. Based on this information, an agent may, at any node, optimize the action by choosing an outgoing link which minimizes the expected travel time to destination. For this purpose, we propose the following distributed self-learning strategy based on the regret matching process in games to converge to Nash equilibrium. The regret matching algorithm is an adaptive learning algorithm based on each agent's payoffs in the past. When arriving at a node $u \in N$, an agent computes the estimated average payoffs of each action (outgoing link) and the difference (*regret*) with the overall average payoffs until day w-1. Hence, the estimated average regret of not choosing an outgoing link $e(u, u') \in Y(u)$ at day w for agent i is

$$R_i^w(e(u, u')) = -[\tilde{U}_i^w(e(u, u')) - \overline{U}_i^w(u)], \forall e(u, u') \in Y(u) \tag{11}$$

where $\overline{U}_i^w(u)$ and $\tilde{U}_i^w(e(u, u'))$ is calculated by eq. 7 and eq. 9, respectively.

Note that a positive regret represents a payoff gain if agent i has chosen that action in the past. In the regret matching process, a player chooses an action with a probability which is proportional to its positive regret value:

$$p_i^{w+1}(e(u, u')) = \frac{[R_i^w(e(u, u'))]_+}{\sum_{e' \in Y(u)} [R_i^w(e')]_+} \tag{12}$$

where $[a]_+ = a$ if $a > 0$, 0 otherwise. $R_i^w(e(u, u'))$ is defined by eq. 11.

Note that if the denominator is 0, uniform choice probability is applied. To guarantee the convergence property, a random term (inertia) $\varepsilon(w) = c / w$ is introduced, where c is a constant with $c \in [0,1]$. Agents may randomly choose their routes with a probability $\varepsilon(w)$ or with a probability $1 - \varepsilon(w)$ based on eq. 12. The convergence proof to Nash equilibrium of the above regret matching process is similar to [11].

4   Numerical Study

4.1   Test Network and Investigated Scenarios

We test the proposed distributed learning algorithm for the dynamic route guidance problem in a small network with 20 nodes and 55 links. The point queue model [19-20] is applied for modeling traffic flow dynamics. It is worth to note that the point queue model gives good estimation of realistic traffic flow dynamics and has been widely applied for urban mobility simulation [20]. The link length is set randomly between [3, 3.5] km. The maximum flow capacity is set at 1,200 vehicles per lane. The free flow speed is set at 50 km/hr. The number of lanes for orthogonal links is set at 2 lanes and 1 lane for negative slopped links, respectively. There are four origin nodes and one destination node (Fig. 1). The departure time is randomly selected within each driver's departure time interval on the initial day and then fixed for the following iterations. Information about travel time delays is updated every 3 minutes. We distinguished two types of users: guided users and non-guided users. The scenarios related to variations in the travel demand and various proportions of guided users are conceived for the numerical study. The simulator is implemented based on discrete event simulation technique by C++.

4.2   Computational Results

In order to verify the performance of the proposed algorithm, we compare the average travel times and delay for guided and non-guided users with respect to different scenarios. The convergence result of the proposed method is shown in Fig. 2. The time-dependent loading profile over 5-minute intervals is shown on the right of Fig. 2. We set a peak hour period to generate a significant level of congestion on the network. The basic case is referred to as no route guidance. When the network is uncongested, increasing the market share of guided users is not beneficial in terms of total travel time delay, computed as the summation of travel time differences compared to free flow travel time. As expected, when all users are guided-users, the resulting routing pattern approaches dynamic user equilibrium, i.e. the average travel times of 100% market penetration is similar to that if non route guidance case (Table 1). However, in congested situations, when the market penetration of guided users increases from 0% to 50%, total system delay reduces accordingly (Table 1). However, when there are too many guided users, the overall system performance may be reduced. When further comparing the average delay of the same quantity of users in each class, we found that the proposed system effectively reduces not only the average travel time, but also the delay experienced by guided users if the market penetration of the system is not too high (Fig. 3 and Table 1).

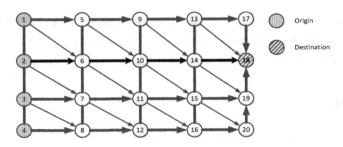

Fig. 1. Dynamic test network

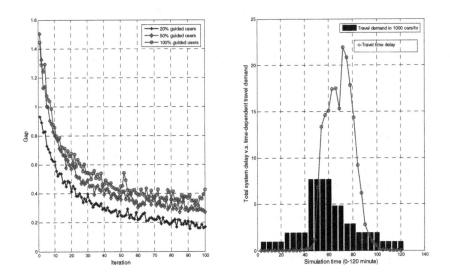

Fig. 2. Convergence result (left) and temporal loading profile and travel time delay (right)

Table 1. Total travel time and delay differences on the network

LF	% of guided users	Total TT (hr)	Avg. TT (min)	Total delay (hr)	Total TT (+/-)	Total delay (+/-)
1	0	898	18.7	15 -	-	
	20%	931	19.4	23	+3.7%	+56.2%
	50%	928	19.3	25	+3.2%	+43.9%
	100%	904	18.8	17	+0.6%	+9.0%
2	0	2047	21.3	199 -	-	
	20%	2097	21.8	173	+2.4%	**-12.9%**
	50%	2019	21.0	146	**-1.4%**	**-30.3%**
	100%	2034	21.2	182	**-0.7%**	**-11.6%**

Remark: 1. LF (loading factor): ratio of the number of users with respect to a reference case (lower congestion) of 2880 users; 2. TT: travel times of users

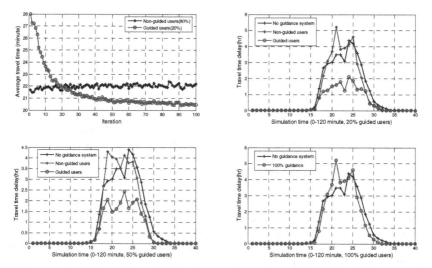

Fig. 3. Average travel times (upper-left) and comparison of travel time delay of the same quantity of users in each class w.r.t to different percentages of guided and non-guided users (loading factor is 2)

5 Conclusions

In this study, a distributed multi-agent self-learning algorithm based on the regret matching process in games is proposed for a dynamic route guidance. We explicitly capture the influence of online information and the users' past routing experiences for an optimal routing guidance. Numerical experiments on various market penetration and different congestion cases were studied and implemented on a point-queue traffic simulation model. The numerical study showed that the proposed method can effectively reduce travel time and delays of guided users in congested situation.

Further extensions might include the impact analysis of the system on unpredicted accident scenarios and also its implementation on a microscopic traffic simulator for realistic application.

References

1. Bottom, J.A.: Consistent anticipatory route guidance. Ph.D. thesis, Massachusetts Institute of Technology (2000)
2. Zuurbier, F.S.: Intelligent Route Guidance. PhD thesis, Technische Universiteit Delft (2010)
3. Kaufman, D.E., Smith, R.L., Wunderlich, K.E.: An iterative routing/assignment method for anticipatory real-time route guidance. In: IEEE Vehicle Navigation and Information Systems Conference, vol. 2, pp. 693–700 (1991)
4. Sarachik, P.E., Ozguner, U.: On Decentralized Dynamic Routing for Congested Traffic Networks. IEEE Transactions on Automatic Control AC-27(6), 1233–1238 (1982)

5. Minciardi, R., Gaetani, F.: A decentralized optimal control scheme for route guidance in urban road networks. In: IEEE Intelligent Transportation Systems Conference, pp. 1195–1199 (2001)
6. Deflorio, F.P.: Evaluation of a reactive dynamic route guidance strategy. Transportation Research Part C 11, 375–388 (2003)
7. Peeta, S., Yu, J.: Adaptability of a hybrid route choice model to incorporating driver behavior dynamics under information provision. IEEE Transactions on Systems, Man, and Cybernetics Part A 34(2), 243–256 (2004)
8. Jha, M., Madanat, S., Peeta, S.: Perception updating and day-to-day travel choice dynamics in traffic networks with information provision. Transportation Research Part C 6, 189–212 (1998)
9. Fudenberg, D., Levine, D.K.: The Theory of Learning in Games. MIT Press, Cambridge (1998)
10. Young, P.H.: Strategic Learning and Its Limit. Oxford University Press, Oxford (2005)
11. Monderer, D., Shapley, L.S.: Fictitious Play Property for Games with Identical Interests. Journal of Economic Theory 68, 258–265 (1996)
12. Hart, S., Mas-Colell, A.: A simple adaptive procedure leading to correlated equilibrium. Econometrica 68, 1127–1150 (2000)
13. Cominetti, R., Melo, E., Sorin, S.: A payoff-based learning procedure and its application to traffic games. Games and Economic Behavior 70(1), 71–83 (2010)
14. Garcia, A., Reaume, D., Smith, R.L.: Fictitious play for finding system optimal routings in dynamic traffic networks. Transportation Research Part B 34, 147–156 (2000)
15. Miyagi, T., Peque Jr., G.C.: Informed-user algorithms that converge to Nash equilibrium in traffic games. Procedia - Social and Behavioral Sciences 54, 438–449 (2012)
16. Friesz, T.L., Bernstein, D., Smith, T., Tobin, R., Wie, B.: A variational inequality formulation of the dynamic network user equilibrium problem. Operations Research 41, 80–91 (1993)
17. Tong, C.O., Wong, S.C.: A predictive dynamic traffic assignment model in congested capacity-constrained road networks. Transportation Research Part B 34(8), 625–644 (2000)
18. Kerner, B.S., Rehborn, H., Aleksic, M., Haug, A.: Traffic Prediction Systems in Vehicles. In: 8th International IEEE Conference on Intelligent Transportation Systems, Vienna, Austria, pp. 72–77 (2005)
19. Kuwahara, M., Akamatsu, T.: Decomposition of the reactive assignments with queues for many-to-many origin-destination pattern. Transportation Research Part B 31(1), 1–10 (1997)
20. Gawron, C.: An iterative algorithm to determine the dynamic user equilibrium in a traffic simulation model. International Journal of Modern Physics C 9(3), 393–408 (1998)

A Uniform Problem Solving in the Cognitive Algebra of Bounded Rational Agents

Eugene Eberbach

Dept. of Engineering and Science, Rensselaer Polytechnic Institute
275 Windsor Street, Hartford, CT 06120, USA
eberbe@rpi.edu

Abstract. The $-calculus cognitive process algebra for problem solving provides the support for automatic problem solving and targets intractable and undecidable problems. Consistent with the ideas of anytime algorithms, $-calculus applies the cost performance measures to converge to optimal solutions with minimal problem solving costs. In the paper, we concentrate on a uniform problem solving and its implementation aspects illustrated on two benchmarks from concurrency and machine learning areas.

1 Introduction

The paper presents a theory of computation for automatic problem solving based on process algebras, utility theory and anytime algorithms. In particular, we try to formalize AI, based on classical (but unformalized) AI textbook by Russell and Norvig on meta-search algorithms and bounded rational agents [8]. Such unifying theory for AI does not exist so far yet. Resource-based reasoning [6, 8] called also anytime algorithms, trading off the quality of solutions for the amount of resources used, seems to be a particularly well suited for such a new AI framework. Process algebras [7], currently the most mature approach to concurrent and distributed systems, seem to be the appropriate way to formalize multiagent systems, and to span AI with the rest of computer science.

In the paper, we describe very briefly the $-calculus algebra of bounded rational agents as a proposal of the unifying framework for AI [3–5]. The $-calculus can be used in the same uniform way for search, planning, evolution, learning, and problem solving under bounded resources in dynamic and uncertain environments. In the paper, we concentrate on automatic problem solving and its implementation aspects illustrated by examples from two divergent areas.

2 The $-calculus Algebra of Bounded Rational Agents

The $-calculus is a mathematical model of processes capturing both the final outcome of problem solving as well as the interactive incremental way how the problems are solved (the main difference compared to other computational theories). The $-calculus is a cognitive process algebra of Bounded Rational

G. Jezic et al. (eds.), *Agent and Multi-Agent Systems: Technologies and Applications*,
Advances in Intelligent Systems and Computing 296,
DOI: 10.1007/978-3-319-07650-8_13, © Springer International Publishing Switzerland 2014

Agents for interactive problem solving targeting intractable and undecidable problems [3–5]. The \$-calculus (pronounced COST calculus) is a formalization of resource-bounded computation (also called anytime algorithms), proposed by Dean, Horvitz, Zilberstein and Russell in the late 1980s and early 1990s [6, 8]. Anytime algorithms are guaranteed to produce better results if more resources (e.g., time, memory) become available. The standard representative of process algebras, the π-calculus [7] is believed to be the most mature approach for concurrent systems. Although being a new approach, the \$-calculus has found already several applications including DSL languages to control NAVY Autonomous Vehicles (AUVs) (for more deatils look at [3–5]).

In \$-calculus everything is a cost expression: agents, environment, communication, interaction links, inference engines, modified structures, data, code, and meta-code. \$-expressions can be simple or composite. Simple \$-expressions α are considered to be executed in one atomic indivisible step. Composite \$-expressions P consist of distinguished components (simple or composite ones) and can be interrupted.

The set $\mathcal{P}$ of \$-calculus process expressions consists of simple \$-expressions α and composite \$-expressions P, and is defined by the following syntax:

$$\alpha ::= (\$_{i \in I} \ P_i) \qquad \text{cost}$$

| $(\rightarrow_{i \in I} \ c \ P_i)$ send P_i with evaluation through channel c
| $(\leftarrow_{i \in I} \ c \ X_i)$ receive X_i from channel c
| $('_{i \in I} \ P_i)$ suppress evaluation of P_i
| $(a_{i \in I} \ P_i)$ defined call of simple \$-expression a with parameters P_i, and and its optional associated definition $(:= (a_{i \in I} \ X_i) \ < R >)$ with body R evaluated atomically
| $(\bar{a}_{i \in I} \ P_i)$ negation of defined call of simple \$-expression a

$$P ::= (\circ_{i \in I} \ \alpha \ P_i) \ \text{sequential composition}$$

| $(\parallel_{i \in I} \ P_i)$ parallel composition
| $(\cup_{i \in I} \ P_i)$ cost choice
| $(\uplus_{i \in I} \ P_i)$ adversary choice
| $(\sqcup_{i \in I} \ P_i)$ general choice
| $(f_{i \in I} \ P_i)$ defined process call f with parameters P_i, and its associated definition $(:= (f_{i \in I} \ X_i) \ R)$ with body R (normally suppressed); $(^1 R)$ will force evaluation of R exactly once

The indexing set I is a possibly countably infinite. In the case when I is empty, we write empty parallel composition, general, cost and adversary choices as $\perp$ (blocking), and empty sequential composition (I empty and $\alpha = \varepsilon$) as ε (invisible transparent action, which is used to mask, make invisible parts of \$-expressions). Adaptation (evolution/upgrade) is an essential part of \$-calculus, and all \$-calculus operators are infinite (an indexing set I is unbounded). The \$-calculus agents interact through send-receive pair as the essential primitives of the model. The \$-calculus rests upon the primitive notion of *cost* in a similar way as the π-calculus was built around a central concept of *interaction* and λ-calculus around a *function*.

Simple cost expressions execute in one atomic step. Cost functions are used for optimization and adaptation. The user is free to define his/her own cost metrics. Send and receive perform handshaking message-passing communication, and inferencing. The suppression operator suppresses evaluation of the underlying $-expressions. Additionally, a user is free to define her/his own simple $-expressions, which may or may not be negated.

Sequential composition is used when $-expressions are evaluated in a textual order. Parallel composition is used when expressions run in parallel and it picks a subset of non-blocked elements at random. Cost choice is used to select the cheapest alternative according to a cost metric. Adversary choice is used to select the most expensive alternative according to a cost metric. General choice picks one non-blocked element at random. General choice is different from cost and adversary choices. It uses guards satisfiability. Cost and adversary choices are based on cost functions. Call and definition encapsulate expressions in a more complex form (like procedure or function definitions in programming languages). In particular, they specify recursive or iterative repetition of $-expressions.

The unique feature of the $-calculus is that it provides a support for problem solving by incrementally searching for solutions and using cost to direct its search. The basic $-calculus search method used for problem solving is called $k\Omega$-optimization. The $k\Omega$-optimization represents this "impossible" to construct, but "possible to approximate indefinitely" universal algorithm. It is a very general search method, allowing the simulation of many other search algorithms, including A*, minimax, dynamic programming, tabu search, or evolutionary algorithms [3–5].

The problem solving works iteratively through select, examine and execute phases. In the select phase the tree of possible solutions is generated up to k steps ahead, and agent identifies its alphabet of interest for optimization Ω. This means that the tree of solutions may be incomplete in width and depth (to deal with complexity). However, incomplete (missing) parts of the tree are modeled by silent $-expressions ε, and their cost is estimated (i.e., not all information is lost). The above means that $k\Omega$-optimization may be (if certain conditions are satisfied) complete and optimal (see [4]). The building trees (or DAGs, in a general case) is done either by using inference rules from LTS (in the style of AI planners, unification from Prolog, or matching from expert systems), or by using random number generators to generate random sequences of simple $-expressions (in the style of genetic programming), or the user is responsible to define the LTS tree. In the examine phase the trees of possible solutions are pruned minimizing cost of solutions, and in the execute phase up to n instructions are executed. Moreover, because the $ operator may capture not only the cost of solutions, but also the cost of resources used to find a solution, we obtain a powerful tool to avoid methods that are too costly, i.e., the $-calculus can directly minimize search cost. This basic feature, inherited from anytime algorithms, is needed to directly tackle hard optimization problems, and allows solving total optimization problems (the best quality solutions with minimal search costs). The variable k refers to the limited horizon for optimization, necessary due to the unpredictable, dynamic nature of the environment. The variable Ω refers to a reduced alphabet of information. The b is the branching factor of the search tree, n - the number of steps selected for execution in the execute phase, and p - the number of agents. No agent ever has reliable information about all factors that influence all agents behavior. To compensate for this, we mask factors where information is not available from consideration; reducing the alphabet of variables used by the \$-function. By using the $k\Omega$-optimization to find the strategy with the lowest \$-function, meta-system finds a "satisficing" (i.e., good enough - term coined by Simon [8]) solution, and sometimes (when appropriate conditions are satisified) - the optimal one. This avoids wasted time trying to optimize behavior beyond the foreseeable future. It also limits consideration to those issues where relevant information is available. Thus the $k\Omega$ optimization provides a flexible approach to local and/or global optimization in time or space. Technically this is done by replacing parts of \$-expressions with invisible \$-expressions ε, which remove part of the world from consideration (however, they are not ignored entirely - the cost of invisible actions is estimated).

The $k\Omega$-optimization meta-search procedure can be used both for single and multiple cooperative or competitive agents working online ($n \neq 0$) or offline ($n = 0$). The \$-calculus programs consist of multiple \$-expressions for several agents. Each agent has its own $k\Omega$-search procedure $k\Omega_i[t]$ used to build the solution $x_i[t]$ that takes into account other agent actions (by selecting its alphabet of interests Ω_i that takes actions of other agents into account). Thus each agent will construct its own view of the whole universe which only sometimes will be the same for all agents (this is analogous to the subjective view of the "objective" world by individuals having possibly different goals and different perception of the universe).

More details on the $k\Omega$-optimization, including the inference rules of the Labeled Transition System, observation and strong bisimulations and congruences, and the standard cost function definition can be found in [3–5].

3   Illustration of Versality and Power of the $k\Omega$-meta Search

3.1   Dining Philosophers - Multi-agent Searching and Planning for a Deadlock-Free and Fair Solution

The Dining Philosophers Problem is a simple abstraction of a typical synchronization problem to allocate mulitple shared reusable resources among several processes in a deadlock and starvation-free manner (for example, an abstraction of the access to I/O devices). It was posed and solved by E. Dijkstra. Scenario: five philosophers are seated around a table. Each philosopher has a plate of spaghetti, which is so slippery that each philosopher needs two forks to eat (sometimes to be more realistic, spaghetti is replaced by rice, and forks by chopsticks). Between each plate is a fork, and if the fork is grabbed, it is not released until a philosopher finishes to eat. Each philosopher does in cycle: taking forks, eating, releasing forks and thinking. The goal is to provide a solution allowing maximum parallelism where philosophers do not deadlock (otherwise all philosophers starve) and sometimes, additionally fairness is required, for a deadlock-free solution to provide a guarantee that no one will starve. Of course, we know how to solve the dining philosophers. We use this problem for illustration of how $k\Omega$-optimization will arrive at a solution that by minimizing costs will avoid deadlocks (having infinite costs) and starvation (by design - philosophers who ate above average will refrain nicely from competing for forks).

Let's consider problem solving (planning + execution) for dining philosophers providing deadlock-free and fair solution, and expressed as a special case of $k\Omega$-search. Note that the solution subsumes hierarchical (user-defined functions Phil, Fork and Count) and partial-order planning (due to concurrency from process algebra).

The system consists of 5 agents-philosophers, i.e., p=5, which are interested and can observe everything, i.e., Ω is an alphabet of all simple $-expressions, with a standard shared by all philosphers cost function $\$ = \$_1(\$_2(k\Omega[t], \$_3(x[t]))$, where $\$_1$ is an aggregating function in the form of addition, $\$_2(k\Omega[t])$ represents costs of the $k\Omega$-search, and $\$_3(x[t])$ represents the quality of solutions. A strong congruence is used. In other words, payoff is associated with empty/invisible actions ε for finding the plan (complete tree) and/or for executing it. The number of steps in the derivation tree selected for optimization in the examine phase $k = \infty$, the branching factor $b = \infty$, and the number of steps selected for execution in the examine phase $n = 0$, i.e., execution is postponed until the plan is found. Flags $gp = reinf = update = 0$ and $strongcong = 1$. The goal of plan is to find the cheapest deadlock-free and fair solution. The solution takes the form of the tree (truly DAGs) of $-expressions that are pruned and passed to execution phase.

Let's assume that user defined functions $Phil_i$, $Fork_i$, $Count$ are given and they are partially designed only, i.e., $\uplus$ represents unsolved alternatives in the design:

$(:= \ grab2_i \ \langle\!\langle \ (\ \| \ (\leftarrow \ p_{i,i} \ f_i) \ (\leftarrow \ p_{(i+4)mod \ 5,i} \ f_{(i+4)mod \ 5}) \rangle\!\rangle$ -atomic $grab2_i$ def.

$(:= \ Fork_i \ (\circ \ // $ allow to grab fork f_i by right/left neighbor and receive it back

$\qquad (\sqcup (\circ (\rightarrow \ p_{i,(i+1)mod \ 5} \ f_i) (\leftarrow \ p_{i,(i+1)mod \ 5} \ f_i))$

$\qquad \quad (\circ (\rightarrow \ p_{i,i} \ f_i) (\leftarrow \ p_{i,i} \ f_i))$
$\qquad)$

$\qquad Fork_i \ // $ call recursively $Fork_i$, $i = 0, 1, ..., 4$ process again

$))$

$(:= \ Count \ (\circ \ (\rightarrow \ ch \ c) \ // $ send and receive global count c of eatings

$\qquad \quad (\leftarrow \ ch \ c)$

$\qquad Count \ // $ call recursively $Count$

$))$

$(:= Phil_i (\circ (\leftarrow ch\ c)$ //grab global count c of eatings; each i-th phil., $i = 0, ..., 4$
// grabs fork $f_{(i+4)mod\ 5}$ through channel $p_{(i+4)mod\ 5,i}$ and fork f_i through channel $p_{i,i}$
$(\sqcup (\circ (\leq 5c_i\ c)$ // for fairness: grab forks if you did not eat above average
$(\uplus (\circ (\leftarrow p_{i,i}\ f_i) (\leftarrow p_{(i+4)mod\ 5,i}\ f_{(i+4)mod\ 5}))$ // grab right next left fork
$(\circ (\leftarrow p_{(i+4)mod\ 5,i}\ f_{(i+4)mod\ 5}) (\leftarrow p_{i,i}\ f_i))$ // grab left next right fork
$grab2_i$ // grab both forks atomically in parallel, def. call of simple \$-expr.
$)$ // in atomic $grab2_i$ all components should not block, else $grab2_i$ will block
eat
$c_i ++$ // increase your private count of eatings; initially all counts 0
$c ++$ // increase global count of eatings
$(\rightarrow p_{i,i}\ f_i) (\rightarrow p_{(i+4)mod\ 5,i}\ f_{(i+4)mod\ 5})$ //return both forks
$think\ )$ // do the job that philosophers supposed to do
$(> 5c_i\ c)$ // for fairness: be nice - do nothing if you ate above average
$)$
$(\rightarrow ch\ c)$ // return global count of eatings through channel ch to $Count$
$Phil_i$ // call recursively $Phil_i$, $i = 0, 1, ..., 4$ process again
$))$

TOTAL OPTIMIZATION: The goal will be to minimize costs for $\$ = \$_2 + \$_3$. The empty actions (representing actions not executed yet) have cost being the sum of payoffs for finding the plan/solutions and for execution of plan (for not starving philosophers). Each action has negative payoffs for action planned and executed (represented by function $\$_3$, and costs for searching for the plan and executing it (costs of running \$-Ruby interpreter - function $\$_2$). Assume that payoff for finding the plan is 500 (negative cost -500) and payoff for executing plan is 1000 (because it is a reactive never terminating program, it will never be reached and we can ignore it). Each action during planning (select and examine phase) and execution has cost 1 and no payoff (payoff will be paid after end of planning). During execution each action has cost 1 and payoff $2 - 0.02m$, where $m = 0, 1, 2, ...$ represents successive uses of the action. This means that payoffs initially will dominate, but after 100 uses actions will incur only costs.

0. t=0, initialization phase $init$:
   $x[0] = (\circ (\parallel Phil_0\ Phil_1\ Phil_2\ Phil_3\ Phil_4\ Fork_0\ Fork_1\ Fork_2\ Fork_3\ Fork_4\ Count)\ \varepsilon)$
   The initial tree consists of the root state $x[0]$ calling in parallel processes $Phil_i$, $Fork_i, i = 0, 1, ..., 4$ and $Count$, and an empty action ε which cost is equal to the estimated payoffs for finding plan and execution, 500. Because $x[0]$ is not the goal state (plan and its execution has not been done yet), interpreter goes to the first loop iteration consisting of select, examine, and execute phases.
1. t =1, first loop iteration:
   select phase sel :   because $k = \infty$ only one loop iteration will be needed and a complete potential tree of solutions is expanded, i.e., user defined functions $Phil_i$, $Fork_i$ and $Count$ are replaced by their body definitions.
   examine phase $exam$:

Execution is postponed ($n = 0$) until pruning is done. Assume that cost of multiset of parallel actions is equal to maximum of its component costs, thus grabbing in parallel two forks will have cost 1, and grabing sequentially 1+1=2. Some sequential orders like grabbing first left and next right fork by all philosophers will result in deadlock (and infinite costs), thus they will be eliminated by the $-calculus optimization mechanism. Parallel grabbing of both forks will be cheaper than sequential (and deadlock-free), thus design will leave only parallel grabbing in ⊎ definition of $Phil_i$.

execute phase $exec$:

Actions are executed, but with gradually decreasing payoffs (corresponding to the span of life of philosophers). Initially it will be cheaper to execute plan, but after 100 cycles for philosphers, costs will decrease payoff for finding plan, and execution should be stopped, because optimum will be lost and it will lead to infinite cost for immortal philosophers. After that the $k\Omega$-search re-initializes for a new problem to solve (e.g., machine learning ID3).

3.2   Learning the Best Decision Tree for a Single Agent

Most algorithms that have been developed for learning decision trees are variations on a core algorithm that employs a top-down, greedy search through the space of possible decision trees, i.e., the ID3 algorithm by Quinlan and its successor C4.5 [8]. ID3 performs a simple hill-climbing search through the hypothesis space of possible decision trees using as an evaluation function the Shannon-based information gain measure. ID3 maintains only a single current hypothesis as it searches through the space of decision trees. The search space is exponential, i.e., the problem is intractable. To alleviate that, ID3 picks up the attribute with the maximum information gain which leads to the shortest tree (i.e., it uses the Occam razor principle). The information gain $Gain(S, A)$ of an attribute A relative to a collection of examples S is defined as $Gain(S, A) = Entropy(S) - \Sigma_{v \in Values(A)} Entropy(S_v)|S_v|/|S|$, where $Entropy(S) = \Sigma_i - p_i log_2 p_i$.

Let's consider problem solving (learning + classification) for ID3 expressed as a special case of $k\Omega$-search finding the shortest classification tree by minimizing the sum of negative gains, i.e., maximizing the sum of positive gains.

The system consists again of one agent only, i.e., p=1, which is interested only in information gain for alphabet $A = \{a_i, a_{ij}\}, i, j = 1, 2$, i.e., $\Omega = A$, i.e., costs of other actions are ignored (being neutral - in this case having zero cost), and it uses a standard cost function $\$ = \$_3$, where $\$_3$ represents the quality of solutions in the form of cummulative negative information gains - payoff in $-calculus. In other words, total optimization is not performed, only regular optimization like in the original ID3 to illustrate a regular optimization, not that the total optimization is not desirable. A weak congruence is used. In other words, empty actions have zero cost. The number of steps in the derivation tree selected for optimization in the examine phase $k = 2$, the branching factor $b = \infty$, and the number of steps selected for execution in the examine phase $n = 0$, i.e., execution is postponed until learning is over. Flags $gp = reinf = update = strongcong = 0$. The goal of learning/classification is to minimize the sum of negative information gains. The machine learning takes the form of the tree of \$-expressions that are built in the select phase, pruned in the examine phase and passed to execution phase for classification work. Data are split into training and test data as usual. Let's assume for simplicity that we have only one decision attribute and two input attributes a_1 and a_2 with data taking two possible values on them denoted by $a_{11}, a_{12}, a_{21}, a_{22}$. Let assume that cost of actions is equal to entropy of data associated with this action, i.e., $\$(a_i) = -Entropy(a_i), \$(a_{ij}) = Entropy(a_{ij}), i, j = 1, 2$.

OPTIMIZATION: The goal will be to minimize the sum of costs (negative gains).

0. t=0, initialization phase $init$: $S_0 = \varepsilon_0$

The initial tree consists of an empty action ε_0 representing a missing classi- fication tree of which cost is ignored (a weak congruence). Because S_0 is not the goal state, the first loop iteration consisting of select, examine, and exe- cute phases replaces an invisible ε_0 two steps deep ($k = 2$) by all offsprings $b = \infty$.

1. t=1, first loop iteration:

select phase sel :

$\varepsilon_0 = (\uplus (\circ a_1(\sqcup (\circ a_{11} \varepsilon_{11})(\circ a_{12} \varepsilon_{12}))) (\circ a_2(\sqcup (\circ a_{21} \varepsilon_{21})(\circ a_{22} \varepsilon_{22}))))$
examine phase $exam$: $\$(S_0) = \$(\varepsilon_0) =$
$= min(\$(a_1) + p_{11}\$(a_{11}) + p_{12}\$(a_{12}), \$(a_2) + p_{21}\$(a_{21}) + p_{22}\$(a_{22}))$
Let's assume that attribute a_1 was selected, i.e., \$-expresion starting from a_1 is cheaper. Note that due to appropriate definition of the standard cost function (for complete definition see [37,38]) this is a negative gain from ID3. This confirms that \$-calculus cost function is defined in a reasonable way (at least from the point of ID3). Note that no estimates of future solutions are used (weak congruence - greedy hill climbing search). Execution is post- poned ($n = 0$), and follow-up ε_{11} and ε_{12} will be selected for expansion in the next loop iteration.

2. t = 2, second loop iteration:

select phase sel : $\varepsilon_{11} = (\circ a_2 (\sqcup (\circ a_{21} \varepsilon_{21})(\circ a_{22} \varepsilon_{22})))$
Let's assume that ε_{12} has data from one class only, thus this is the leaf node - no further splitting of training data is required.
examine phase $exam$:   nothing to optimize/prune - all attributes were used in the path or the leaf node contained sample data from one class of the decision attribute. Thus the end of the learning phase and the shortest decision tree is designated for the execution:
execute phase $exec$:
Test data are classified by the decision tree left from the select/examine phases. After that the $k\Omega$-search re-initializes for the new problem to solve.

Note that we can change for example values of k (considering a few attributes in parallel), b, n and optimization to total optimization, then this will be related, but not ID3 algorithm any more. This is the biggest advantage and flexibility of \$-calculus problem solving. It can modify "on fly" existing algorithms and design new algorithms, and not simulation of ID3 alone (as obviously ID3 can be expressed in any powerful enough programming language).

Other examples of problem solving by search simulated by $k\Omega$-optimization, including A*, minimax, dynamic programming, TSP, the halting problem of UTM, neural networks, cellular automata, simulation of λ-calculus or π-calculus can be found in [3–5]. This illustrates the power and versatility of the $k\Omega$-search meta-algorithm.

4 Implementation in $-cala and $-calculisp

We started to implement $-calculus framework in $-cala [1] and $-Calculisp [9] to gain necessary experience, to solve some theoretical problems, and as a preliminary step in the main implementation in Ruby. Ruby is a powerful and versatile object-oriented language that has at the same time a flexibility to that of LISP in dealing with code and meta-code management. We selected Scala, because of very concise implementation of related π-calculus process algebra in Scala [2]. On the other hand, Common Lisp seems be a natural contrcandidate to Ruby. Initially, we implemented in $-cala and $-Calculisp the same examples as $k\Omega$-search: n-puzzle, A* and TSP, and next added dining philosphers and ID3. It looks that both Scala and Lisp have similar metaprogramming capabilities as Ruby, however they are less popular at this moment (and, most likely, in the future) and seem to have less powerful programming environment.

$-Calculisp is implemented in Common Lisp using a combination of functional programming, the powerful Common Lisp macro meta-programming system, and the Common Lisp Object System. It implements an object tree of $-nodes, and $-calculus operators for manipulating the $-node object tree. $-nodes are sub-classed and their methods extended to match a given problem domain or algorithm. Once the problem domain is defined, $-operators and the $k\Omega$ construct are used to implement the bulk of the algorithm.

$-cala is implemented in Scala - an object functional language with an extensible syntax that makes it an ideal host language [2]. $-cala attempts to provide all the primitives from $-calculus as objects. The programmer uses syntax similar to $-calculus to create an abstract syntax tree (AST) and then the AST is interpreted as $k\Omega$-search algorithm. As a consequence a $-cala program corresponds closely to a $-calculus program.

5 Conclusions and Further Work

In the paper, we presented a uniform problem solving by the same $k\Omega$-meta-search algorithm, and its preliminary implementation in $-cala and $-Calculisp. Using the same $k\Omega$-search we can investigate completeness, optimality, search optimality and total optimality of uninformed and informed search methods for adversary and cooperating agents (see [3–5]). Both sequential and concurrent (partial-order) planning, conditional planning, hierarchical planning can be expressed and investigated (their completeness and optimality) in the same uniform way as $-calculus search. We can quite easily express Dynamic Bayesian Networks - DBNs (using sequential composition for conditional probabilities, and general choice operator combining varius choices weighted by probabilities). Bayesian learning (e.g., MAP/ML [8]) has a natural mapping by $k\Omega$-search tree.

We want to use $-calculus to experiment and combine various approaches. Even simple modifications with simulation of evolutionary algorithms, like changing for example values of n, k, b will lead to algorithms that are not *sensu stricto* evolutionary algorithms any more. Similarly, we would be able to experiment with algorithms based on existing algorithms, but they will not be classical A* or minimax any more. For example, A* with n!=0 leads to "on-line A*", or A* using general choice (besides cost choice) will lead to "probabilistic, fuzzy set or rough set A*" [5]; fixing b leads to SMA*. Additionally, using total optimization leads to algorithms that stop to be classical A*, minimax or evolutionary algorithms (they even do not have corresponding names in the literature). Of course, you can try to do that in any language, but it will be much easier to do in a language designed for it.

The author, besides earlier work on GPS by Simon and Newell and Koza's GPPS is not aware about any related approach so far. Thus the approach is totally novel and if successful can be compared only in its significance with the role of the Turing Machine model in computer science.

References

1. Ansari, B.: $-cala: An Embedded Programming Language Based on $-Calculus and Scala, Master Thesis, Rensselaer Polytechnic Institute at Hartford (2013)
2. Cremet, V., Odersky, M.: PiLib: A Hosted Language for Pi-Calculus Style Concurrency. In: Lengauer, C., Batory, D., Blum, A., Odersky, M. (eds.) Domain-Specific Program Generation. LNCS, vol. 3016, pp. 180–195. Springer, Heidelberg (2004)
3. Eberbach, E.: $-Calculus of Bounded Rational Agents: Flexible Optimization as Search under Bounded Resources in Interactive Systems. Fundamenta Informaticae 68(1-2), 47–102 (2005)
4. Eberbach, E.: The $-Calculus Process Algebra for Problem Solving: A Paradigmatic Shift in Handling Hard Computational Problems. Theoretical Computer Science 383(2-3), 200–243 (2007), doi:dx.doi.org/10.1016/j.tcs.2007.04.012
5. Eberbach, E.: Approximate Reasoning in the Algebra of Bounded Rational Agents. Intern. Journal of Approximate Reasoning 49(2), 316–330 (2006), doi:dx.doi.org/10.1016/j.ijar.2006.09.014
6. Horvitz, E., Zilberstein, S.: Computational Tradeoffs under Bounded Resources. Artificial Intelligence 126, 1–196 (2001)
7. Milner, R., Parrow, J., Walker, D.: A Calculus of Mobile Processes, I & II. Information and Computation 100, 1–77 (1992)
8. Russell, S., Norvig, P.: Artificial Intelligence: A Modern Approach, 3rd edn. Prentice-Hall (1995) (2010)
9. Smith, J.: $-Calculisp: an Implementation of $-Calculus Process Algebra in Common Lisp. Master Project, Rensselaer Polytechnic Institute at Hartford (2013)

CAUMEL: A Temporal Logic Based Language for Causal Maps to Explain Agent Behaviors

Aroua Hedhili Sbaï and Wided Lejouad Chaari

Strategies of Optimization and Intelligent Computing Laboratory (SOIE),
National School of Computer Studies (ENSI)-University of Manouba, Tunisia
{aroua.hedhili,wided.chaari}@ensi.rnu.tn
http://www.soie.isg.rnu.tn

Abstract. Causal maps are a powerful tools, used to deal with causal relations between events. They are frequently developed for specific issues such as decision analysis and problems diagnostic. The approach described in this paper underlines their novel utility providing a foundation to explain how agents have done actions. In fact, Multi-Agent Systems (MAS) are considered as complex systems, in which agent actions are affected by several factors as uncertain beliefs, intentions of other agents, high interaction, and the dynamic aspect of the environment. Thus, we believe that it is crucial to elucidate the agent system's behavior. To address the explanation of agent behaviors, this research presents, summarily, our method to build the causal map that corresponds to observed events during agent activities. Then, it focuses on a formal logic theory to interpret the built causal map, which includes causation between temporally ordered actions.

Keywords: causal map, explanation, agent behavior, temporal logic.

1 Introduction

Agents present a promising approach for dealing with complex system development. During its execution, an agent may have several alternatives to attend an objective and it chooses the most appropriate one based on the context. This makes agents flexible and robust, however it makes explanation and comprehension of agents reasoning challenging. Consequently, the description of agent behaviors has been investigated through different works. We note that the existent researches in MAS are based on adjusting the structure, the design and the architecture of the agent in order to be comprehensible or to produce a description of its behaviors [1], [2], [3]. So the explanation process is considered in the design phase as a behavior description rather than a reasoning explanation. This methodology may have severe limits: (i) it requires that the behavior description starts with the design and the development of the MAS, we can not use it to explain the reasoning in systems that have been already developed, (ii) the cited approaches are concerned with specific types of agents; the one illustrated by [1] requires an agent who is able to learn while the solution highlighted in [2]

G. Jezic et al. (eds.), *Agent and Multi-Agent Systems: Technologies and Applications*,
Advances in Intelligent Systems and Computing 296,
DOI: 10.1007/978-3-319-07650-8_14, © Springer International Publishing Switzerland 2014

is only applied for believable agents, (iii) it overburdens the agents' behaviors which reduce the system performance; the agent should resolve a problem and at the same time deal with the explanation process.

Our aim is to provide an explanation system independent of agent architecture and MAS design. So, we propose an explanation system divided into three modules: an observation module presented in [8], [10], a modeling module described in [9], [10] and an interpretation module. The role of the observation module consists in visualizing the agents' activities and collecting the context of the agent reasoning as explanatory knowledge. The modeling module relates the acquired explanatory knowledge with causal relationships in different levels of causal maps. Causal Maps or Cognitive Maps (CM) are represented as a directed graph where the basic elements are simple. The concepts are represented as points, and the causal links between these concepts are represented as arrows between these points [4]. Eventually, the interpretation module analyzes the built causal maps to generate new knowledge for explanation purpose. The expected explanations are produced during the execution of the MAS. We intend an on-line explanation. The main issue underlined in this paper focuses on the interpretation module. This module concerns the analysis of the represented explanatory knowledge in the provided causal maps. However, the existent works as [4], [5] and [6] are based on the intuitive view of causal maps with ad hoc rules, no precise semantic or formal treatment is provided to analyze concepts and relations between them. In our case, the concepts and relations translate the agents behaviors. These behaviors are considered as *intra-agent* reasoning that deals with agent attitude and *inter-agent* reasoning that focuses on the entire system attitude. So, the highlighted interpretation module in this paper presents an *intra-agent* explanation (individual explanation) and an *inter-agent* explanation (social explanation) that generate new knowledge according to the different levels of the built causal maps.

This paper is organized as follows. In section 2, we present briefly our method to construct causal maps in which the output of the observation module is represented as concepts and causal assertions. In section 3, we quote some works related to agent knowledge representation and causal reasoning. In section 4, we elucidate our contribution to analyze the built causal maps. In our solution, we define a specific language based on first order logic, designated to the generated causal maps and formulated with agent context.

2   Causal Map Construction for Explanation

As mentioned, our explanation process starts with an observation phase. An observed action could be: a communication act, a perception, a resource acquisition, a removal/addition of an agent, or any action performed by the agent to achieve its goal. The observation module collects the context of the execution of detected actions as explanatory knowledge, in a tuple $< K, A, G, R >$ (Knowledge, Action, Goal, Relation). Thus, the reasoning state of an agent i detected at the moment t is described as the tuple $< K_i(t), A_i(t), G_i(t), R_i(t) >$ where:

- $K_i(t) = \bigwedge_{m=0}^{n} k_m$; k_m is a set of concepts defined based on the knowledge structure used to represent agent knowledge during the conception of the MAS as ontology, semantic network, set of production rules, etc.
- $A_i(t) = (agent\ identifier, a_l, execution\ time)$; the action value, a_l, executed by an agent, $agent\ identifier$, at the moment, $execution\ time$. The action value is always different from zero, it could be a communication act, a perception of the environment, an external action, a resource holding, etc.
- $G_i(t) = goal_i$; the goal value and it could have the value null.
- $R_i(t) = \bigwedge_{m=0}^{n} r_m$ where r_m indicates information about the communication act and it is defined using the communication language structure.

These acquired explanatory knowledge could not reflect an explanation of performed actions by agents, they are deprived of a clear and a semantic explanation. Consequently, we identify the causal links between these knowledge to elucidate the cause/effect relationships among visualized events [9], [10]. So, we proceed according to different levels of construction in order to build connected causal maps in what we call an Extended Causal Map (ECM). These levels denote several types of causal links. The first level presents a **temporal** one. It describes a causal graph where the concepts are the reasoning states, RS_i. These concepts express the detected agents' behaviors while the graph arrows (t_i, t_{i+1}) shows the alteration between behaviors. Each observed behavior depicted in the graph concept is represented in our case with the explanatory knowledge collected in the tuple $< K, A, G, R >$. Therefore, from this level we generate a second level, labeled an **horizontal level**, that indicates the causal relations between the tuple attributes. Furthermore, we recognize additional causal links via the temporal relation depicted between RS_i and RS_{i+1} that refer to the detected behaviors of, on the one hand, the same agent, and on the other hand, of different agents. So, we consider that there are causal relations between explanatory knowledge collected at the moment t_i in the tuple $< K(t_i), A(t_i), G(t_i), R(t_i) >$ and the ones collected at the moment t_{i+1} in the tuple $< K(t_{i+1}), A(t_{i+1}), G(t_{i+1}), R(t_{i+1}) >$ for the same agent. These relationships outline an **internal vertical level** in the ECM. This level indicates the causal relations between the actions performed by the agent and its satisfied goals. Then, we note that the relation between the behaviors of different agents of the MAS depends on the interaction process elaborated between these entities. We remind that each performed communication act is detected by the observation module, and its execution parameters are stored in the attributes R and A. So, we focus on these attributes for each agent at different moments to describe and analyze the established interactions. We create these causal maps in what we call an **external vertical level** based on the performatives of the agent communication language:

CM_RELATIONS: In this map, we connect the attribute $R_i(t_x)$ that presents the details of the sending action performed by the sender agent i at the moment t_x to the one accomplished by the addressee, an agent j, of which the detail of the sent message is stored in the attribute $R_j(t_y)$. The causal relation between

$R_i(t_x)$ and $R_j(t_y)$ refers to the provided interaction protocol between the agents i and j. It has the positive value to show that the sent message at the moment t_y is the cause of the sent one at t_x.

CM_RELATIONK: An agent i could interact with an agent j to require or to verify the value of some knowledge. This fact could not be visualized explicitly but it could be deduced via a set of actions accomplished by these agents. We decide to relate the essential performed actions to conclude this fact as a causal map. This latter is composed of the attributes $R_j(t_x)$ and $K_i(t_y)$. The causal relation has the positive value to indicate that the sent message at the moment t_x is the cause of the used knowledge at t_y. Notice that this link is established after a comparison between the content of the message, $R_j(t_x)$, and the used knowledge, $K_i(t_y)$.

CM_RELATIONA: Another reason makes an agent i interact with an agent j is to require the execution of an action. In this case, a causal link is created between the attribute $R_i(t_x)$ that expresses an acceptation of a previously submitted proposal to perform an action and the attribute $A_j(t_y)$ which presents the required action, executed by the agent j at the moment t_y.

We note that the presented causal maps in this section are explicitly described and illustrated in [10].

Further, a knowledge representation model as a graph could provide a visual interpretation in terms of a set of relationships between concepts. Such interpretation is easier if the graph has a limited number of nodes and arrows. This is not the case of our obtained ECM elaborated for complex systems like MAS. So, we propose to associate this causal map with an interpretation formalism to deduce a novel knowledge for explanation.

3   Related Works

We introduce in a previous work [13] a first vision to interpret the built causal maps that represent the explanatory knowledge of each agent. This interpretation deals with the intra-agent explanation. So, we were inspired by theories and frameworks that have been developed in order to describe agent reasoning as BDI [11], and KARO [12]. Besides, the represented explanatory knowledge in the causal map refer to agent actions that express the change between the agent behaviors. Or, the concepts of changes and time are closely related. This idea is also represented in the built causal map (the internal vertical level and the external vertical one). In this context, there are different formalisms to represent the causal and the temporal relationship between events and actions as McDermott's logic [14] and Allen's logic [15] that describe predicates to show the causality between certain events. In addition, the representation of knowledge in dynamic world was described as an action language based on the situation calculus in several research works as [16], [17], etc. However, these works deal with the representation of the expected knowledge related to agents and they have limits to present temporal and causal reasoning as frame and ramification problems. Through the previous state of the literature study, we note that there is no formalism that investigates at once the inter-agent reasoning concepts, time in causal theory and causal maps. Our original contribution is essentially oriented toward providing a semantic interpretation of represented explanatory knowledge in causal maps provided in both horizontal level and external vertical level. For this purpose, we improve the work presented in [13] to introduce the dynamic aspect of the MAS and we define the CAUsal Map Explanation Language (CAUMEL).

4   Causal Map Explanation Language: CAUMEL

The highlighted language in this paper, CAUMEL, aims to provide a new knowledge from the concepts and cause/effect relations between them. For this purpose, we were inspired by notations and terminologies presented in the literature to define our language. In fact, the proposed language CAUMEL is defined at two levels:

- The first level, L1, consists of a first order language to depict static information.
- The second level, L2, is a temporal logic described as extended first order logic with temporal argument to represent dynamic information. It is composed of predicates, and functions with temporal variables. For instance, if the predicate $p(x_1, x_2, ..., x_n)$ is a well formed formula of L1, T a set of temporal elements and $t \in$ T then $p(t, x_1, x_2, ..., x_n)$ is a formula of L2.

4.1   Alphabet of CAUMEL

The alphabet of the presented language consists of the following symbols:

- Connectives: $\wedge, \vee, \neg, \Longleftrightarrow$
- Quantification symbols: $\exists, \forall$
- Collection of variables V that contains the following sets:
  - $Ag = \{ag1, ag_2, ..., ag_n\}$: set of active agents of the MAS.
  - $Ac = \{a_1, a_2, ..., a_n\}$: set of performed actions by agents.
  - $K = \{k_1, k_2, ..., k_n\}$: set of used knowledge by agents.
  - $R = \{r_1, r_2, ..., r_n\}$: set of exchanged messages between agents.
  - $G = \{g_1, g_2, ..., g_n\}$: set of agents' goals.
  - $N = \{K, A, G, R\}$: set of the nodes of the built causal maps.
- Set of predicates P that involves the temporal predicates, Pt, and the static ones, Ps.
- Set of temporal elements T that indicates the temporal intervals and the temporal points.
- Set of functions F.

4.2 Functions of CAUMEL

The set F describes three functions. First, we present the correspondence between the static level and the dynamic one of the language CAUMEL through the following functions:

The function f_p associates a static knowledge, represented as a predicate, that carried out at a temporal element t with a temporal predicate:

$$f_p : Ps \longrightarrow Pt$$
$$p(x_1, x_2, ..., x_n) \longmapsto f_p(p) = p(t, x_1, x_2, ..., x_n)$$

The function f_v attributes values to some variables depending on time:

$$f_v : V \longrightarrow V \times T$$
$$v \longmapsto f_v(v) = v(t)$$

Then, we define a function $Period(p(t, x_1, x_2, ..., x_n))$ that distinguishes a temporal interval from a temporal point. It determines for each temporal predicate the occurrence period.

$$Period : Pt \longrightarrow \mathbb{R}_{>=0}$$
$$p(t, x_1, x_2, ..., x_n) \longmapsto Period(p(t, x_1, x_2, ..., x_n)) = 0 \text{ or } > 0$$

-$Period(p(t, x_1, x_2, ..., x_n)) = 0$ when the predicate p presents a description that appears immediately.
-$Period(p(t, x_1, x_2, ..., x_n)) > 0$ when the predicate p presents a description that appears at a period.

4.3 Well-formed formulas

The well-formed formulas are defined as follows:

- if $p \in$ P, $t \in$ T and $x_1, x_2, ..., x_n$ are individuals of the set V then $p(t, x_1, x_2, ..., x_n)$ is a well-formed formula.
- if p, q are well-formed formulas then $p \wedge q$ is a well formed formula.
- if p is a well-formed formula then $\neg p$ is a well formed one.
- if p is a well-formed formula, $x \in$ V or $x \in$ T then $(\forall x p)$ and $(\exists x p)$ are well formed formulas.
- if t_1 and t_2 are temporal points of the set T then the operators of Allen $EQUALS(t_1, t_2)$ and $BEFORE(t_1, t_2)$ are a well formed formulas.
- if t_1 and t_2 are temporal intervals of the set T then the operators of Allen $IN(t_1, t_2)$, $BEFORE(t_1, t_2)$, $MEETS(t_1, t_2)$, $OVERLAPS(t_1, t_2)$, and $EQUALS(t_1, t_2)$ are well formed formulas.

4.4 Formulas of CAUMEL

In this section, we present the formulas of the proposed language CAUMEL, used to interpret knowledge depicted in the concepts and the cause/effect relationships of the built causal maps.

A Causal Map Description in CAUMEL. Our provided causal maps in the modeling module are composed of the following concepts $< K, A, G, R >$ that represent the explanatory knowledge. Besides, these maps are built at fixed moments. Thus, we define a set of formulas that refer to the occurrence of the map concepts at a temporal moment t for an agent i as follows:

- $Exists(t, A) \iff f_v(A) = A_i(t)$
- $Exists(t, K) \iff f_v(K) = K_i(t)$
- $Exists(t, R) \iff f_v(R) = R_i(t)$
- $Exists(t, G) \iff f_v(G) = G_i(t)$

The values $A_i(t)$, $K_i(t)$, $R_i(t)$, and $G_i(t)$ are already detailed in section 2. In addition to concepts, the causal map is constituted of causal assertions as the positive and the negative relationships in our case. We interpret a causal relation that occurs in a temporal element t between two concepts x and y as:

- $R^+(t, x, y) \iff \exists t'|(Exists(t', x) \wedge Exists(t, y)) \wedge (IN(t', t) \vee BEFORE(t', t) \vee MEETS(t', t) \vee OVERLAPS(t', t) \vee EQUALS(t', t)) \wedge (inflH^+(x, y) \vee inflVI^+(x, y) \vee inflVE^+(x, y))$
- $R^-(t, x, y) \iff \exists t'|(Exists(t', x) \wedge Exists(t, y)) \wedge (IN(t', t) \vee BEFORE(t', t) \vee MEETS(t', t) \vee OVERLAPS(t', t) \vee EQUALS(t', t)) \wedge (inflH^-(x, y) \vee inflVI^-(x, y) \vee inflVE^-(x, y))$

The first parenthesis that includes the predicate $Exists$ mentions the occurrence of the map concepts, x and y, at respectively the temporal elements t and t'. Then, the second parenthesis shows the temporal relation between t and t' according to Allen operators. Ultimately, the third parenthesis reveals a static information about the value of the causal relation between the concepts x and y. It indicates as well the type of the causal map considered in this interpretation:

- $inflH^{+|-}(x, y)$: A causal influence between the concepts x and y of the built causal map at the horizontal level.
- $inflVI^{+|-}(x, y)$: A causal influence between the concepts x and y of the built causal map at the vertical internal level.
- $inflVE^{+|-}(x, y)$: A causal influence between the concepts x and y of the built causal map at the vertical external level.

Intra-agent Behavior in CAUMEL. In our approach, we explain agent reasoning using acquired explanatory knowledge, represented as cause, effect, and causal relation in causal maps. This explanation is divided into an intra-agent explanation which focuses on the reasoning explanation of each agent separately in the system and an inter-agent explanation which uses the interaction between agents and the change of the agents' behaviors in the time to clarify the entire system reasoning process. This section is devoted to the intra-agent explanation. We interpret each built causal map that represents an agent activity in order to translate it to comprehensible predicates according to the notions and terminology of CAUMEL already presented. We note that the concepts of the generated causal map belong to the set $\{K, A, G, R\}$ and the causal relationship has the positive or negative value. So, the first causal map illustrated in the figure 1 (a) is interpreted as an interaction act performed successfully by an agent ag_i at a temporal element t based on the message details saved in the concept R. This is expressed with the following predicate:

- $communicate\_success(t, ag_i, R) \Longleftrightarrow \exists A, R, G | R^+(t, R, A) \wedge R^+(t, A, G)$

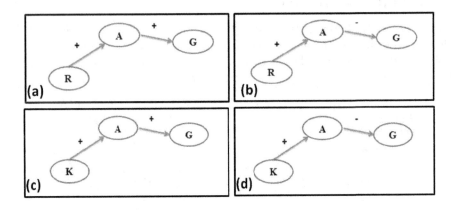

Fig. 1. Intra-agent causal maps

However, the causal map depicted in the figure 1 (b) shows a failure interaction act. This is expressed as follows:

- $communicate\_failure(t, ag_i, R) \Longleftrightarrow \exists A, R, G | R^+(t, R, A) \wedge R^-(t, A, G)$

In addition to a communication act, an agent performs tasks. A task could be executed to achieve goals or to prevent them. These knowledge are represented in the causal maps depicted in the figure 1 (c) and (d) and they are interpreted as follows:

- $act(t, ag_i, K) \Longleftrightarrow \exists A, K | R^+(t, K, A)$
- $perform(t, ag_i, A, G) \Longleftrightarrow \exists A, K, G | R^+(t, K, A) \wedge R^+(t, A, G)$
- $prevent(t, ag_i, A, G) \Longleftrightarrow \exists A, K, G | R^+(t, K, A) \wedge R^-(t, A, G)$

The first predicate shows the used knowledge by the agent to execute an action. The second one translates the figure 1 (c), it shows an executed action to reach a goal. Nevertheless, the third predicate indicates an executed action to prevent a goal, this is illustrated in the graph (d).

So far, our study focuses on the horizontal level of the generated causal maps. It interprets the intra-agent level without considering the inter-agent level that includes the agent society. The analysis of this level is based on the maps provided at the external vertical level which is deduced from the transition between agents' behaviors according to the temporal causal graph. An interpretation of this temporal transition is needed in order to investigate the social reasoning of the agent.

Agents' Behaviors Change in CAUMEL. We consider the temporal causal graph presented in section 2. We represent the graph concepts RS_1, RS_2,..., RS_i respectively with the predicates $p1(t_1, x_1, x_2, ..., x_n)$, $p2(t_2, x_1, x_2, ..., x_n)$,..., $pi(t_i, x_1, x_2, ..., x_n)$ which express agents' behaviors. We introduce the following formula to indicate a temporal relationship between pi and pj:

$$- Tchange(pi(t_1, x_1, x_2, ..., x_n), pj(t_2, x_1, x_2, ..., x_n)) \iff IN(t_1, t_2) \lor$$
$$BEFORE$$
$$(t_1, t_2) \lor MEETS(t_1, t_2) \lor OVERLAPS(t_1, t_2) \lor EQUALS(t_1, t_2)$$

This formula shows the dynamic aspect of MAS. It describes the temporal change at agents' behaviors level.

Inter-agent Behavior in CAUMEL. In complex systems such MAS, the agents' goals are not always achievable. We cannot ensure that the developed plans will be properly executed. An expected result implies that all resources are available and all action plans' are successfully accomplished. We believe that it exists two alternatives to achieve an objective (1) the agent is completely autonomous, it is considered as an active entity that perceives its environment, reasons and executes appropriate actions to accomplish its goal without the intervention of another agent or user (2) the agent is dependent of another agent if the latter can help or prevent him to achieve its goal. This latest point concerns the inter-agent explanation. In order to generate the inter-agent explanation, we focus on the causal map, CM, generated at the current temporal element t and the one, CM', obtained at a previous temporal element t' and which has at least one concept related to CM concepts with a causal relation. In this case, we present the following causal maps and its interpretation according to CAUMEL:

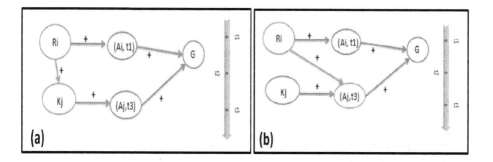

Fig. 2. Inter-agent causal maps

- $social\_knowledge(t3, ag_j) \iff \exists K, R|Tchange(communicate\_success(ag_i, t1, R), act(ag_j, t3, K)) \wedge R^+(t3, R, K)$

The map in figure 2 (a) indicates that the agent ag_j uses a knowledge, $act(ag_j, t3, K)$, deduced from an interaction process with the agent ag_i, $communicate\_success$ $(ag_i, t1, R)$. So, the agent ag_j enhances its knowledge base with this new information. Otherwise, the concept K_j is not linked with a concept R, in this case, we determine that the agent uses its own knowledge. This could be exploited to differentiate between a dependent agent and an autonomous one.

- $delegate(t1, ag_i) \iff \exists K, R|Tchange(communicate\_success(ag_i, t1, R), perform(ag_j, t3, K)) \wedge R^+(t3, R, A)$

The figure 2 (b) shows that the agent ag_i delegates an action to the agent ag_j. In fact, the agent ag_i interacts with ag_j and this latter performs the required action. This required action appears in the content of the exchanged message. This knowledge is expressed by the predicate $delegate(t, x)$.

We have shown the explanatory knowledge of an agent using extended causal map and we have encoded this model based on an explanation language CAUMEL. This language contains predicates that refer to the agent reasoning, expressed according to causal map concepts, influences in the paths, and temporal relation between actions. By doing so, we retrieve from the causal relations a semantic interpretation of the represented explanatory knowledge.

4.5   Semantics

The semantic of CAUMEL is defined using the interpretation structure M=<W, O, D>, where W is a set of temporal elements, O presents the ordering relations defined on W, and D is a set of individuals disjoint from W. The model-structure M is associated with an interpretation function I that fulfills the links between temporal terms of W and elements of D, non-temporal terms and elements of D, the Allen operators and relations of O. Consequently, we consider the following interpretations:

- The function I assigns a member of D to every variable in V, a member of W to each member of T, for each predicate having as parameters $(t, x_1, x_2, .., x_n)$, a member of W to the parameter t and a member of D to each x_i.
- $I : Pt \longrightarrow T$
  $p_i(t, x_1, x_2, .., x_n) \longmapsto I(p_i(t, x_1, x_2, .., x_n)) = T_i = \{t_i/p_i(x_1, x_2, .., x_n)$ true in $t_i\}$
- if $p_i \in Pt$ then $I(\neg p_i) = T - I(p_i) = T - T_i$
- if $p, q \in P$ then $I(p \wedge q) =$ true iff $I(p)=$ true and $I(q) =$ true.
- if t_1, t_2 are temporal points then $I(EQUALS(t_1, t_2))$ is true iff $I(t_1.start) = I(t_2.start)$ and $I(t_1.finish) = I(t_2.finish)$.

- if t_1, t_2 are temporal elements then $I(BEFORE(t_1, t_2))$ is true iff $I(t_1.start) < I(t_2.start)$.
- if t_1, t_2 are temporal elements then $I(Meets(t_1, t_2))$ is true iff $I(t_2.start) = I(t_1.finish)$.
- if t_1, t_2 are temporal elements then $I(Overlaps(t_1, t_2))$ is true iff $I(t_1.start) < I(t_2.start)$ and $I(t_1.finish) > I(t_2.start)$ and $I(t_1.finish) < I(t_2.finish)$.
- For the quantification, we consider $I(\forall xp)$ is true iff $I(p)$ is true for each evaluation, $I(\exists xp)$ is true iff $I(p)$ is true for this element.

5   Conclusion

We believe that causal maps present a suitable tool for event explanation since they describe the cause/effect relations between concepts related to this event. In this paper, we highlight the use of this structure to explain agent behaviors during the MAS execution (on-line explanation). Our contribution consists in first observing the agent activity, detected at such temporal element, and collecting the necessary explanatory knowledge in the tuple $< K, A, G, R >$ at each temporal element. Then, we propose to identify the causal relationships between the acquired explanatory knowledge in causal maps. The method used to build the causal map is divided into different levels; temporal level, horizontal level, internal vertical level, and external vertical level. Finally, we define an explanation language CAUMEL to analyze the built causal maps. This language is based on the classic temporal logic that extends the first order logic to include the dynamic information using the method of temporal arguments. We note that the presented approach provides an explanatory Knowledge Based System (KBS) for MAS (knowledge acquisition phase, knowledge representation phase, and knowledge interpretation phase). However, this approach is not applied on MAS developed for virtual environments because of the emotional aspect involved in such environments and not considered in our work. In addition, our solution is not validated in Robotics since the developed observation module uses software probes that do not detect robots behavior. Currently, we aim to produce agent behaviors explanations using a storytelling tool, to test the underlined approach in a complex MAS, and to evaluate the quality of the produced explanations.

References

1. Johnson, W.L.: Agents that explain their own actions. In: Proceedings of the Fourth Conference on Computer Generated Forces and Behavioral Representation (1994)
2. Sengers, P.: Designing comprehensible agents. In: Proceedings of the Sixteenth International Joint Conference on Artificial Intelligence, IJCAI 1999, pp. 1227–1232 (1999)
3. Görz, G., Ludwig, B., Reib, P., Schiemann, B., Seutter, T.: Self-describing agents. In: Multikonferenz Wirtschaftsinformatik. GITO-Verlag, Berlin (2008)

4. Chaib-draa, B.: Causal Maps: Theory, Implementation and Practical Applications in Multiagent Environments. IEEE Transactions on Knowledge and Data Engineering 14, 1201–1217 (2002)
5. El Fallah Seghrouchni, A., Haddad, S., Mazouzi, H.: A Formal Study of Interactions in Multi-Agent Systems. In: The Proceeding of the 14th International Conference on Computers and Their Applications (1999)
6. Guillermo, V., Jorge, J., Gómez, S., Juan, A., Botía, B., Juan, P.: Using Semantic Causality Graphs to Validate MAS Models. In: Innovations in Hybrid Intelligent Systems. Advances in Soft Computing, pp. 9–16. Springer, Heidelberg (2007)
7. Vasseur, A.: Dynamic AOP and runtime weaving for Java-How does AspectWerkz address it? In: DAW: Dynamic Aspects Workshop (2004)
8. Hedhili, A., Lejouad Chaari, W., Ghédira, K.: Explanation Issue in Multi-Agent Systems. In: The International Conference on Computational Science and Information Management (2012)
9. Hedhili, A., Chaari, W.L., Ghédira, K.: Causal Maps for Explanation in Multi-Agent System. In: Abraham, A., Thampi, S.M. (eds.) Intelligent Informatics. AISC, vol. 182, pp. 183–191. Springer, Heidelberg (2013)
10. Hedhili Sbaï, A., Lejouad Chaari, W.: Extended Causal Map for Explanation in Multi-Agent Systems. Special Issue on: Intelligent Systems and Applications Using Knowledge and Agent-Based Technologies of the Int. Journal of Intelligent Systems Technologies and Applications 12(3/4), 301–315 (2013)
11. Rao, A.S., Georgeff, M.P.: Modeling rational agents within a BDI-architecture. In: Allen, J., Fikes, R., Sandewal, E. (eds.) Proceedings of the Second International Conference on Principles of Knowledge Representation and Reasoning, pp. 473–484. Morgan Kaufman (1991)
12. Van Linder, B., Van der Hoek, W., Meyer, J.-J.: Ch: Formalising motivational attitudes of agents: On preferences, goals and commitments. In: Tambe, M., Müller, J., Wooldridge, M.J. (eds.) IJCAI-WS 1995 and ATAL 1995. LNCS, vol. 1037, pp. 17–32. Springer, Heidelberg (1996)
13. Hedhili Sbaï, A., Lejouad Chaari, W., Ghédira, K.: Intra-agent Explanation Using Temporal and Extended Causal Maps. In: Proceeding of the International Conference Knowledge-Based and Intelligent Information and Engineering Systems, KES 2013, Procedia Computer Science, vol. 22, pp. 241–249 (2013)
14. McDermott, D.V.: A temporal logic for reasoning about processes and plans. Process and Cognitive Science 6, 101–155 (1982)
15. Allen, J.: Towards a general theory of actions and time. Artificial Intelligence, 123–154 (1984)
16. Lesprance, Y., Levesque, H., Lin, F., Lin, M.D., Reiter, R., Scherl, R.: Foundations of logical approach to agent programming. In: Tambe, M., Müller, J., Wooldridge, M.J. (eds.) IJCAI-WS 1995 and ATAL 1995. LNCS, vol. 1037, pp. 331–347. Springer, Heidelberg (1996)
17. Baral, C., Geffond, M.: Reasoning about effects of concurrent actions. Journal of Logic Programming (1997)

Communication Security Prognosis Realized as the Parallel Dynamic Auditing Intelligent System

Henryk Piech, Grzegorz Grodzki, and Piotr Borowik

Czestochowa University of Technology,
Dabrowskiego 73, 42201 Czestochowa, Poland
h.piech@adm.pcz.czest.pl, ggrodzki@icis.pcz.pl, piotrborowikii@gmail.com

Abstract. Communication operations are realized according to cryptography protocols in a typical network. Such communication among users takes place on the basis of public keys, secrets, supplies, encrypted messages and nonces. The investigation of the communication run gives security information about forthcoming threats. The main goal of the research consists in the elaboration of a useful and simple (in the sense of complexity) prognosis algorithm adapted to an auditing form of investigation. The results of the prognosis presented in the time parameter about impending threats are dynamically changed, operation by operation. Therefore, users can prepare a strategy of avoiding the closest (in the sense of time) or the most dangerous threat. The proposed approach is based on probability counting rules that guarantee fast realization at the cost of accuracy. The large scale of parallelization possibilities is also worth noticing. This follows from the independent module structure of fundamental security elements which are associated with dynamically designated counting threads.

Keywords: protocol logic, probabilistic timed automata, communication security prognosis.

1 Introduction

The auditing investigation program is treated as an intelligent system [1], [2] for dynamic generation of warnings about possible forthcoming threats. There are many approaches to elaborate convenient and useful information about the security of communication protocols [3]. Usually, proposed solutions refer to a single protocol [4] and independently give short deterministic or approximate information about the security of a protocol during the moment of realization [5]. Therefore, they have static character but it is obvious that network communication, in practice, consists of the operations of interleaving protocols. Consequently, the full process is realized on time [6], [7]. As a result our proposition concerning the algorithm tries to regard the practice situation. Such convention does not decrease the number of possibilities regarded in the exploitation of timed automata TA, probability timed automata or Petri nets [8],[9],[10]. On the contrary, in this

G. Jezic et al. (eds.), *Agent and Multi-Agent Systems: Technologies and Applications*,   139
Advances in Intelligent Systems and Computing 296,
DOI: 10.1007/978-3-319-07650-8_15, © Springer International Publishing Switzerland 2014

case possibilities that appear pertain to the creation of a complex structure of automaton nodes consisting of a set of security attributes. The auxiliary task consists in finding the connections among actions from protocol operations and predefined security attributes. To resolve this problem, we exploit the system of communication logic rules [11]. Actions in these rules are treated as a condition and security attributes are treated as conclusions. Therefore, the set of actions leads to choosing attributes, which will be corrected. Then the next problem appears: how to define methods and evaluate parameters for correction procedures. This problem has been resolved on the basis of experiments. We use two methods of attribute correction [12] based on the attribute value exchange (for time and the number of user attribute parameters) and on attribute multiplication by the correction coefficient (for probabilistic attribute parameters).

2   The Structure of Security Auditing Intelligent Systems in Reference to Parallel Realization Predispositions

Let us not forget the communication security investigation algorithm based on the probability timed automaton (PTA) model [13], [14] or on an adequate colored Petri net (CPN) structure [10]. The general configuration is presented in Fig.1. This algorithm characterized a multi stage structure and a simple conversion in particular stages. Up till now, almost all stages are described in detail and implemented. The largest complexity characterizes the procedure that combines the stages of rule activation and attribute correction. The complexity of this procedure is equal to $O(n^5)$. These parameters are only established from nested conversion cycles. As it is depicted in the algorithm block scheme, the possibility concerning the designation of a parallel thread appears only after the recognition of actions [15]. On the other hand, the observation of network communication may be divided into parts (segments). This creates added possibilities regarding parallelization. Generally, let us not forget that threads described and defined security situations with reference to keys, messages, users, secrets, etc. (the so called main security factors). Simplified theoretical analysis permits us to estimate the acceleration of calculations due to the parallel realization that is presented in the following way:

$$Accel = \frac{lo \cdot la \left(2lr \cdot lat + 2lm + lm \cdot lpr\right)}{lo \left(lat + lr \cdot lat + 2lat + lpr\right)} = \frac{la \left(2lr \cdot lat + 2lm + lm \cdot lpr\right)}{lat + lr \cdot lat + 2lat + lpr} \quad (1)$$

where:
   lo - the number of operations in a run,
   la - the number of actions,
   lat - the number of security attributes,
   lr - the number of communication rules,
   lm - the number of security modules (the main security factors),
   lpr - he number of generated types of prognosis.

In practice, the acceleration parameter is approximately equal to $0, 51 \cdot Accel$ on average. Example 1 illustrates the change of the acceleration parameter for the following data: $lo = 30$ (irrelevant parameter due to (1)), $la=8$, $lat =[4,8]$ lr $=[8,12]$, $lm =3$, $lpr =2$ (Fig.2). Example 2 illustrates the change of the acceleration parameter for the following data: $lo = 30$ (irrelevant parameter due to (1)), $la=[8,10]$, $lat =[4,8]$ $lr =10$, $lm =3$, $lpr =2$ (Fig.3) Example 3 illustrates the change of the acceleration parameter for the following data: $lo = 30$ (irrelevant parameter due to (1)), $la=8$, $lat =[4,8]$, $lr =10$, $lm =[2,4]$, $lpr =2$ (Fig.4).

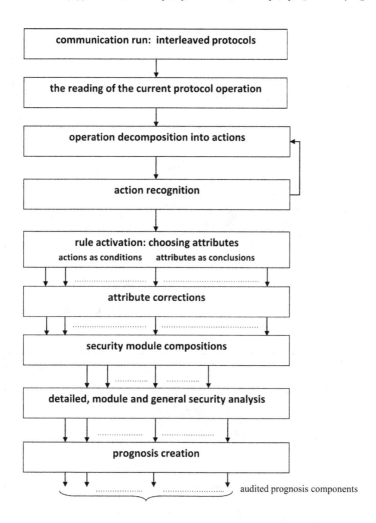

Fig. 1. The block scheme of the security investigation algorithm

The nonlinear and linear dependencies reflect the simple form of conversion and the easiness of parallelization regarding calculation that generally consists in thread decomposition. Each thread may be connected with: protocols, messages, actions, rules, users and will be subjected to parallel execution. These elements

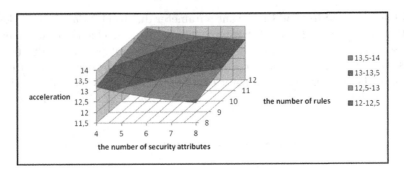

Fig. 2. Acceleration dependence deriving from the number of security attributes and the number of rules (example 1)

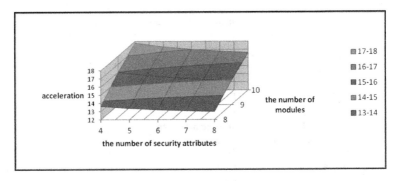

Fig. 3. Acceleration dependence deriving from the number of security attributes and the number of modules (example 2)

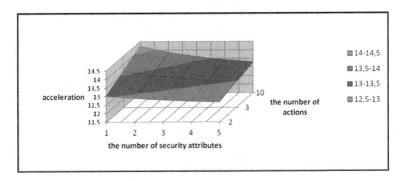

Fig. 4. Acceleration dependence deriving from the number of security attributes and the number of actions (example 3)

create security modules. Parallelization possibilities still appear after operation reading. In the designated thread, we may recognize one kind of action, and we should also exploit several threads if *la* actions are searched up. When the full set of actions recognizes the next group of threads it is activated for choosing security attributes, which will be corrected on the basis of rules. In this case we exploit lr threads. Attribute corrections should be sequentially realized due to a specific action depending on the character of attribute influence. The calculation parallelization possibilities reappear after attribute modifications. It is connected with creation security modules. Only then, we may start the prognosis analysis.

3   Searching for the Useful Communication Prognosis Form

Firstly, we approve the decision concerning the sensibility of longtime and short-time prognosis preparation. This prognosis refers to main security factors (analyzed as security modules). The prognosis may have a general and a detailed character. The prognosis creation is based on current attribute values $at(i)$, threshold attribute levels $th(i)$ (the minimal accepted attribute value), the probability of attribute corrections and the attribute structure of a given security module $sm(k) = atp(1, k), atp(2, k), ..., atp(lat, k)$, where $atp(i, k) = 0, 1$ - binary participation index referring to the i-th attribute in the k-th security module structure. It is useful to introduce the following types of prognosis:

- detailed, referring to security attributes,
- module, referring to security modules,
- general, referring to all or chosen sets of security modules.

Another prognosis classification refers to the way of probability estimation regarding attribute corrections. In this case we propose the following classification structure:

- with intruders,
- without intruders.

In both of these estimation variants, we may use a different approach (Fig.5):

- according to past communication operations,
- according to predicated future operations regarding the communication protocol structure.

The full prognosis analysis is realized on the basis of probability and binary variables.

The analyzing system will define and predicate the threat zone (expressed in time or probability) on the basis of $pp(i)$ $(pf(i))$, according to single attributes, single modules or the set of modules (Fig.6).

A more detailed presentation requires the formation of formal definitions and grammar fixing of security prognosis.

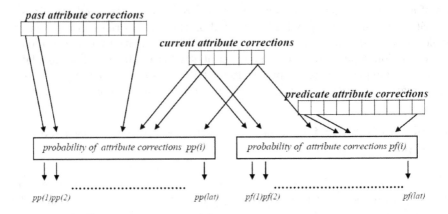

Fig. 5. The dependence diagram about attribute correction connected with probability estimation procedures

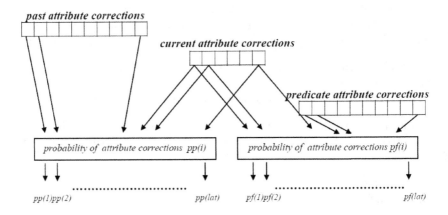

Fig. 6. Diagram regarding a different type of threat zone creation

4   Security Prognosis Formalisms and Evaluation System

Let us start from atomic security elements, i.e. security attributes. The security zone and the threat zone are defined as follows:

Definition 1. *Security attribute zone is the difference between the current attribute value and the threshold attribute value $sz(i) = at(i) - th(i)$. The threat attribute zone is the difference between the threshold attribute value and the current attribute value $fz(i) = th(i) - at(i)$.*

The next definition is connected with the so called tokens, which appoint the attribute security state.

Definition 2. *Token delivers information about the location of the current at-tribute value in relation to the security threshold (attribute state): $tk(i) = 1$, if $at(i) \geq th(i)$, $tk(i) = 0$, otherwise.*

Definition 3. *Security module state is defined by the model set of the attribute state and the security token state: $ams(k) = pat(1, k) \cdot pat(2, k) \cdot ... \cdot pat(lat, k)$, where:*

    $p(i, k)$- probability security factor referring to the k-th model,
    $pat(i, k) = at(i)$ if $atp(i, k) = 1$,
    $pat(i, k) = 1$ if $atp(i, k) = 0$,
and
    $tms(k) = btk(1, k) \cdot btk(2, k) \cdot ... \cdot btk(lat, k)$,
where:
    $btk(1, k)$ - binary security factor referring to the k-th model,
    $btk(i, k) = tk(i)$ if $atp(i, k) = 1$,
    $btk(i, k) = 1$ if $atp(i, k) = 0$.

Definition 4. *Security i-th attribute (k-th module) approbation is connected with fulfilling the condition $tk(i)=1$ ($tms(k)=1$).*

Definition 5. *Attribute security spectrum is defined by the sets of $at(1), at(2), ..., at(lat)$ and $tk(1), tk(2), ..., tk(lat)$.*
K-th module security spectrum is defined by the sets of $at(1), at(2), ..., at(lat)$, $tk(1), tk(2), ..., tk(lat)$ and $atp(1, k), atp(2, k), ..., atp(lat)$.

Now, the protocol structure will be considered according to attribute engag-ing. The communication protocol consists of operations which activate actions and those lead to chosen (by rules) attribute corrections. Therefore, it is sensible to define the number of given attribute corrections in the investigated protocol.

The protocol structure indicates the set of corrected attributes. The approach of these attributes cannot be ordered in our prognosis. The distribution of at-tribute corrections is the only relevant element. For each standard protocol, the number of attribute corrections should be predefined, e.g. in the following form: for protocol Pi distribution of attribute corrections:
$dpa(i) = nat(1, i), nat(2, i), ..., nat(lat, i)$,

At the beginning of the prognosis analysis, a problem appears consisting in small quantity of data. Generally, the situation, which takes place after the part concerning the communication run reading, can be depicted as in Fig.7.

Both the set of corrected and the set of not corrected attributes (black rings and white rings in Fig.7) can be exploited for the prognosis creation:

$$nca\,(i) = \sum_{j=1}^{k} cat\,(i, c\,(j)) \tag{2}$$

$$nnca\,(i) = \sum_{j=1}^{k} nat\,(i, c\,(j)) - cat\,(i, c\,(j)) \tag{3}$$

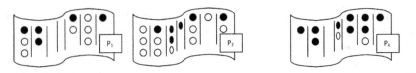

Fig. 7. The situation after the start of k protocol reading and the set of attribute corrections (black rings). For different protocols we define different distributions $dpa(i) \neq dpa(j)$.

where:

i - attribute number (code),

j - activated protocol number,

$c(j)$ - the code of the j-th activated protocol,

k - the current number of activated protocols,

$nca(i)$ - the current number of corrected attributes in all recognized protocols,

$cat(i, j)$ - the current number of corrected attributes in the j-th protocol,

$nnca(i)$ - the current number of not corrected attributes.

Obviously, the kind of protocols, which appear in the next part of the communication run, is unknown. Therefore, only activated protocols will be regarded. We propose two variants of prognosis, according to data exploitation:

- *ex post*, where the probability of attribute corrections $pexp(i)$ is calculated on the basis of an action that takes place after reading up the run operation:

$$pexp(1) = \frac{nca(1)}{\sum_{j=1}^{k} nat(i, c(j))} \qquad (4)$$

- *ex ante*, where the probability of attribute corrections $pexa(i)$ is calculated on the basis of an action in the future realization of protocols activated till now:

$$pexa(1) = \frac{nnca(1)}{\sum_{j=1}^{k} nat(i, c(j))} \qquad (5)$$

The result of prognosis analysis consists of:

- $ta(i, o)$ - the predicated time of the i-th attribute life (with its level above a given threshold $th(i)$) counted from the current o-th operation reading,

- $tm(j, o)$ - the predicated time of the j-th module life (with all included in module attributes above given security thresholds, e.g. the protocol module. It is also counted from the current o-th operation reading.

The main drawback of the prognosis implementation consists in the approximated character of base data:

- $pexp(i)$ and $pexa(i)$, $i=1, 2,..., lat$,
- $sz(i)$,

- $po(i)$ - the probability of the i-th attribute correction after a single operation.

The last parameter contributes especially strong simplification. On the other hand, such approach essentially decreases the complexity of the prognosis algorithm. The $pexp(i)$ and $pexa(i)$ refer to the action and $po(i)$ refers to the operation, therefore, the conjunction of these elements will be represented by multiplication: $pexp(i) \cdot po(i)$ or $pexa(i) \cdot po(i)$. Finally, the time prognosis parameter, which refers to attributes, can be defined in the following way:

- *ex post* prognosis variant:

$$ta'(i,o) = \left[\frac{sz(i)}{ccr(i) \cdot pexp(i) \cdot po(i)} \right] [the\ steps\ of\ operating\ reading] \quad (6)$$

- *ex ante* prognosis variant:

$$ta''(i,o) = \left[\frac{sz(i)}{ccr(i) \cdot pexa(i) \cdot po(i)} \right] \quad (7)$$

where:
[$*$] - round function,
$ccr(i)$ the given correction coefficient for the i-th attribute.

The time prognosis parameter, which refers to modules, can be defined in the following way:

- *ex post* prognosis variant:

$$ta'(i,o) = \left[\min_{\substack{i=1,...,lat \\ atp(i)=1}} \frac{sz(i)}{ccr(i) \cdot pexp(i) \cdot po(i)} \right] \quad (8)$$

or

- *ex ante* prognosis variant:

$$ta''(i,o) = \left[\min_{\substack{i=1,...,lat \\ atp(i)=1}} \frac{sz(i)}{ccr(i) \cdot pexa(i) \cdot po(i)} \right] \quad (9)$$

Sometimes, we adapt the more complex data about protocols realized in the future and it is possible to simply regard this information in calculation parameters $pexa(i)$.

5   Conclusions

The result of security communication prognosis permits us to send warnings about the necessity of key or secret exchanging, the appearance of additional users (intruders), the necessity of refreshing the nonce, etc. By supplying time distances concerning the impending threats it is possible to choose the most dangerous and the closest one and make the adequate preventions. The proposed intelligent system works in real time and investigates the continuously changing situation in the auditing convection. We realize the simple prognosis analysis regarding 16 attribute corrections after a single communication operation by program implementation in part of a second (0.6-0.92 sec.). Generally, the reaction to the current communication situation is quick enough to stop protocol realization and to create a prevention strategy.

References

1. Tadeusiewicz, R.: Introduction to Inteligent Systems. In: Wilamowski, B.M., Irvin, J.D. (eds.) The Industrial Electronic Handbook, ch. 1, pp. 1-1 – 1-12. CRC Press, Boca Raton (2011)
2. Tadeusiewicz, R.: Place and role of Intelligence Systems in Computer Science. Computer Methodsin Material Science 10(4), 193–206, 13. Tadeusiewicz, R.: Place and role of Intelligence Systems in Computer Science. Computer Methodsin Material Science 10(4), 193–206 (2010)
3. Kwiatkowska, M., Norman, R., Sproston, J.: Symbolic Model Checking of Probabilistic Timed Automata Using Backwards Reachability. Tech. rep. CSR-03-10, University of Birmingham (2003)
4. Kwiatkowska, M., Norman, G., Segala, R., Sproston, J.: Automatic Verification of Real-time Systems with Discrete Probability Distribution. Theoretical Computer Science 282, 101–150 (2002)
5. Evans, N., Schneider, S.: Analysing Time Dependent Security Properties in CSP Using PVS. In: Cuppens, F., Deswarte, Y., Gollmann, D., Waidner, M. (eds.) ESORICS 2000. LNCS, vol. 1895, pp. 222–237. Springer, Heidelberg (2000)
6. Focardi, R., Gorrieri, R., Martinelli, F.: Information Flow Analysis in a Discrete -Time Process Algebra. In: Proc. of 13th CSFW, pp. 170–184. IEEE CS Press (2000)
7. Gray III, J.W.: Toward a Mathematical Foundation for Information Flow Security. Journal of Computer Security 1, 255–294 (1992)
8. Alur, R., Courcoubetis, C., Dill, D.L.: Verifying Automata Specifications of Probabilistic Real- Time Systems. In: Huizing, C., de Bakker, J.W., Rozenberg, G., de Roever, W.-P. (eds.) REX 1991. LNCS, vol. 600, pp. 28–44. Springer, Heidelberg (1992)
9. Alur, R., Dill, D.L.: A Theory of Timed Automata. Theoretical Computer Science 126, 183–235 (1994)
10. Szpyrka, M.: Fast and exible modeling of real-time systems with RTCP- nets. Computer Science, 81–94 (2004)
11. Burrows, M., Abadi, M., Needham, R.: A Logic of Authentication. In: Harper, R. (ed.) Logics and Languages for Security, pp. 815–819 (2007), Di Pierro, A., Hankin, C., Wiklicky, H.: Approximate Non-Interference. Journal of Computer Security 12, 37–82 (2004)
12. Piech, H., Grodzki, G.: The system conception of investigation of the communication security level in networks. In: Abramowicz, W. (ed.) BIS Workshops 2013. LNBIP, vol. 160, pp. 148–159. Springer, Heidelberg (2013)

13. Beauquier, D.: On Probabilistic Timed Automata. Theoretical Computer Science 292, 65–84 (2003)
14. Focardi, R., Gorrieri, R.: A Classification of Security Properties. Journal of Computer Security 3, 5–33 (1995)
15. Tudruj, M., Masko, L.: Toward Massively Parallel Computation based on Dynamic Clasters with Communication on the Fly. In: IS on Parallel and Distributed Computing, Lille, France, pp. 155–162 (2005)

....

Simulation and Statistical Model Checking of Logic-Based Multi-Agent System Models

Christian Kroiß

Department for Informatics,
Ludwig-Maximilians-Universität, Munich, Germany
kroiss@pst.ifi.lmu.de

Abstract. In this paper we introduce a new approach for multi-agent simulation and statistical model checking that allows the use of very generic logical models based on the well-established situation calculus to describe the behavior of agents in the context of their environment. A consequent logic-based framework is achieved by combining the situation calculus with a first order version of bounded linear time logic (BLTL) as a property specification language. This creates a much more expressive and flexible modeling-verification workflow than existing solutions.

Keywords: statistical model checking, multi agent systems, situation calculus, discrete event simulation.

1 Introduction

In the analysis of multi-agent systems, one of the most common general approaches is the use of *simulations* to generate *execution traces* for a given system model that can then be examined with the most appropriate methods. Statistical model checking (SMC) can be regarded as a variant of this idea. Similar to traditional model checking approaches, SMC allows the use of temporal logics like linear temporal logic (LTL) [1] to specify certain requirements for the system (aka. *properties*). However, instead of using exact model checking algorithms, a statistical hypothesis test is performed on the recorded simulation traces to assure that the property is violated only with a certain maximal probability. This approach overcomes the critical state-space explosion problem of exact model checking. Statistical model checking has already been applied to various types of system models and with several different temporal logics. However, in existing solutions the monitored properties are in a way separated from the system model because they refer to symbolic logical *propositions* whose connection to the model have to be defined explicitly.

In order to achieve a better integration, we introduce an approach that uses the *situation calculus* to define system models in a first-order logic language. We combine this with an extended first-order version of bounded LTL (BLTL) that makes it possible to directly use concepts and entities of the system model in the property formulas. This leads to a very intuitive and expressive modeling approach with wide applicability.

G. Jezic et al. (eds.), *Agent and Multi-Agent Systems: Technologies and Applications,*
Advances in Intelligent Systems and Computing 296,
DOI: 10.1007/978-3-319-07650-8_16, © Springer International Publishing Switzerland 2014

In the next sections, we first briefly mention some related work and their relevance to our solution. Section 3 shortly introduces the situation calculus together with some necessary extensions for the description of multi-agent systems. Then, the core simulation mechanism is described in section 4. After that, we describe the use of our approach for statistical model checking in section 5, first by describing syntax and semantics of our property specification language, and then by summarizing how the classical sequential probability ratio test of A. Wald [2] is integrated to perform hypothesis tests. Finally, the use of the approach is demonstrated by means of a small example in section 6 before we conclude in section 7.

2   Related Work

A rather recent overview about previous work in the field of statistical model checking can be found in [3]. Especially relevant to our approach is [4] where the authors also suggest the use of the sequential probability ratio test (SPRT). Instead of BLTL they use a variant of continuous stochastic logic (CSL) [5], which also has a bounded until-operator but additionally supports nested probabilistic formulas. As in most work on SMC, the authors do not cover the topic of how the simulated model is defined in detail but rather point out that any discrete event model is supported. The situation calculus has already been used intensively as basis for a agent programming languages, particularly in GoLog [6] and its many descendants (e.g. [7,8]). While many of these systems can technically be used for discrete event simulation, no attempt has been made (to our knowledge) to integrate them directly with statistical model checking. Finally, there is a overwhelming number of systems for discrete event multi-agent simulation. Some of them (e.g. MASON [9]) are very generic and in this respect similar to our system. While these solutions offer more or less declarative approaches to modeling, they are still not directly combinable with existing statistical model checkers.

3   Modeling Actions and Their Effects

Modeling and simulating the interaction of agents with each other and with their environment is traditionally done in a very problem-dependent fashion, ranging from simple models with a few variables to full-fledged 3D physics engines. As we show in the following sections, a very generic modeling approach can be realized by using an extended version of the well-known situation calculus.

3.1   The Classical Situation Calculus

The *situation calculus* [10,6] is a first order logic (FOL) language whose main ingredients are *actions* and *fluents*, which are variables whose values depend on the system's *situation*. Actions are encoded in the natural way by FOL functions, and fluents by functions or predicates that take a situation argument as

an additional last formal parameter. A situation itself is represented as a sequence of actions, encoded as a recursive FOL term with the special function
$do(Action, Situation)$. Based on these ingredients, the modeler specifies for each
fluent exactly one *successor state axiom (SSA)*, e.g.

$$broken(x, do(a, s)) \equiv a = drop(x) \vee (a \neq repair(x) \wedge broken(x, s)).$$

Each SSA exactly defines the value of a fluent at the situation that results from
performing any action, i.e. it captures all ways a fluent's value might change.
This is complimented by *action precondition axioms* that define action executed
are possible, e.g. $Poss(drop(x), s) \equiv holding(x, s)$. Starting from there, one can
use well-established logical methods to reason about the modeled system. In
the context of this paper, this basically means solving the *projection problem*:
checking whether a given formula $G(s)$ holds in the situation that results from
performing a given action sequence starting from the initial situation S_0. The
basic computational mechanism that is used there is *regression*. In short, regression recursively replaces fluents by the right-hand side of their SSA. This finally
yields a sentence that only mentions the *initial situation* S_0 and therefore can
be treated like any FOL sentence. In spite of its elegance, this schema is not
well-suited for the purpose of simulating long-running systems. The problem is
that the evaluation would gain complexity for each step. To avoid this, our simulation engine uses an alternative mechanism that is called *progression* (see [11]),
which basically means in our context that regression is only used on rather short
action sequences to update the complete *database* of fluent values. This updated
database is then used as the initial situation for the next step.

3.2 Processes and Concurrency

When a more realistic description of the world is desired, it is necessary to
distinguish between *instantaneous actions* and *processes* or *activities* with a
certain duration. In the situation calculus this can be achieved easily by modeling
processes as fluents that are started and stopped by actions, which on their part
are assumed to be timeless, e.g.

$$moving(do(a, s)) \equiv a = start \vee (moving(s) \wedge a \neq stop)$$

This already allows expressing multiple simultaneous process and therefore *concurrency* to some degree. In order to handle actions by multiple agents, the
simulation engine additionally uses an *interleaving concurrency model* that is
established by gathering executed actions from all agents in each simulation step
and then compiling them into a sequence with randomized order (cf. sec. 4). This
model is usually sufficient for the kind of analyses we are interested in. Mainly
when a detailed physical view is intended (see [6, sec. 7.2.2]), other models of
concurrency are needed.

3.3 Stochastic and Exogenous Actions

Another step towards realism is to consider the inherent uncertainty that is
imposed by the agents' *environment*. First of all, there might be *exogenous*

actions that are not initiated by the agent but by nature or an external actor (see [6, chap. 8]) and hence occur in a stochastic fashion. Additionally, actions performed by the agents can have stochastic outcomes, e.g. imposed by a failure (see [6, chap.12]). Each of these additional influences can easily be integrated in the simulation control loop. When a stochastic action is executed (exogenous or intentional), the simulation engine compiles a deterministic action by stochastically selecting one of the declared possible outcomes. Our simulation engine provides a very flexible extensible infrastructure for the realization of this probabilistic sampling and selection process, including for instance the ability to define time-dependent distributions by referring to timestamps of actions and condition changes.

3.4 Quantitative Time and Clocks

For the kind of simulation and statistical model checking we have in mind, time has to be represented explicitly in a quantitative way. A detailed discussion of the treatment of time in the situation calculus can be found in [6, chap. 7], where the *occurrence time* of each action is added as an additional temporal argument, e.g. *explode*(*bomb*, 12.32). The advantage of this modeling style is that time becomes an integral part of the situation term and can therefore be elegantly integrated in regression. However, we pursue a *discrete event simulation* (DES) approach, which allows a simpler treatment. Instead of a continuous time model, we use a regular integer fluent *time* that holds a counter for the current simulation step. Time advances only through the action *tick*, which is executed (using progression) in each step by the simulation engine. We extend this basic model by relative time measurements we call *clocks* that mark the most recent occurrence of actions or changes of conditions. As mentioned in the last section, clocks can be used for example to define time-dependent probability distributions.

4   Modeling and Simulating Multi-Agent Behavior

In addition to a model of the features and mechanisms of the system, the behavior of each agent has to be defined by means of a *control procedure*. In our approach, these procedures are executed by a Python module that is equipped with an interface to a Prolog (ECL^{i}PS^e) runtime engine, which stores and manages the situation calculus model. A simple example for an agent control procedure can be found in section 6. The simulation engine interprets such procedures for all agents, and in each step it compiles the joined action sequence that should be performed. This is summarized in Algorithm 1. The inner loop executes the `step()` method for each agent, which proceeds its control procedure until an action is executed or the agent has to wait for some condition. Actions are not performed directly but instead stored in a list for later execution. In the case of stochastic actions, their *outcomes* and parameter values are generated first by sampling from the corresponding distributions that were defined in the model.

The engine distinguishes between actions that take observable time, and *immediate actions* whose duration can be neglected, e.g. when manipulating a fluent that acts as a status flag. The engine continues each agent's control flow and progresses all immediate actions within the current time step until each agent either has generated a non-immediate action, is waiting, or has finished. Then the engine chooses which exogenous actions should take place in this simulation step by evaluating their action precondition axioms and probability distributions. The selected exogenous actions are interwoven with the non-immediate actions by the agents and progressed together. Finally, all registered property queries (see below) are evaluated, and the simulation step is ended by performing a time-step (i.e. by running a progression with the action *tick*, see section 3.4).

Algorithm 1. Simulation Procedure

```
verdict ← undefined ;
repeat
    actions ← ∅ ;
    repeat
        immediateActions ← ∅ ;
        foreach agent in World.agents do
            action ← agent.step() ;
            if action.immediate then
            |   immediateActions ← immediateActions ∪ {action};
            else
            |   actions ← actions ∪ {action};
        end
        shuffle(immediateActions) ;
        progress(immediateActions) ;
    until immediateActions = ∅;
    exoActions ← generateExogenousActions() ;
    actions ← actions ∪ exoActions ;
    shuffle(actions) ;
    progress(actions) ;
    verdict ← evaluateProperties() ;
    progress({tick}) ;
until verdict = true ∨ verdict = false ∨ world.finished;
```

As the algorithm shows, concurrency between agents is simulated by interleaving their actions. Additionally, the order of the actions is randomized (shuffled) every time before a progression is performed to avoid unfair and therefore unrealistic preference of individual agents.

5 From Simulation to Statistical Model Checking

In order to use the described simulation mechanism for statistical model checking, three main components have to be provided: (1) a suitable property specification language, (2) an evaluation engine that is able to check for each simulation step whether the declared properties hold, and (3) an adequate statistical sampling plan that is able to produce statistically significant results with a reasonable number of samples. This section discusses each of these topics shortly and sketches how they are tackled in our approach.

5.1   Defining and Evaluating First-Order Bounded LTL Properties

The predominant general method of reasoning about *execution traces* in computer science is to use *temporal logics*. In agreement with that, we based our property specification language upon a first-order version of *bounded linear temporal logic (BLTL)* (see [1]), a variant of LTL that adds an upper time bound for the temporal operators. The structure of the language is given by the following grammar:

$$\phi ::= \top \mid \bot \mid \psi_1 \sim \psi_2 \mid \phi_1 \wedge \phi_2 \mid \phi_1 \vee \phi_2 \mid \neg\phi \mid \phi_1 \rightarrow \phi_2 \mid$$
$$\forall x \in \mathcal{E}.\phi \mid \exists x \in \mathcal{E}.\phi \mid \phi_1 \mathsf{U}^{\leq T}\phi_2 \mid occur(a(\psi_1, \ldots, \psi_n)) \mid$$
$$start(\phi) \mid end(\phi) \mid f(\psi_1, \ldots, \psi_n)$$
$$\psi ::= C \mid x \mid g(\psi_1, \ldots, \psi_n) \mid \psi_1 \odot \psi_2$$

Here $\sim$ denotes one of the comparison operators $<, \leq, =, \geq, >$, x is a variable, $\mathcal{E}$ is a finite domain, T is an integer constant, and C is either a number or a constant of any defined finite domain. Furthermore, a is an action, f is a predicate or relational (boolean) fluent, and g is a functional fluent or situation-independent function that returns a number or a value of one of the defined domains. Lastly, $\odot$ stands for one of the arithmetic operators $+, -, *, /, \%$. For fluents, the otherwise mandatory situation argument at the last position is suppressed in this language, i.e. the current situation is assumed implicitly. Besides the common temporal until-operator U, there are also the special temporal predicates *occur*, *start*, and *end*. These predicates mark states where an action occurs or a property starts or stops being true, which is particularly important for modeling the temporal boundaries of processes or activities.

In order to describe the semantics of the language, we first introduce the concept of a *simulation trace*, denoted by σ, that represents a sequence of *situations* $s_0, \ldots, s_n$, where s_0 is the initial situation at the simulation start and s_k is the situation after k actions have been performed. Let further be σ^k the suffix of the simulation trace that starts at situation s_k and $[\![\psi]\!]_{s_k}$ be the semantic denotation of ψ at situation s_k. We also write $\models \phi$ as a shortcut meaning that ϕ is entailed by the situation calculus considering only the axioms of the *basic action theory* [6, sec. 4.4] established by the system model. Finally, we write $\sigma^k \models \phi$ to express that the simulation trace suffix starting at s_k satisfies the property ϕ. Then the semantics of properties specified in the language described above can be defined by means of the situation calculus (omitting the well-known defintions for the logical connectives $\wedge$, $\vee$, $\neg$ and $\rightarrow$):

1. $[\![v]\!]_s = v$ iff v is a numeric or boolean literal, or a constant of a defined finite domain.
2. $[\![g(t_1, \ldots, t_n)]\!]_s = v$ iff (1) g is a situation-independent function and the evaluation of $g([\![t_1]\!]_s, \ldots, [\![t_n]\!]_s)$ yields the result v, or (2) g is a functional fluent and $\models g([\![t_1]\!]_s, \ldots, [\![t_n]\!]_s, s) = v$.
3. $[\![t_1 \odot t_2]\!]_s = v$ iff $\odot$ is any arithmetic operator and the evaluation of $[\![t_1]\!]_s \odot [\![t_2]\!]_s$ yields v.
4. $\sigma^k \models f(t_1, \ldots, t_n)$ iff (1) f is a relational fluent and $\models f([\![t_1]\!]_{s_k}, \ldots, [\![t_n]\!]_{s_k}, s_k)$, or (2) f is a situation-independent predicate and $\models f([\![t_1]\!]_{s_k}, \ldots, [\![t_n]\!]_{s_k})$.
5. $\sigma^k \models g(t_1, \ldots, t_n) \sim \psi$ iff g is a functional fluent or a situation-independent function, and $\models [\![g(t_1, \ldots, t_n)]\!]_{s_k} \sim [\![\psi]\!]_{s_k}$.
6. $\sigma^k \models \phi_1 \mathsf{U}^{\leq T} \phi_2$ iff there exists a situation s_l in σ with $k \leq l$ so that (a) $\sigma^l \models \phi_2$, (b) $\sigma^i \models \phi_1$ for each $k \leq i < l$, (c) $[\![time]\!]_{s_l} - [\![time]\!]_{s_k} \leq T$, and (d) $[\![time]\!]_{s_i} \leq [\![time]\!]_{s_l}$ for each $k \leq i < l$.
7. $\sigma^k \models occur(a(t_1, \ldots, t_n))$ iff $s_{k+1} = do(a([\![t_1]\!]_{s_k}, \ldots, [\![t_n]\!]_{s_k}), s_k)$.
8. $\sigma^k \models start(\phi)$ iff $\sigma^{k-1} \not\models \phi$ and $\sigma^k \models \phi$.
9. $\sigma^k \models end(\phi)$ iff $\sigma^{k-1} \models \phi$ and $\sigma^k \not\models \phi$.
10. $\sigma^k \models \forall x \in \mathcal{E}.\phi$ iff $\bigwedge_{e \in \mathcal{E}} (\sigma^k \models \phi[e/x])$.
11. $\sigma^k \models \exists x \in \mathcal{E}.\phi$ iff $\bigvee_{e \in \mathcal{E}} (\sigma^k \models \phi[e/x])$.

During a simulation run, the evaluation of the property is not performed for every executed action but at the end of every simulation step. Due to the until-operator, the satisfaction of a property cannot always be determined conclusively in the current step. In this case a *property schedule* is used to keep track of such properties and to re-evaluate them until eventually a definite verdict is for ϕ. Since all uses of the until-operator are limited by time bounds, and time is guaranteed to advance in each step, this is always possible in a finite number of steps.

5.2 Sequential Hypothesis Tests

If the simulated model contains stochastic actions or exogenous events, then a property ϕ will be violated in a simulation run σ (i.e. $\sigma \not\models \phi$) with some probability p. Consequently, when N simulation runs are performed, then the number of violations (defects) follows a binomial distribution $B(N, p)$. The basic approach in *statistical model checking* is now to perform a hypothesis test to show that the probability of a property violation is at most a given *marginal probability* p_{max}, i.e. the null hypothesis is $H_0 : p \leq p_{max}$. Since it is possible to repeat the simulation arbitrarily often, the most important question is how many runs are enough. In order to solve this, our tool implements the widely used *sequential probability ration test (SPRT)*, introduced by Wald in 1945 [2]. Here, an *indifference region* is specified by two probabilities p_0 and p_1 around p_{max}. During the test, the actual value of p is estimated by the ratio of defects, and a test decision will only be considered an error if (a) H_0 is rejected and $p \leq p_0$ (type I error), or (b) H_0 is accepted and $p \geq p_1$ (type II error). Additionally, two parameters α and β have to be defined for the maximum probability of type I and type II errors, respectively. Then, after each simulation run, a *probability ratio* is calculated based on the observations so far. Depending on this value and the chosen parameters, the tool decides if the null hypothesis H_0 should be accepted, rejected, or if a further simulation run is necessary.

The expected number of required simulations N_{req} depends on the values chosen for the parameters p_0, p_1, α, and β, but also on the actual probability p. If p is very close to the center of the indifference region $[p_0, p_1]$, then it is expected that the highest number of simulation runs will be required to find a significant result (cf. [2, sec. 5.2.46]).

6   Example and First Experimental Results

A prototype of the tool has been developed that is already able to provide all functionality described above. As a first thorough system test for the whole approach, a simple example scenario was defined with a property for which the violation (defect) probability can be calculated analytically. This example is described here to demonstrate the use of the most important modeling concepts. Furthermore, some experimental results are presented that confirm the correctness of the approach by comparison with results from Wald's sequential hypothesis test (cf. section 5.2).

The narrative of the chosen example scenario can be described as follows: (1) The simulated world consists of a number N_R of *robots* and one *item* for each robot. (2) When the simulation starts, all robots will simultaneously grab their item and start moving to the right. (3) Each robot is placed in a separate row so a collision is impossible. (4) At each (simulation) step in time, each robot may accidentally drop its item with the probability p_d. From this informal description, a situation calculus model has been derived and implemented in Prolog. Here, only the successor state axiom for the fluent *xpos* is shown:

```
xpos(0, Pos, do(A,S))  :- xpos(0, OldPos, S),
( A = move_right(0), !, Pos is OldPos+1 ;
A = move_right(R), carrying(R, 0, S), !,
Pos is OldPos+1 ; Pos = OldPos ).
```

This example demonstrates a concept called *ramification*: the axiom *xpos* models the fact that a move action of a robot not only changes the robot's position but also the position of a carried item. Besides the axiomatization, the user also has to provide some Python code to configure the simulated scenario. Here this mostly means to declare the probability distribution for the exogenous action *accidental_drop*, and to provide a simple agent control procedure, e.g.:

```
main = Procedure("main",[], [ Act("grab", [SELF, "item1"]),
    While(PYTHON_EXPRESSION, "True", [],
        Act("move_right", [SELF])) ])
```

Furthermore, the following property is evaluated in this example:

$$\phi \equiv \forall r \in dom(robot).\forall i \in dom(item).$$

$$occur(grab(r,i)) \rightarrow carrying(r,i) \mathsf{U}^{\leq 12} xpos(i) > 20$$

This formula asserts that, whenever some robot grabs an item, within 12 time units from that moment, the item's horizontal position must exceed 20 and the robot must keep carrying the item until then. Encoded for our system, this is:

```
forall([r,robot], forall([i,item],
    implies(occur(grab(r,i)),
        until(12, carrying(r,i), xpos(i) > 20)))))
```

If the simulated world is initialized with all robots and items at the same position $xpos = 10$, then all robots reach the goal after 10 steps, unless an accidental_drop event occurs. Let p_v be the probability that the property is violated during a simulation run σ. Then, since the probability for an accidental drop is known (p_d), and all events are assumed to be independent, p_v can be calculated as follows: $p_v = P(\sigma \not\models \phi) = 1 - ((1 - p_d)^{10})^{N_R}$.

For the experiment, we simulated a world with $N_R = 3$ and $p_d = 0.01$. In this case the formula above yields $p_v \approx 0.260$. We further set both the parameters α and β to 0.05 and performed sequential probability ratio tests as described in section 5.2, effectively testing the null hypothesis $H_0 : p'_v \leq p_{max}$. We repeated the test 200 times while moving the value for p_{max} from 0 to 1 and used an indifference region of size 0.05 around p_{max}, i.e $p_0 = p_{max} - 0.25$ and $p_1 = p_{max} + 0.25$. The decision for each test was correct in our experiment, i.e. neither errors of type I nor type II occurred, considering the indifference region. As predicted in section 5.2, the results also showed a significant difference in the number of required simulation runs, depending on how close the marginal probability of the hypothesis p_{max} was to the actual violation probability p_v. This can be seen in Figure 1, where the number of required simulation runs m is plotted against p_{max}. The dotted vertical line marks the calculated probability p_v, and a very strong peak can be recognized at this area.

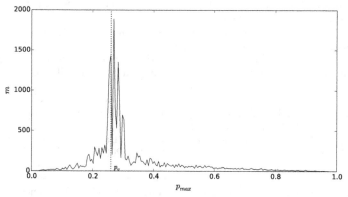

Fig. 1. Number of required runs m by marginal probability p_{max}

7 Conclusion

In this paper, we presented a new approach for simulation and statistical model checking of multi-agent system models that is mainly based on the idea of combining extended and specialized versions of the situation calculus and bounded linear time logic (BLTL). The main advantage of the introduced approach is that it allows a very flexible and generic way of modeling while maintaining the clear representation of logics. The presented combination of the situation calculus with the Python-based simulation engine allows to capture the interaction of agents with their environment and with each other with an almost arbitrary level of detail. This fosters a modeling style that endorses the surfacing of *emergent effects* which might not have been foreseen by the modeler. By using a first-order version of BLTL that provides direct access to the model concepts like actions, fluents and events, this flexibility also arises for the specification of properties. Altogether this seems to create a much more generic and holistic way of modeling than in other SMC approaches. In order to confirm this last claim, we plan to apply the approach to several case studies that are currently conducted as shared effort of he EU research project ASCENS (www.ascens-ist.eu).

Acknowledgments. This work has been partially sponsored by the EU project ASCENS, FP7 257414.

References

1. Pnueli, A.: The temporal logic of programs. In: 18th Annual Symposium on Foundations of Computer Science, pp. 46–57. IEEE (1977)
2. Wald, A., et al.: Sequential tests of statistical hypotheses. Annals of Mathematical Statistics 16, 117–186 (1945)
3. Legay, A., Delahaye, B., Bensalem, S.: Statistical model checking: An overview. In: Barringer, H., et al. (eds.) RV 2010. LNCS, vol. 6418, pp. 122–135. Springer, Heidelberg (2010)
4. Younes, H.L.S., Simmons, R.G.: Probabilistic verification of discrete event systems using acceptance sampling. In: Brinksma, E., Larsen, K.G. (eds.) CAV 2002. LNCS, vol. 2404, pp. 223–235. Springer, Heidelberg (2002)
5. Aziz, A., Sanwal, K., Singhal, V., Brayton, R.: Verifying continuous time markov chains. In: Alur, R., Henzinger, T.A. (eds.) CAV 1996. LNCS, vol. 1102, pp. 269–276. Springer, Heidelberg (1996)
6. Reiter, R.: Knowledge in action: logical foundations for specifying and implementing dynamical systems. MIT Press (2001)
7. De Giacomo, G., Lespérance, Y., Levesque, H.J.: Congolog, a concurrent programming language based on the situation calculus. Artificial Intelligence 121, 109–169 (2000)
8. De Giacomo, G., Lespérance, Y., Levesque, H.J., Sardina, S.: Indigolog: A high-level programming language for embedded reasoning agents. In: Multi-Agent Programming, pp. 31–72. Springer (2009)
9. Luke, S., Cioffi-Revilla, C., Panait, L., Sullivan, K., Balan, G.: Mason: A multiagent simulation environment. Simulation 81, 517–527 (2005)
10. Levesque, H., Pirri, F., Reiter, R.: Foundations for the situation calculus. Linköping Electronic Articles in Computer and Information Science 3 (1998)
11. Lin, F., Reiter, R.: How to progress a database. Artificial Intelligence 92, 131–167 (1997)

Multi-agent Based Execution Environment for Task Allocation via Coalition Formation

Merve Özbey and Nadia Erdoğan

Istanbul Technical University, Faculty of Computer and Informatics, Computer Engineering
Department 34469, Maslak Istanbul, Turkey
merve.ozbey@live.com,
nerdogan@itu.edu.tr

Abstract. This paper focuses on a solution to the problem of task allocation through coalition formation and presents the design and implementation of a framework for this purpose. The framework serves as a multi-agent execution environment where worker agents follow the Shehory-Kraus's coalition formation algorithm to negotiate and form coalitions. Worker agents prefer to form an optimal coalition to maximize the joint utility of the system as a whole, rather than to maximize their own utilities. There is no central authority for distribution of tasks among the agents. The framework provides the infrastructure, which, allows the formation of coalitions, assigns a task to each coalition and then monitors task execution, taking the necessary steps for rescheduling of tasks in case of noncompletion due to agent errors, generating and recording of task execution reports, handling cyclic execution of coalition formation process. In addition, an efficient and fast messaging infrastructure has been developed for effective agent communication.

Keywords: Multi-agent system, negotiation, coalition formation.

1 Introduction

Multi-agent systems are systems of intelligent autonomous agents which communicate and collaborate with each other. Agents cooperate when the capability and knowledge of each agent alone is not sufficient to solve a problem. Coalition formation is an important method for cooperation in a multi-agent environment. It is a process where agents form coalitions and work together to solve a joint problem via cooperating or coordinating their actions within each coalition.

In this paper, we present a framework which serves as a multi-agent execution environment for coalition formation. Agents (employee) are autonomous. They follow the Shehory-Kraus's algorithm [1] to negotiate and decide on coalitions, with no user interaction. The reason behind the choice of this algorithm is that, instead of having a single agent calculate all possible coalitions, the algorithm distributes the coalitional value calculations among agents, thus both improving execution time and preventing the existence of a single point of failure. They prefer to form an optimal coalition to maximize the joint utility of the system as a whole, rather than to maximize their own

G. Jezic et al. (eds.), *Agent and Multi-Agent Systems: Technologies and Applications,*
Advances in Intelligent Systems and Computing 296,
DOI: 10.1007/978-3-319-07650-8_17, © Springer International Publishing Switzerland 2014

utilities. There is no central authority for distribution of tasks among the agents. Each agent has a predefined task execution capability and agents negotiate with each other concurrently until they form a grand coalition to perform a predefined task. Agents cannot be members of multiple coalitions, as the algorithm enforces disjoint coalitions. After a task is completed, the members of the coalition can join new coalitions. Furthermore, agents cannot join an already formed coalition due to extra agent addition cost. It is assumed that tasks have no precedence order and inter dependency.

The framework we present provides an infrastructure that supports multiple agents with different roles that cooperate to present the following functionalities:

- definition of employee agent capabilities and task requirements
- formation of coalitions
- assignment of tasks to each coalitions
- monitoring of task execution
- rescheduling of tasks in case of error and noncompletion
- generation and recording of task execution reports
- dynamic inclusion of new tasks
- cyclic execution of coalition formation process

The framework builds on various types of agents which are defined and implemented to handle different issues. In the design phase, we have determined the actors of the system, specifying their tasks and responsibilities. After associating each actor with an agent type, e.g. employee agents, a controller agent, coalition manager agents and a database agent, protocols that define in detail the interaction and coordination between agents were developed. In addition, an efficient and fast messaging infrastructure has been developed for effective agent communication. The framework will be used for further research on coalition formation.

2   Related Work

In existing multi-agent literature there are several works about negotiation and coalition formation. Shehory and Kraus's work [1] proposes several methods for task allocation through coalition formation. These methods are disjoint coalitions, overlapping coalitions and coalitions for tasks that have precedence orders. In this work, we focus on disjoint coalitions for tasks with no precedence orders, leaving the addition of precedence to tasks as future work.

Shehory and Kraus's another study with Taase [2] proposes a protocol that enables agents to negotiate and form coalitions with time constraints and incomplete information. Test results show that their solution is close to optimal solutions. Continuation of this study [3] also shows that their solution gives good results with incomplete information under time constraints.

Ferber, Gutknecht and Michel's study [4] proposes an organizational view of multi-agent systems. They define roles for agents and assign a group manager to each agent groups. Group managers are responsible from agents in their group.

[5] presents a MAS based infrastructure for the specification of a negotiation framework in multi-agent systems. They present role definitions for agents, and a negotiation environment with design details. Another study [6] also proposes an application of coalition formation in multi-agent systems particularly for electricity markets. The framework presented is generic, appropriate for any type of application.

A current study [7] presents coalition formation process in detail. They present definition of equilibrium process in coalition formation and describe basic concepts of classical game theory which has coalitional constraints. They show how to cover some of the coalitional stability's standard notions using equilibrium process concept.

3   Shehory-Kraus Coalition Formation Algorithm

Employee agents in the execution environment follow the Shehory-Kraus's algorithm [1] to negotiate and decide on coalitions. There exist several variations of the algorithm; our implementation focuses on disjoint coalitions. We first state the underlying assumptions and definitions, and then describe the algorithm.

3.1   Assumptions

- Agents can communicate, negotiate and make agreements.
- Coalition enlargement is not considered, due to the cost of addition of agents to an already formed coalition.
- Agents cannot participate multiple coalitions.
- There exists no explicit dependency between tasks.
- Agent population does not change during coalition formation.
- Complete information is not required. However, all of the agents know about all the tasks and all other agents, and coalition members know all the details necessary for fulfilling their task.
- In this work, it is assumed that agents are group rational. Agents are not concerned with their own profits, but aim to maximize the total outcome of the system.

3.2   Definitions

- Agent set consists of n agents, $N = \{A_1, A_2, \dots, A_n\}$. For each agent A_i, a capability vector $B_i = \langle b_1^i, b_2^i, \dots, b_r^i \rangle$ has been defined. Each capability is a property of an agent which quantifies its ability to perform a specific action.
- Task set consists of m independent tasks, $T = \{t_1, t_2, \dots, t_m\}$. For the fulfillment of each task t_l, a vector of capabilities $B_l = \langle b_1^l, b_2^l, \dots, b_r^l \rangle$ is necessary. The utility gained from performing the task is a linear function of resource amount.
- Expected outcome after executing a task is selected as a linear function for decreasing complexity of calculations.
- A coalition C has a vector of capabilities B_c which is the sum of capability vectors of the coalition member agents. A coalition C can perform a task t only if all capabilities in the t's capability vector B_t are satisfied by $b_i^t \le b_i^C$ for all $0 \le i \le r$.

- The joint utility of the coalition members by contributing to a coalition is defined as the coalitional value, V.
- Coalitional cost, c, is calculated as the reciprocal of the coalitional value.

- *Coalition formation problem:*

Given a set of m independent tasks, $T = \{t_1, t_2, \dots, t_m\}$ and a set of n agents, $N = \{A_1, A_2, \dots, A_n\}$ with their capabilities, the coalition formation problem is assigning tasks $t_j \in T$ to coalitions $C_i \subseteq N$ such that $\Sigma i V i$ (the total outcome) is maximal.

3.3   Shehory-Kraus Coalition Formation Algorithm

The algorithm proceeds in three stages. The agents that negotiate execute the steps in all three stages concurrently, contributing to the process in distributed manner.

- **Stage 1:** All possible coalitions are calculated.
- **Stage 2:** Coalitional values of possible coalitions are calculated.
- **Stage 3:** Agents decide upon preferred coalitions and form them through iterations.

First Stage of the Algorithm

The first stage entails the calculation of the coalitional values in a distributed manner. The maximum coalition size, k, is initialized. Each agent communicates with others and eliminates duplicate coalitions from its potential coalition list L.

1. Calculate all of the possible coalitions up to size k in which A_i is a member and form P_i, the set of the potential coalitions of agent A_i.
2. For each coalition in P_i, contact each member A_j and retrieve its capabilities B_j.
3. Commit to the calculation of the values of a subset S_{ij} of the common potential coalitions (i.e., a subset of the coalitions in P_i in which both A_i and A_j are members).
4. Subtract S_{ij} from P_i. Add S_{ij} to long-term commitment list L_i.
5. For each agent A_k that has contacted A_i, subtract from P_i the set S_{ki} of the potential coalitions A_k had committed to compute values for.
6. Repeat contacting of other agents until $Pi = \{A_{ij}\}$.

Second Stage of the Algorithm

After the first stage, the agent has a list of coalitions for which it had committed to calculate the values. It also has all of the necessary information about the capabilities of the members of these coalitions. In the second stage each agent checks whether a coalition's capability vector is sufficient for the tasks in the task list. Coalition's capability vector is the sum of capability vectors of the agents in that coalition. For sufficient coalitions; agent calculates expected values. Then, agent selects the maximum expected value as the coalition value for each coalition, calculates coalitional cost.

1. Calculate the coalitional potential capabilities vector B_c^{pc} by summing up the unused capabilities of the members of the coalition.
2. Form a list E_c of the expected outcomes of the tasks in T when coalition C performs them. For each task $tj \in T$, perform:

(a) Check what capabilities B_j are necessary for the satisfaction of t_j.
(b) Compare B_j to the sum of the unused capabilities of the members of the coalition B_c^{pc}, thus finding the tasks that can be satisfied by coalition C.
3. Calculate the expected outcome of the tasks that can be performed by the coalition.
4. Among all of the expected outcomes, on list Ec, choose the maximal. This will be the coalitional value V_c.
5. Calculate the coalitional cost which is $c_c = 1/V_c$.

Third Stage of the Algorithm
In this stage, each agent calculates coalitional weights w, as the cost divided by coalitional size, the number of agents in that coalition. Each agent chooses the best coalition from among its list, i.e., the coalition Ci that has the smallest Wi. Next, each agent announces the coalitional weight that it has chosen, and the lowest among these is chosen by all agents. The members of the coalition that is chosen are deleted from the list of candidates for new coalitions. In addition, any possible coalition from the lists of any agent, that includes any of the deleted agents, is deleted from its list. These steps are iteratively repeated until all agents join a coalition or all tasks are assigned to a coalition. Each agent Ai iteratively performs the following steps:

1. Locate in L_i the coalition C_j with the smallest w_j.
2. Announce the coalitional weight w_j that it has located.
3. Choose the lowest among all of the announced coalitional weights. This w_{low} will be chosen by all agents. Choose corresponding coalition C_{low} and task t_{low} as well.
4. Delete the members of the chosen coalition C_{low} from the list of candidates.
5. If A_i is a member of the chosen coalition C_{low}, join its members and form C_{low}.
6. Delete from L_i the possible coalitions that include deleted agents.
7. Delete from T the chosen task t_{low}.
8. Assign to L_i^{cr} the coalitions in L_i for which values should be re-calculated.
9. Execute second stage of the algorithm.

4   System Architecture

Four different types of agents cooperate to provide an infrastructure with the functionalities listed in Section 1. Every type of agent is specialized in its actions, following well defined protocols. The agent types and their roles are as follows:

- *employee agents* which execute the coalition formation algorithm to form coalitions and, after the coalitions are determined, perform the tasks assigned to them.
- *a controller agent* which is in charge of system's overall coordination.
- *coalition manager agents(s)* which reduce the controller agent's heavy work load via coordinating the interaction and the communication between employee agents and the controller agent.
- *a database agent* which stores and maintains information associated with employee agents and tasks in a database, and meets query requests of all types of agents.

Fig. 1 depicts the architectural design of the framework, reflecting agent interactions during the coalition formation phase. Employee agents can communicate with each other and the controller agent. After a coalition is formed, this interaction slightly changes, as described in Section 4.3.

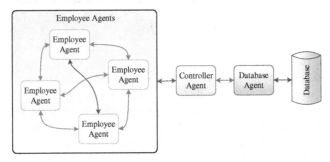

Fig. 1. System architecture displaying agent interaction during coalition formation phase

4.1   Database Agent

The database agent is the only agent in the system which has the capability of accessing the database. It accepts database requests, executes queries and delivers results. Only, the controller agent has direct communication with the database agent. If other agents need to query the database, they transmit their requests through the controller agent to database agent and receive results in the same way.

4.2   Controller Agent

The framework presents an execution environment where coalitions are formed in a distributed manner with no central control. However, the framework itself has a central management system, the controller agent being in charge of system's overall coordination. Central management usually becomes a bottleneck; therefore the control agent conveys some of its tasks to coalition manager agents as to decrease its work load. The controller agent monitors agents' status and task control period. When a predefined task control period ends, it restarts the coalition formation phase, thus eliminating the need to reset the system after new tasks have been defined.

Controller agent monitors the coalition formation phase. After coalition formation is completed, the controller agent starts a new coalition manager agent for each coalition, sends them the employee agents' identifications and task details. Next, the controller agent waits for messages from coalition manager agents. When a coalition manager reports, the controller agent prepares an execution report and sends it to database agent to be inserted into the database.

Furthermore, the controller agent is responsible for error handling. When an error message is received from a coalition manager agent, the controller agent updates the status of employee agents and also changes the task state to "failed" on the database. The failed task is scheduled on the next round of coalition formation phase. With this approach, the system becomes fault tolerant against agent failures.

4.3   Employee Agents

Employee agents execute the coalition formation algorithm to form coalitions and, after deciding on the coalitions, they perform the tasks which have been assigned to the coalition. They cannot be members of multiple coalitions at a time because of the disjoint coalition assumption. When activated, these agents firstly contact the controller agent and register themselves on the active agent list. Next, they wait for the arrival of information related to currently active agents and tasks from controller agent. After all employee agents receive the required information, they start executing the coalition formation algorithm and determine the coalitions they will join. Employee agents inform the controller about the result and wait until they are notified by the coalition manager agents assigned to their coalition. At this point in time, employee agents lose contact with the controller agent and carry on all further interactions with their manager agents. Fig 2 depicts agent interactions after coalitions are formed. Employee agents join their coalitions, execute their tasks and inform coalition manager agents that they have completed subtasks. If an error occurs during subtask execution, it is reported to the coalition manager agent.

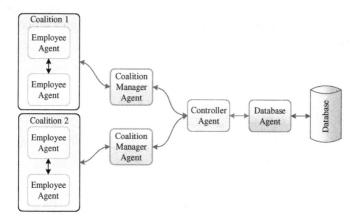

Fig. 2. Agent interaction after coalition formation

4.4   Coalition Manager Agents

For each coalition formed, a new coalition manager is created and remains in existence during the lifetime of the coalition. The coalition manager agent is representative of its coalition. It checks whether employee agents have completed subtasks; after a task is completed, the manager agent informs the controller and shuts down itself.

Coalition manager agent is also responsible for error checking, informing the controller agent about the error. The coalition manager also checks employee agents at predefined time periods to see if they are alive. If any employee agent does not respond in certain duration, the coalition manager assumes the agent has failed and informs the controller agent about the error. In both error cases, coalition manager stops execution of all subtasks and shuts down itself as the coalition is broken down.

5   Experiments and Evaluation

We have performed a comprehensive experimental analysis of the framework we have presented. The implementation is coded in Java 7 and runs on a notebook computer with 1.73 GHz Intel Core i7 processor, with 8.0 GB memory on a 64-bit Windows 7 Professional operating system. Experiments were performed on agents which were developed using JADE 4.2.0 version [8].

We tested our system approximately with 3000 runs to observe the effect of increasing values of agents, tasks and coalition sizes on coalition formation time. Test results reflect the time employee agents have consumed during the execution of the coalition formation algorithm and exclude the time taken by other activities, such as employee agent registration or task list sharing. We have applied the coalition formation algorithm to transportation problem and have carried out the test runs on the described environment over this application.

5.1   Effect of the Number of Employee Agents on Coalition Formation Time

In order to see the effect of varying numbers of employee agents on coalition formation time, we have carried out two sets of experiment, each with a different of tasks. Fig. 3 shows the results.

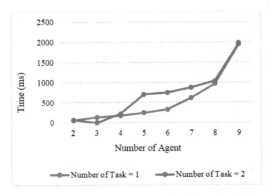

Fig. 3. Effect of number of employee agents on coalition time for different numbers of tasks

In both experiments, the maximum coalition size, k, was set to 2. The first set of experiments were executed with the number of tasks equal to 1 (blue graph), and the second with the number of tasks equal to 2 (red graph). In both cases, the number of employee agents is increased from 2 to 9 and the time taken for coalition formation is recorded. We have observed that increase in the number of employee agents directly effects coalition formation time. While the increase is acceptable up to 8 agents, further increases in the number of agents result in a steep rise. This is due to the heavy communication load between agents. We also observe that an increase in the number of tasks also effect coalition formation time, with larger values of tasks resulting in longer time periods, as seen in Figure 3.

Next, we have carried out experiments to observe how varying numbers of employee agents effect coalition formation time in systems with different maximum coalition sizes. The results are depicted in Fig. 4. In the experiments, the number of tasks was to 1. The experiments were executed for two maximum coalition size, one with k equal to 2 (blue graph), and another with k equal to 3 (green graph).

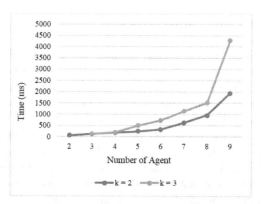

Fig. 4. Effect of number of employee agents on coalition time for different coalition sizes

We have observed that formation of larger coalitions take more time, compared to smaller coalitions. In conclusion, we can say that increasing the number of employee agents results in longer coalition formation time due to increase in the iterative calculations and inter agent communication.

5.2   Effect of Number of Tasks on Coalition Formation Time

We have also carried out experiments to observe how varying the number of tasks effects coalition formation time in systems with different numbers of employee agents. The results are depicted in Fig. 5.

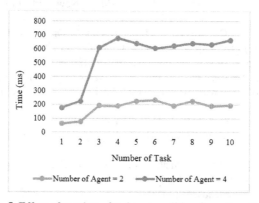

Fig. 5. Effect of number of tasks on coalition formation time

In these experiments, we have set the maximum coalition size, k, to 2. Two sets of experiments were executed, one with 2 (orange graph) and another with 4 (blue graph) employee agents. The results conform with the experiments; coalition formation time is directly affected by the number of employee agents and increasing the number of tasks does not create a significant difference in the time required.

6   Conclusions and Future Work

In this paper, we focus on a particular solution to the problem of task allocation through coalition formation. First, we shortly describe the Shehory-Kraus coalition formation algorithm, and next present a multi-agent based execution environment where employee agents can negotiate to form coalitions and execute tasks in accordance with the Shehory-Kraus algorithm. We describe the system architecture of the framework and discuss its various components in detail. For an assessment of the framework, we present experiments and report the effects of several factors, such as numbers of employee agents, tasks, and coalition sizes on coalition formation time.

Experimental results show that the framework fully provides the requirements for a negotiation environment for agents to form coalitions and execute tasks. The framework can be used as a test bed for further research as well. We plan to enhance the coalition formation algorithm by allowing agents to join multiple coalitions and adding precedence constraints to tasks as future work.

References

1. Shehory, O., Kraus, S.: Methods for task allocation via agent coalition formation. Artificial Intelligence 101(1-2), 165–200 (1998)
2. Kraus, S., Shehory, O., Taase, G.: Coalition formation with uncertain heterogeneous information. In: Proceedings of the Second International Joint Conference on Autonomous Agents and Multiagent Systems, AAMAS 2003, Melbourne, Australia, July 14-18 (2003)
3. Kraus, S., Shehory, O., Taase, G.: The Advantages of Compromising in Coalition Formation with Incomplete Information. In: Proc. AAMAS, New York, July 19-23, pp. 588–595 (2004)
4. Ferber, J., Gutknecht, O., Michel, F.: From Agents to Organizations: an Organizational View of Multi-Agent Systems. In: Giorgini, P., Müller, J.P., Odell, J.J. (eds.) AOSE 2003. LNCS, vol. 2935, pp. 214–230. Springer, Heidelberg (2004)
5. Alfonso, B., Botti, V., Garrido, A., Giret, A.: A MAS-based Infrastructure for Negotiation and its Application to a Water-Right Market. In: Infrastructures and Tools for Multiagent Systems, Valencia, Spain (2012)
6. Pinto, T., Morais, H., Oliveira, P., Vale, Z., Praça, I., Ramos, C.: A new approach for multi-agent coalition formation and management in the scope of electricity markets. Energy 36(8), 5004–5015 (2011)
7. Ray, D., Vohra, R.: Coalition Formation. In: Young, P., Zamir, S. (eds.) Preliminary draft, prepared as a chapter for Handbook of Game Theory, vol. 4. North-Holland, The MIT Press, Cambridge, Massachusetts (2013)
8. JADE Home Page, http://jade.tilab.com

Multi Agent Based Dynamic Task Allocation

Arambam James Singh*, Poulami Dalapati, and Animesh Dutta

Multi Agent and Distributed Computing Lab, Department of Information
Technology, National Institute of Technology Durgapur, India
{jamesastrick,dalapati89,animeshrec}@gmail.com

Abstract. Multi-Agent system, which is relatively a recent term, can be
viewed as a sub-field of Distributed AI. In order to construct and solve
complex real world problems like task allocation in an organization, a
number of agents can perform work cooperatively and collaboratively.
To achieve the maximum system utility through the proper task allocation among agents, there is a notion of dynamic team formation. In this
paper we focus on task allocation mechanism to agents with dependencies
among various tasks to agents either individually or by forming a team
whenever needed according to their capability to do a particular task.
We propose few algorithms which dynamically filter the capable agents
for performing task and hence allocating them with the help of synergy
value and their individual utility. We also simulate the task allocation
mechanism done by our proposed algorithms and show the performance
of the system by defining some metrics like, agent utility, team utility,
system utility.

Keywords: Multi Agent System, Synergy, Dynamic Team Formation,
Task Allocation.

1 Introduction

As we see in the human society, in order to organize an event, we need a group
of individuals who perform the task associated with the event. For effective
completion of this event, the compatibility among the individuals in the group
is also important to consider. We extend this notion to an agent based system.

As the field of multi agent system is advancing, this event organization problem could be an emerging trend, in this regard. Here the event can be described
as a goal that is to be achieved or performed with the help of an organization
or rather an agency. An event may consists of several tasks which are either
dependent or independent. So, event organization means successful completion
of all the tasks in an efficient way. In this scenario, the allocation of tasks among
agents is a major challenge as there are various agents with various capabilities
which may or may not be matched with the required characteristics to complete
the task. So, here comes the concept of team formation with compatible and
capable agents. There is a common term *Synergy* [1] which is generally used in

---

* Corresponding author.

G. Jezic et al. (eds.), *Agent and Multi-Agent Systems: Technologies and Applications,*   171
Advances in Intelligent Systems and Computing 296,
DOI: 10.1007/978-3-319-07650-8_18, © Springer International Publishing Switzerland 2014

team formation in human society as well as in multi agent system. It means how well the group members in a team can work together.

Prior researches in agent system and task allocation [2] [3] have dealt with single agent's capability and mostly the task allocation through team formation [1] in undependable environment [4], i.e. all the tasks and subtasks are independent of each other, though there are some algorithms which supports dynamic task allocation scenario [5]. But here we have tried to introduce specifically the event organization through task allocation in the atomic level after decomposition of tasks into subtasks [6] to number of agents either individually or in a team, where some tasks are dependent on other, in terms of priority, as well as for efficient completion of tasks to maximize the overall system utility.

The structure of the paper is organized as follows: In section 2 we dicuss about some previous works in related domain, Section 3 present scope of our work. Section 4 & 5, model our proposed system where we define some metrics. Section 6 highlights some problems as well as its solution following the algorithms of these solutions. In section 7 we discuss about our experiments and results along with its practical application and finally section 8 concludes.

2   Related Work

In MAS(Multi Agent System) many prior works have been focused on the agents as homogeneous environment [7], taking into account that all the agents have same identical capabilities and there are no dependencies in the environment. Various algorithms are proposed on the basis of number of practical problems in this domain where all the agents are continuously adopting environmental inputs and accordingly forming teams to handle problems. But being in heterogeneous environment [2] agents have different capabilities and some of them have the goal to maximize system's utility where others are dedicated to their own work. There are various task allocation algorithm [4] [8] [9] [10] [3] which allocate tasks to agents in optimized way. Yichuan Jiang *et.al.* [4], in their paper, have given the solution of task allocation in undependable multi agent environment. They proposed the idea of deceptive agents who cannot provide desired resources and take more time in seeking those and also the idea of contractor agents. They have also provided a solution in this regard by giving the guarantee of access of dependable resources and also minimizing access time. Sherief Abdallah *et.al.* [8] come up with a generalized semi-MDP($SMDP$) model for efficient representation. They have referred a term *mediation* for decomposition of tasks and also for picking up the right agent negotiating with other agents. In adhoc system [11] all the capable agents collaborate with other agents to complete a task. With the increasing number of autonomous agents, the need of their effective interaction is also growing gradually. Agent has different perspective of the world. So, to achieve a common goal agents should have capability to adopt new team-mates and learn to cooperate dynamically as a part of the adhoc team [12] [13]. An important aspect of adhoc team formation is the idea of leading team-mates which influence a team to alter their decision in the notion of maximizing system utility. In some recent paper, Katie Genter *et.al.* [12] have mentioned the idea of flocking agents where the main challenge is that whether more than one agent can lead the team to a desired objective and if yes, what would be the policy to do that efficiently as there will be no explicit control over the behavior of the flocking agents, where influence over the adhoc agents is implicit. Agents compute their best action based on their most recent observations of their team-mates' actions. The forming of team [1] highly depends on the composition or the combination of the team members whether they are compatible enough to work in a group to achieve a goal or not. And for checking compatibility between them a team synergy is used in multi agent system. The synergy of two agents can be formulated and calculated from synergy graph where distance between them signifies how well they can work together in a team.

3   Scope of the Work

The work done so far this domain is concerned with tasks assignment problems i.e. assigning a set of tasks to a set of agents in MAS. But there are very few remarkable mechanism, which deal with the task allocation dynamically considering some disastrous condition [14] [15]. There is another issue like optimized allocation of tasks here [8] [9]. An automated agent based system should address all these practical phenomena in an efficient way. Here we address such issues and also take some practical scenario to implement the same. In our proposed technique, the agency perform tasks either by forming team or by decomposing tasks into atomic subtasks according to the need. We also simulate our work to achieve better performance than previous works.

4   System Model

The task of event organization can be formally modeled by some parameters. An event can be described as a goal that is to be achieved and the agency should perform the event in such a manner that the allocation of tasks to the agency will be the optimal allocation to maximize the total system utility. For the sake of clarity here the whole system can be defined by two attributes,

$$Sys = <E,A>$$

where E is the *event* which can be represented as a task graph [16] and A is the agency which will perform the tasks to complete the event.

An *agency* is composed of number of agents as,

$$A = \{A_i | i \in [1, n]\}$$

Here we define some terms which we use in modeling the task allocation problem.

- *Attributes* : Attributes are some properties which should be available in the characteristics set of tasks and capability set of agents so that tasks can be allocated to agents or in the other way agents can perform some task successfully.

$$C = \{C_i | i \in [1, k]\}$$

- *Characteristics or Attributes of a task* : Characteristics of a task T_i which is denoted by char(T_i) represents some attributes of the individual task which are required in agents' capability set in order to complete the task.
  $char(T_i) \subseteq C$
- *Capability of an agent* : Capability of an gent A_i is the set of skills or attributes an agent must have to perform tasks and is denoted by cap(A_i).
  $cap(A_i) \subseteq C$
- *Priority of a task* : Priority of any task can be described in terms of dependencies i.e if T_j is dependent on T_i then T_i needs to be executed first. So T_i has the higher priority than T_j. In our scenario we have represented priority as

  $P = \{P_i | i \in [1, l]\}$

  Please note that more than one task can have same priority and these tasks can be executed concurrently provided system have enough agents. A task T_i can only be assigned to an agent A_i iff A_i has the capability to perform the task T_i provided A_i is available at that time instant. i.e. $char(T_i) \cap cap(A_i) = char(T_i)$. If we are not able to find a single agent who possesses all the attributes required to complete the task, then the task T_i can either be decomposed into further subtasks as,

  $T_i = \{T_{ij} | j \in [1, k]\}$

  or, T_i can be solved by forming a team of agents whose collective capability set matches to the required attributes set of T_i, i.e $char(T_i) = \forall i, A_i \in T_m \cup cap(A_i)$, where T_m means team of agents.
- *Synergy between agents* : In every agent community, there is a synergy graph associated with it. Synergy graph provides the relationship among agents. Mathematically, A synergy graph S_{syn} is a connected graph, $S_{syn} = (V_s, E_s)$,

  $V_s = \{A_i : A_i \in A\}$
  $E_s = \{E_{ij}, \text{ where } E_{ij} \text{ represents the compatibility between } A_i \text{ and } A_j\}$
  $S_{syn}(A_i, A_j) = 1 / d(A_i, A_j), i \neq j \text{ and } \forall i, j \in [1, n]$
  $d(A_i, A_j) = $ Distance or Compatibility between any two agents A_i, A_j in S_{syn}.
  For example, In Fig. 1 we have a synergy graph.
  $S_{syn}(A_1, A_2) = 1 / d(A_1, A_2) = 1/1 = 1$
  $S_{syn}(A_1, A_3) = 1 / d(A_1, A_3) = 1/2 = 0.5$
  The synergy values among agents may change overtime as in when the synergy network or graph change.

Fig. 1. Synergy (Source: [1], Sec 3.1)

5   Metric Definition

Here we propose some of the metrics to analyse the whole system of task allocation to agents.

- *Agent Utility* : Utility of an agent A_j is defined w.r.t a task T_i , i.e. how well an agent can execute that task, which will depend on how much capability the agent has for that task. So mathematically can be defined as :

$$U(T_i, A_j) = \frac{|char(T_i) \cap cap(A_i)|}{|cap(A_i)|}$$

- *Team Utility* : Utility of a team can be defined as, how Effectively members of the team, perform the task in a cooperative manner.
  Mathematically,
  $U_t = U(T_m) + S_{syn} (T_m)$
  where,
  $U(T_m) = \Sigma_{\forall j, j \in [1,n]} U(T_i, A_j)$,
  $S_{syn}(T_m) = \Sigma_{\forall i,j, A_i, A_j \in T_m} Syn(A_i, A_j)$, $T_i \in E$ and $A_i \in A$
- *System Efficiency* : In our scenario efficiency of the system can be described from various perspectives as, we allocate a task to a single agent if the agent has all the necessary capabilities, we also allocate task to a team of agents if we are not able to find a single agent which has all the necessary capabilities. Now here, even after using various techniques to allocate tasks, one question remains as to how much we are able to execute those tasks which can be executed concurrently i.e tasks which are in same priority. So, if our system is able to execute all the concurrent tasks concurrently then it means we have no task in waiting state and thus it should account for the efficiency of the system.
  Efficiency can be mathematically represented as: $Eff = \Sigma_{\forall i, P_i \in P}$
  $(\frac{\lambda}{|same-prio|})$
  where,
  $same - prio$: Set of tasks with same priority,
  λ : Total Number of Concurrently Executed Tasks in same-prio

6   Problem Formulation

In order to decide which task should be allocated first and which agent should do the task so that the maximum system utility or the efficiency will be achieved, we introduce the notion of priority value P for each task, i.e. all the task T_i has a priority value from the set $P = \{ P_i | i \in [1,l] \}$, where $P_i > P_j$ if i<j, means task T_i with priority value P_j can only be allocated to an agent $A_j \in A$ iff task T_i with priority value P_i is completed, i.e. T_j dependent on T_i or T_i has the higher priority to complete the event efficiently.

Problem 1: *If more than one task have same priority value, then there may arise an ambiguity that in which order tasks should be performed.*
When number of tasks have the same priority value, i.e.

$P(T_i) = P_i, P(T_j) = P_j$ and $P_i = P_j$

where $P(T_i)$ = the priority value of the task T_i, $T_i \in E$, then agent can pick any one of them as the tasks have no dependencies. But agent should calculate the utility value for the order of picking a single task so that the total utility of the system be the maximum. And if there are sufficient number of capable agents are available for doing all those tasks, then the tasks can be done parallely. i.e. for any task T_i and T_j, ($i \neq j$), if $P_i = P_j$, then those tasks can be done concurrently provided,

$char(T_i) \cap cap(A_i) = char(T_i)$ and $char(T_j) \cap cap(A_j) = char(T_j)$

Problem 2: *If there are number of tasks T_i with certain characteristics and number of available agents with all the capability to perform a task or with partial capability, then which agent(s) should do which task so that the system utility get maximized.*
There are two possible cases in this scenario.

Case 1 : When there are more than one agents with same capability set, then any agent can perform the task. But only one agent will do the task.

$char(T_i) \cap cap(A_i) = char(T_i)$ and $char(T_i) \cap cap(A_j) = char(T_i)$

where $T_i \in E$ and $A_i, A_j \in A$.

To maximize the total system efficiency every agent will calculate their individual utility value to perform a particular task. Among all the capable agents who has the maximum utility to complete the task, will perform that one.

$allocate(T_i, A_i) = 1$, iff $(T_i, A_i) = max\{U(T_i, A_i)\}$

Case 2 : Among all the available agents no single agent has all the required capabilities or attributes to perform any particular task completely, i.e. $char(T_i) \cap cap(A_i) \neq char(T_i)$, where $T_i \in E$ and $A_i \in A$. So, in this case either the tasks should be decomposed into number of subtasks T_{ij}.

$T_i = \{ T_{ij} \mid j \in [1, k] \}$

or some teams of agents are to be formed in order to complete the whole task.

$\forall i, A_i \in T_m \cup cap(A_i)$,

where T_m means team of agents.

Now, whether to form any team or to divide a task further into subtasks can be decided using the synergy graph S_{syn} of the available agents. Here, to take the decision of team formation, we have selected a threshold value for the synergy T_h between agents. If for any two agents in the synergy graph, $\{S_{syn}(A_i, A_j) \forall i, j \in [1, n]\} < T_h$, then the agents are not compatible to work together in a team. In this situation decomposition of task T_i into subtasks T_{ij} is must. After decomposition if the agent A_i has the full capability to perform T_{ij} and also it has the maximum utility value then directly allocate the task to that agent.

$(char(T_{ij}) = cap(A_i)) \bigwedge (U(T_{ij}, A_i) = max\{U(T_i, A_i)\}) = TRUE$,
then $allocate(T_{ij}, A_i) = 1$

Again, if the attributes required to complete any particular task match to the united capability set of number of available agents and the calculated synergy value between agents are greater or equal to the threshold value, then they are compatible to work together and hence to complete any particular task a team of those agents can be formed without any decomposition of task T_i.

$$(char(T_i) = (cap(A_i)) \cup cap(A_j)))\forall i, j \in [1, n]) \bigwedge$$
$$(\{S_{syn}(A_i, A_j)\forall i, j \in [1, n]\} > T_h) = TRUE,$$
then, $T_m = \{A_i, A_j\}, \forall A_i, A_j \in A$ and $allocate(T_i, T_m) = 1$

Algorithm 1. Selects Agents with all the characteristics of a tasks, kept in full[] and also select agents with atleast one characteristic and kept in partial[]

---

full[]   /*Array containing agents with all the capabilities of any task*
partial[]   /*Array containing agents with partial capability*

```
1: k=0, l=0
2: for  all tasks T_i, T_i ∈ E  do
3:     for  all agents A_j, A_j ∈ A do
4:        if  char( T_i ) ∩ cap( A_j ) == char( T_i ) then
5:           T_i.full[k++] ← A_j
6:        end if
7:        if  char(T_i) ∩ cap( A_j ) ⊂ char( T_i ) then
8:           T_i.partial[l++] ← A_j
9:        end if
10:    end for
11: end for
```

---

7   Experiments and Results

In our system we use three metrics Agent Utility, Team Utility and System Efficiency. System performance is measured using the expression of System efficiency given in section 5. Efficiency is the ratio of total number of concurrently executed task out of total number of task present in same-prio. Now in order for a task to execute we need to find the agent or team of agents. Finding agent or team of agents is done using the metrics Agent Utility and Team Utility. So, ultimately all the three metrics are accounted for System Performance.

Now to simulate our proposed methodology to organize an event, consisting of tasks, i.e. to efficiently allocate the tasks to an appropriate agent or to a team of agents and check how effectively our system responds, we consider an event E which may consists of various number of tasks. We run our algorithm for different number of tasks an event can have also for different priorities(3,6,9) of the task set. Fig. 2. represents the Efficiency Graph 1. Here we fix number of agents in the agency to 8 and test the efficiency for number of tasks varying from 5 to 20. We found an effective result as we can see from the Efficiency graph 1 the average efficiency is 85.13%.

Algorithm 2. Filters the partial[] column of each task, i.e find the best combination or a team of agents which has maximum synergy value and also their collective capabilities are sufficient enough to perform the task

```
1: for  i = 1 to | E | do
2:     Evaluate(Tᵢ.partial[])
3: end for
```

```
1: Evaluate(Tᵢ.partial[])
2: Group = φ, U_max = 0
3: for  j = 1 to Tᵢ.partial.length  do
4:     for  k = j to Tᵢ.partial.length  do
5:         if char(Tᵢ) == char(Group) ∪ cap( Tᵢ.partial[k])  then
6:             Group ← Tᵢ.partial[k]
7:             U ← 1/|Group| ∑_{∀aᵢ,aⱼ∈Group∧i≠j} Sy(aᵢ, aⱼ)
8:             break
9:         end if
10:        if cap(Tᵢ.partial[k])∉(cap(Group)∩ char(Tᵢ)) then
11:            Group ← Tᵢ.partial[k]
12:        end if
13:    end for
14: end for
```

In second experiment, we fix the number of tasks of the event E to 10 and run our algorithm for different number of agents in the agency A varying from 5 to 19, also for different priorities(3,6,9) of tasks set. Fig. 3. represents Efficiency Graph 2. Here also we obtained a positive result with average efficiency 83.68%.

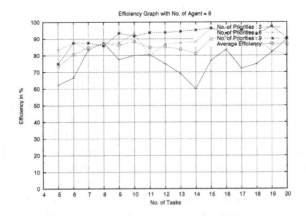

Fig. 2. Efficiency Graph 1 (Source: Experiments and Results)

Algorithm 3. Allocates tasks to a agent or to a team of agents based on priority and then tasks are also executed

```
1: if A_i ∉ T_k.partial[],∀T_k ∈ E ∧ k ≠ j then
2:     Allocate(T_j, A_i)
3: end if
4: if A_i ∈ T_k.partial[] and A_i ∈ T_i.full[] ,∀T_k,T_j ∈ E ∧ k ≠ j  then
5:     Allocate(T_k, A_i)
6: end if
7: for all task T_i ∈ E, T_i.full[] ≠ φ and T_i.partial ≠ φ do
8:     if Utility(T_i.Group) < U_threshold then
9:         decompose(T_i)
10:    else
11:        form-team(T_i)
12:    end if
13: end for
14: for all task T_k ∈ E, T_k.full[] ≠ φ do
15:    for all agent A_i ∈ A ∧ A_i ∈ T_i.full[] do
16:        A_i.Utility = |char(T_k) ∩ cap(A_i)| / |cap(A_i)|
17:    end for
18: end for
19: for all priorities P_i ∈ P, i=1 to P.length do
20:    Same-Prio ← Task(P_i) {//Task(P_i) returns tasks with same priority P_i}
21:    Exclusive-full = φ ; Flag-full = True;
22:    for all tasks T_i, T_i ∈ Same − Prio do
23:        if Exclusive-full ∩ max(A_i.Utility, A_i ∈ T_i.full[]) == φ then
24:            Exclusive-full ← max(A_i.Utility, A_i ∈ T_i.full[])
25:        else
26:            Flag-full ← False
27:            Break
28:        end if
29:    end for
30:    if Flag-full == True  then
31:        for all tasks T_i, T_i ∈ Same − Prio do
32:            Allocate(T_i,max(A_i.Utility, A_i ∈ T_i.full[]))
33:            Execute(T_i)
34:        end for
35:    end if
36:    Exclusive-partial = φ
37:    Flag-partial = True
38:    for all tasks T_i, T_i ∈ Same − Prio do
39:        if Exclusive-partial ∩ T_i.partial[]) == φ then
40:            Exclusive-partial ← T_i.partial[])
41:        else
42:            Flag-partial ← False; Break
43:        end if
44:    end for
45:    if Flag-full == False and Flag-partial == True  then
46:        for all tasks T_i, T_i ∈ Same − Prio do
47:            Allocate(T_i,T_i.Group); Execute(T_i);
48:        end for
49:    end if
50:    if Flag-full == False and Flag-partial == False then
51:        for all tasks, T_i ∈ Same − Prio do
52:            if A_k ∈ ∩_{∀i,T_i∈Same−Prio}T_i.full[] and A_l ∉ T_j.full[],∀T_j ∈ Same − Prio and cap(A_k) ∩
               cap(A_l) == cap(A_k) then
53:                Replace(A_k, A_l)
54:            end if
55:            if A_k ∈ ∩_{∀i,T_i∈Same−Prio}T_i.partial[] and A_l ∉ T_j.partial[],∀T_j ∈ Same − Prio and
               cap(A_k) ∩ cap(A_l) == cap(A_k) then
56:                Replace(A_k, A_l)
57:            end if
58:            Execute(T_i)
59:        end for
60:    end if
61: end for
```

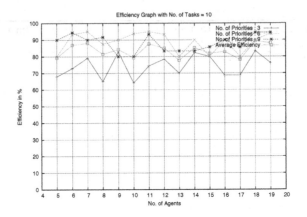

Fig. 3. Efficiency Graph 2 (Source: Experiments and Results)

7.1 Practical Application to RoboCup Rescue

We have tested our system for a practical scenario of disaster rescue, the scenario chosen is that of a Robocup Rescue Problem [14] [15].

In a city after a disaster like an earthquake suffers from devastation. There are chaos everywhere, road blocks, Fires, people stuck in bulidings, seriously injured, urgent need of medical attentions, failure of communication infrastructures.

Now in order to save the civilians the rescue teams need to perform various tasks. The rescue agents that we have considered are : Ambulance Agents, Police Agents and Fire-Fighter Agents. Each of these agents are different, they perform different tasks. But there are few tasks that any of those agents can also perform.

Followings are the tasks that we have considered for our system :

- Task 1 : Providing First Aid to Injured civilians and taking them to near by hospitals
  Agent Required : Ambulance Agents

- Task 2 : To extinguish fire if any
  Agent Required : Fire Fighter Agent or Police Agents

- Task 3 : To save stucked civilians from a multi storey buildings
  Agent Required : Fire Fighter Agent or Police Agent

- Task 4 : To evacuate civilians from the affected areas
  Agent Required : Police Agent or Fire Fighters Agent

- Task 5 : To clear the blocked roads
  Agent Required : Police Agent

– Task 6 : Coordinating civilians
  Agent Required : Police Agent

Our aim here is to rescue all the civilians, which means completion of all the 6 tasks. This further raises the question of task allocation among the agents. Here we want all the tasks to be performed in an efficient manner. A task can be performed by any agent if it has all the necessary capabilities or characteristics. Here it is obvious that though the instances of tasks are predefined, the total number of tasks are much more than the number of available agents in practical. The priority and dependencies between those works change over time. As an example, say at a point of time rescue team rescue an injured people and decide to send them by ambulance to the hospitals. In the mean time another injured person comes into notice whose condition is worse. So, we have to send this people to the hospitals more urgently than the previous rescued people. And as per the dependencies are concerned, such scenario may happen that untill the road blockage is removed, the ambulance cannot move forward.

We calculate the efficiency of the system for this scenario using the system utility formula given in Section 5 and we got an efficiency of 73.25%.

8   Conclusions

In this paper we present three algorithms for dynamic task allocation, where all the tasks have different priorities which determine the execution order of the tasks. We able to achieve approximately 85.13% of efficiency by varying number of tasks while number of agents are fixed. Again, solution finds 83.68% of efficiency when number of agents in agency are varying.

Here we only propose the mechanism of an agent based dynamic task allocation with interdependencies and priorities among tasks. But there can be another important issue of *fault tolerance*, i.e. when any number of capable agents for doing any particular task fail. So, the future direction of our work is to extend this mechanism by developing another dynamic algorithm which takes into account this practical issue, i.e. how other agents can manage to do the task in absence of its previously allocated capable agents.

References

1. Liemhetcharat, S., Veloso, M.: Modeling and Learning Synergy for Team Formation with Heterogeneous Agents. In: Conitzer, Winikoff, Padgham, van der Hoek (eds.) Proceedings of the 11th International Conference on Autonomous Agents and Multiagent Systems (AAMAS 2012), Valencia, Spain (2012)
2. Anders, G., Hinrichs, C., Siefert, F., Behrmann, P., Reif, W., Sonnenschein, M.: On the Influence of Inter-Agent Variation on Multi-Agent Algorithms Solving a Dynamic Task Allocation Problem under Uncertainty. In: IEEE Sixth International Conference on Self-Adaptive and Self-Organizing Systems, pp. 29–38 (2012)

3. Shehory, O., Kraus, S.: Methods for task allocation via agent coalition formation. Artificial Intelligence 101, 165–200 (1998)
4. Jiang, Y., Zhouand, Y., Wang, W.: Task Allocation for Undependable Multiagent Systems in Social Networks. Advan. IEEE Transaction on Parallel and Distributed Systems 24(8), 1671–1681 (2013)
5. Macarthur, K.S., Stranders, R., Ramchurn, S.D., Jennings, N.R.: A Distributed Anytime Algorithm for Dynamic Task Allocation in Multi-Agent Systems. In: Association for the Advancement of Artificial Intelligence (2011)
6. Skowron, A., Nguyen, H.S.: Task Decomposition Problem in Multi-Agent System. Artificial Intelligence 101
7. Albrecht, S.V., Ramamoorthy, S.: Comparative Evaluation of MAL Algorithms in a Diverse Set of Ad Hoc Team Problems. In: Conitzer, Winikoff, Padgham, van der Hoek (eds.) Proceedings of the 11th International Conference on Autonomous Agents and Multiagent Systems (AAMAS 2012), Valencia, Spain (2012)
8. Abdallah, S., Lesser, V.: Modeling Task Allocation Using a Decision Theoretic Model. National Science Foundation Engineering Research Centers Program
9. Weerdt, M., Zhang, Y., Klos, T.: Multiagent task allocation in social networks. Autonomous Agent Multi-Agent System 25, 46–86 (2012), doi:10.1007/s10458-011-9168-3.
10. Jiang, L., Zhan, R.: An Autonomous Task Allocation for Multi-robot System. Journal of Computational Information Systems 7: 11 25, 3747–3753 (2011)
11. Barrett, S., Stone, P.: An Analysis Framework for Ad Hoc Teamwork Tasks. In: Conitzer, Winikoff, Padgham, van der Hoek (eds.) Proceedings of the 11th International Conference on Autonomous Agents and Multiagent Systems (AAMAS 2012), Valencia, Spain (2012)
12. Genter, K., Agmon, N., Ston, P.: Ad Hoc Teamwork for Leading a Floc. In: Ito, Jonker, Gini, Shehory (eds.) Proceedings of the 12th International Conference on Autonomous Agents and Multiagent Systems (AAMAS 2013). Saint Paul, Minnesota (2013)
13. Agmon, N., Stone, P.: Leading Ad Hoc Agents in Joint Action Settings with Multiple Teammates. In: Conitzer, Winikoff, Padgham, van der Hoek (eds.) Proceedings of the 11th International Conference on Autonomous Agents and Multiagent Systems (AAMAS 2012), Valencia, Spain (2012)
14. Santos, F.D., Bazzan, A.: Towards efficient multiagent task allocation in the RoboCup Rescue: a biologically-inspired approach, pp. 465–486. Springer (2010)
15. Chapman, A.C., Micillo, R.A., Kota, R., Jennings, N.R.: Decentralised Dynamic Task Allocation:A Practical Game Theoretic Approach. In: Decker, Sichman, Sierra, Castelfranchi (eds.) Proc. of 8th Int. Conf. on Autonomous Agents and Multiagent Systems (AAMAS 2009), Budapest, Hungary (2009)
16. Upadhyay, P., Acharya, S., Dutta, A.: Task Petri Nets for Agent based computing. INFOCOMP Journal of Computer Science (2013)

Microeconomic Demand Functions Implementation in Java Experiments

Roman Šperka and Marek Spišák

Silesian University in Opava, School of Business Administration in Karviná,
Department of Informatics, Univerzitní nám. 1934/3a, 733 40, Karviná, Czech Republic
{sperka,spisak}@opf.slu.cz

Abstract. The aim of this paper is to introduce microeconomic demand functions (Marshallian demand function and Cobb-Douglas utility function) in Java simulation experiments. The motivation is to use these function as a core element in a seller-to-customer price negotiation in an agent-based simulations. Furthermore, multi-agent model is proposed and implemented in Java to serve as a simulation framework to support the virtual company trading processes. The main background of this framework is to be integrated in management information systems as a decision support module. The paper firstly presents some of the existing principles about consumer behavior, agent-based modeling and simulation in the same area and demand function theory. Secondly, presents multi-agent model and demand functions negotiations. Lastly, depicts some of the simulation results in a trading processes throughout one year of selling commodities to consumers. The results obtained show that in some metrics the demand functions could be used to predict the trading results of a company.

Keywords: seller-to-customer, negotiation, Marshallian demand function, virtual company, simulation, agent-based.

1 Introduction

In the contemporary, dynamic, global and competitive market environment, consumer behavior depends on many different types of factors, which are difficult to grasp. The understanding of consumers could overcome some of the problems contemporary businesses are dealing with [1]. We concentrate on the use of some economic models and theories in our research to build advanced decision making tools for the trading companies. Previously, we presented partial research results using the decision function [2,3,4,5,6].

The approach introduced in this paper uses an agent-based model in the form of multi-agent system to serve as a simulation platform for the seller-to-customer negotiation in a virtual trading company. The main idea concentrates around the negotiated price establishment. Here we used a demand function. The overall scenario comes from the research of Barnett [7]. He proposed the integration of the real system models with the management models to work together in real-time. The real system (e.g. ERP system) outputs proceed to the management system (e.g. simulation framework)

G. Jezic et al. (eds.), *Agent and Multi-Agent Systems: Technologies and Applications*,
Advances in Intelligent Systems and Computing 296,
DOI: 10.1007/978-3-319-07650-8_19, © Springer International Publishing Switzerland 2014

to be used to investigate and to predict important company's metrics (KPIs – Key Performance Indicators). Actual and simulated metrics are compared and evaluated in a management model that identifies the steps to take to respond in a manner that drives the system metrics towards their desired values. We used a generic control loop model of a company and implemented multi-agent simulation framework, which represents the management system. This task was rather complex, therefore we took only a part of the model – trading processes and the seller-to-customer negotiation.

Implemented simulation framework will be a basic part of a future management system simulating business metrics of a real company's system. The paper is structured as follows. Section 2 represents some of the theoretical incomes. In the section 3 the multi-agent model is described. In the section 4 the seller-to-customer negotiation is introduced. The core of this section is the demand function definition. The simulation results are presented in section 5.

2   Literature Review

With personal and social factors deals e.g. Enis [8]. With physical factors deal e.g. McCarthy and Perreault [9]. More complex view on the social, economic, geography and culture factors gave Keegan et al. [10]. Schiffman [11] brought marketing mix and environment into the types of factors mentioned herein above. Previous discussions have so far either relied on an objectivist (complete information of customers, constant decision mechanism, constant consumer preferences) or a constructivist view (consumption discourses, consumption as a crucial aspect in the construction of identity). However, both have failed to integrate the consumers' interactions with their social behavior and physical environment as well as the materiality of consumption [12]. The complexity of the factors influencing consumer behavior and their changes in the time shows relations between external stimuli, consumer's features, the course of decision-making process and reaction expressed in his choices. As a result, the investigation of consumer behavior seems to be too complicated for traditional analytical approaches [13].

Agent-based modeling and simulation (ABMS) provides some opportunities and benefits resulting from using multi-agent systems as a platform for simulations with the aim to investigate the consumers' behavior. Agent-based models are able to integrate individually differentiated types of consumer behavior. They are characterized by a distributed control and data organisation, which enables to represent complex decision processes with only few specifications. In the recent past there were published many scientific works in this area. They concern in the analysis of companies positioning and the impact on the consumer behavior [14,15,16]. Often discussed is the reception of the product by the market [17,19], innovation diffusion [19,20,21]. More general deliberations on the ABMS in the investigating of consumer behavior show e.g. [22,23].

The core problem to be solved in the business process of selling the commodities to consumers while using the simulation is the price negotiation. We used in this partial research some functions from economic theory. We built our experimental research on a demand function. In microeconomics, a consumer's Marshallian demand function (named after Alfred Marshall) specifies what the consumer would buy in each price and wealth situation [24], assuming it perfectly solves the utility maximization problem[1]. Marshallian demand is sometimes called Walrasian demand (named after Léon Walras) or uncompensated demand function instead, because the original Marshallian analysis ignored wealth effects[2] [25,26]. We also used a Cobb-Douglas utility function and preferences saying that quantity demanded of each commodity does not depend on income, in fact quantity demanded of each commodity is proportional to income [27]. We based the seller-to-customer negotiation in our virtual company Java simulations on these two approaches. In the next section the agent-based model is introduced.

3   Agent-Based Model

To ensure the outputs of business processes simulations a simulation framework was implemented and used to trigger the simulation experiments. The framework covers business processes supporting the selling of commodities by company sales representatives to the customers – seller-to-customer negotiation (Fig. 1).

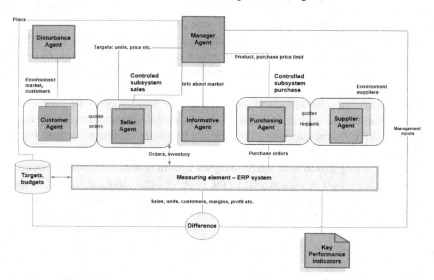

Fig. 1. Generic model of a business company. (Source: adapted from [4])

It consists of the following types of agents: sales representative agents (representing sellers, seller agents), customer agents, informative agent (measures time, informs agents about period passing), and manager agent (manages the seller agents,

---

[1] In microeconomics, the utility maximization problem is the problem, which consumers face: "How should I spend my money in order to maximize my utility?" It is a type of optimal decision problem.

[2] The wealth effect is an economic term, referring to an increase (decrease) in spending that accompanies an increase (decrease) in perceived wealth.

calculates KPI). Disturbance agent is responsible for the historical trend analysis of sold amount (using his influence on customer agent). All the agent types are developed according to the multi-agent approach. The interaction between agents is based on the FIPA contract-net protocol [28].

The number of customer agents is significantly higher than the number of seller agents in the model because the reality of the market is the same. The behavior of agents is influenced by two randomly generated parameters using the normal distribution (an amount of requested goods and a sellers' ability to sell the goods). In the lack of real information about the business company, there is a possibility to randomly generate different parameters (e.g. utility ratio of the current commodity, or an income of the customer). The influence of randomly generated parameters on the simulation outputs while using different types of distributions was previously described in [29].

4   Demand Functions in Negotiation

In this section, the seller-to-customer negotiation workflow is described and the definition of the Marshallian demand curve is proposed. Marshallian demand function is used during the contracting phase of agents' interaction. It serves to set up the limit price of the customer agent as an internal private parameter.

Only a part of the company's generic structure, defined earlier, was implemented. This part consists of the sellers and the customers trading with commodities (e.g. tables, chairs). One stock item simplification is used in the implementation. Participants of the contracting business process in our multi-agent system are represented by the software agents - the seller and customer agents interacting in the course of the quotation, negotiation and contracting. There is an interaction between them. The behavior of the customer agent is characterized by the Marshallian demand function based on the Cobb-Douglas utility function.

At the beginning disturbance agent analyses historical data – calculates average of sold amounts for whole historical year as the base for percentage calculation. Each period turn (here we assume a week), the customer agent decides whether to buy something. His decision is defined randomly. If the customer agent decides not to buy anything, his turn is over; otherwise he creates a sales request and sends it to his seller agent. Requested amount (generation is based on a normal distribution) is multiplied by disturbance percentage. Each turn disturbance agent calculates the percentage based on historical data and sends the average amount values to the customer agent. The seller agent answers with a proposal message (a certain quote starting with his maximal price: *limit price * 1.25*). This quote can be accepted by the customer agent or not.

The customer agents evaluate the quotes according to the demand function by calculating his maximal price. The Marshallian demand function was derived from Cobb-Douglas utility function and represents the quantity of the traded commodity as the relationship between customer's income and the price of the demanded commodity. If the price quoted is lower than the customer's price obtained as a result of the demand function, the quote is accepted. In the opposite case, the customer rejects the quote and a negotiation is started. The seller agent decreases the price to the average of the minimal limit price and the current price (in every iteration is getting effectively closer and closer to the minimal limit price), and resends the quote back to the customer. The message exchange repeats until there is an agreement or a reserved time passes.

Marshallian function specifies what would consumer buy at each specific price and income, assuming it perfectly solves utility maximization problem. For example: If there are two commodities and the specific consumer's utility function[3] is:

$$U(x_1, x_2) = x_1^{0.5} x_2^{0.5} U(x_1, x_2) = x_1^{0.5} x_2^{0.5} \tag{1}$$

Then the Marshallian demand function is a function of income and prices of commodities:

$$x(p_1, p_2, I) = \left(\frac{I}{2p_1}, \frac{I}{2p_2}\right) \tag{2}$$

Where I represents income and p_1 and p_2 are the prices of the commodities. In general, Cobb-Douglas utility function can be defined as:

$$U(x_1, x_2) = x_1^{\alpha} x_2^{1-\alpha} \tag{3}$$

The corresponding Marshallian demand function is:

$$x(p_1, p_2, I) = \left(\frac{\alpha I}{p_1}, \frac{(1-\alpha)I}{p_2}\right) \tag{4}$$

In the model there is calculated only one commodity (which is traded by the simulated company). In this case – using the Marshallian demand function there are two commodity baskets, where one is represented by company traded one and the rest represents all alternative commodities that customer can buy. So only x_1 is used supposing that utility ratio α is known and that for the rest of commodities the utility ratio is *(1-α)*. Therefore the demand function looks like this:

$$X = \frac{\alpha I}{p} \tag{5}$$

Where X represents amount of commodity, α is utility ratio, I is income and p is the price of the commodity. Customer's decision is described by retrieving the price from the demand function. We also include here the ability of the seller for increasing/decreasing the price according to his skills:

$$p = S \frac{\alpha I}{X} \tag{6}$$

This is the core formula, by which the customer decides if the quote is acceptable.

The aforementioned parameters represent global simulation parameters set for each simulation experiment. Other global simulation parameters are:

I – customer's income – it's normal distributed value generated at the beginning and not being changed during the generation;

---

[3] Ratio of the same utility, consuming the commodities – Cobb-Douglas utility function.

α – utility ratio – normal distributed value, which is generated for each customer each turn (week, while customers' preferences can change rapidly);

p – commodity price;

S – seller skills (ability to change price);

X – amount of commodity – normal distributed value generated, when customer decides to buy something.

Customer agents are organized in groups and each group is being served by specific seller agent. Their relationship is given; none of them can change the counterpart. Seller agent is responsible to the manager agent. Each turn, the manager agent gathers data from all seller agents and stores KPIs of the company. The data is the result of the simulation and serves to understand the company behavior in a time – depending on the agents' decisions and behavior. The customer agents need to know some information about the market. This information is given by the informative agent. This agent is also responsible for the turn management and represents outside or controllable phenomena from the agents' perspective.

5   Simulation Results

At the start of simulation experiments phase some parameters were set. Agent count and their parameterization are listed in Table 1. The purpose of the simulations is to prove if the demand functions could serve as a core element in the seller-to-customer negotiation.

Table 1. Multi-agent system parameterization. Source: own.

Agent Type	Agent Count	Parameter Name	Parameter Value
Customer	500	Maximum Disc. Turns	10
		Mean Quantity	40 m
		Quantity S. Deviation	32
		Mean Income	600 EUR
		Income St. Deviation	10
		Mean Utility Ratio	1.15
		Utility St. Deviation	0.2
Seller	25	Mean Ability	1
		Ability St. Deviation	0.03
		Minimal Price	0.36 EUR
Manager	1	Purchase Price	0.17 EUR
Market Info	1	Iterations count	52 weeks
Disturbance	1		

Agents were simulating one year – 52 weeks of interactions. As mentioned above – manager agent was calculating the KPIs. The results of the simulation are shown in graph (Fig. 2). The results are depicted in four categories frequently used to describe the company's trading balance. The categories are: sold amount, income, costs, and gross profit.

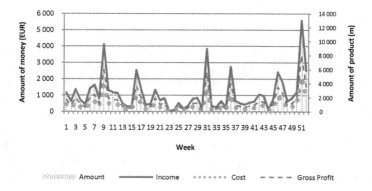

Fig. 2. KPIs results for 1 year generation (Source: own)

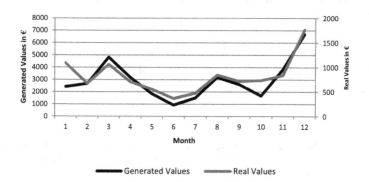

Fig. 3. The generation values graph (gross profit) – monthly (Source: own)

The commodity to be traded with was a UTP cable. Of course, companies are dealing with a whole portfolio of products. In our simplification we concentrated only on one product and this was a UTP cable. Further, total gross profit was chosen as a representative KPI. Figure 3 contains the month sums of total gross profit for real and generated data. As can be seen from this figure, the result of simulation represents trend, which is quite similar to the real data.

Real data was taken from a Slovak anonymous company trading with PC components and supplies. The time series was discovered for the 2012 and the parameters of the simulations were set to mirror the situation on the market in that time.

To see the correlation between the real and generated total month profit the correlation analysis was done. Correlation coefficient for total gross profit amount was 0.894, which represents very strong correlation between real and generated data. These results show that the demand functions could be used in further experiments to support the predictive purposes of decision making tools based on it.

Conclusion

The paper introduces an agent-based simulation approach dealing with trading processes within a virtual company. The experiments were set to prove the idea, that microeconomic demand functions could be used as a core element in seller-to-customer price negotiation. The overall idea is to use this approach to implement decision support models that could be connected to real management information systems in order to serve as a prediction modules. We obtained successful results in some of the KPIs of a virtual company. This supports our motivation to proceed with the experiments, to enhance our approach to extend the results on the rest of the KPIs.

In our future research we will concentrate on the enhancement of the model proposed in second section. We will make a comparison of the simulation results with other microeconomic models such as a decision functions, and we will try to implement a Monte Carlo simulation to compare agent-based models with classical approach.

Acknowledgement. This work was supported by grant of Silesian University no. SGS/6/2013 "Advanced Modeling and Simulation of Economic Systems".

References

1. Chramcov, B., Bucki, R., Suchánek, P.: Logistic Optimization of the Complex Manufacturing System with Parallel Production Lines. Journal of Applied Economic Sciences-VIII(3(25), 17–23 (2013) ISSN: 1843-6110
2. Šperka, R.: Application of a Simulation Framework for Decision Support Sys-tems. MITTEILUNGEN KLOSTERNEUBURG, Hoehere Bundeslehranstalt und Bundesamt fuer Wein- und Obstbau, Klosterneuburg, Austria 64(1) (2014) ISSN 0007-5922
3. Šperka, R., Vymětal, D., Spišák, M.: Towards the Validation of Agent-based BPM Simulation. In: Proc. Frontiers in Artificial Intelligence and Applications, 7th International Conference KES-AMSTA 2013, Hue city, Vietnam, vol. 252, pp. 276–283. IOS Press BV, Amsterdam (2013) ISBN 978-1-61499-253-0 (print)
4. Šperka, R., Spišák, M., Slaninová, K., Martinovič, J., Dráždilová, P.: Control Loop Model of Virtual Company in BPM Simulation. In: Snasel, V., Abraham, A., Corchado, E.S. (eds.) SOCO Models in Industrial & Environmental Appl. AISC, vol. 188, pp. 515–524. Springer, Heidelberg (2013)
5. Šperka, R., Spišák, M.: Transaction Costs Influence on the Stability of Financial Market: Agent-based Simulation. Journal of Business Economics and Management 14(suppl. 1), S1–S12 (2013), doi:10.3846/16111699.2012.701227 ISSN 1611-1699

6. Šperka, R., Vymětal, D.: MAREA - an Education Application for Trading Company Simulation based on REA Principles. In: Proc. Advances in Education Research, USA. Information, Communication and Education Application, vol. 30, pp. 140–147 (2013) ISBN 978-1-61275-056-9

7. Barnett, M.: Modeling & Simulation in Business Process Management', Gensym Corporation, pp. 6–7 (2003),
http://news.bptrends.com/publicationfiles/1103%20WP%20Mod%20 Simulation%20of%20BPM%20-%20Barnett-1.pdf (accessed January 16, 2012)

8. Enis, B.M.: Marketing principles: the management process. Goodyear Pub. Co. (Pacific Palisades, California), 608 p. (1974) ISBN 0876205503

9. McCarthy, E.J., Perreault, W.D.: Basic marketing: a global-managerial approach. Irwin, 792 p. (1993) ISBN 025610509X

10. Keegan, W., Moriarty, S., Duncan, T.: Marketing, 193 p. Prentice-Hall, Englewood Cliffs (1992)

11. Schiffman, L.G., Kanuk, L.L.: Purchasing Behavior, 9th edn. Pearson Prentice Hall, Upper Saddle River (2007)

12. Gregson, N., Crewe, L., Brooks, K.: Shopping, space, and practice. Environment and Planning D 20(5), 597–617 (2002), doi:10.1068/d270t

13. Challet, D., Krause, A.: What questions to ask in order to validate an agent-based model. In: Report of the 56th European Study Group with Industry, pp. J1–J9 (2006)

14. Tay, N., Lusch, R.: Agent-Based Modeling of Ambidextrous Organizations: Virtualizing Competitive Strategy. IEEE Transactions on Intelligent Systems 22(5), 50–57 (2002)

15. Wilkinson, I., Young, L.: On cooperating: Firms. Relations. Networks. Journal of Business Research (55), 123–132 (2002)

16. Casti, J.: Would-be Worlds. How Simulation is Changing the World of Science. Wiley (1997)

17. Goldenberg, J., Libai, B., Muller, E.: The Chilling effect of network external-ities. International Journal of Research in Marketing 27(1), 4–15 (2010)

18. Heath, B., Hill, R., Ciarallo, F.: A survey of agent-based modeling practices (January 1998 to July 2008). Journal of Artificial Societies and Social Simulation 12(4), 5–32 (2009)

19. Rahmandad, H., Sterman, J.: Heterogeneity and network structure in the dynamics of diffusion: Comparing agent-based and differential equation models. Management Science 54(5), 998–1014 (2008)

20. Shaikh, N., Ragaswamy, A., Balakrishnan, A.: Modelling the Diffusion of In-novations Using Small World Networks. Working Paper. Penn State University. Philadelphia (2005)

21. Toubia, O., Goldenberg, J., Garcia, R.: A New approach to modeling the adoption of new products: Aggregated Diffusion Models. MSI Reports: Working Papers Series 8(1), 65–76 (2008)

22. Adjali, I., Dias, B., Hurling, R.: Agent based modeling of consumer behavior. In: Proceedings of the North American Association for Computational Social and Organizational Science Annual Conference. University of Notre Dame, Notre Dame (2005)

23. Ben, L., Bouron, T., Drogoul, A.: Agent-based interaction analysis of con-sumer behavior. In: Proceedings of the First International Joint Conference on Autonomous Agents and Multiagent Systems: Part 1, pp. 184–190. ACM, New York (2002)

24. Marshall, A.: Principle of Economics, 8th edn. MacMillan, London (1920)

25. Mas-Colell, A., Whinston, M., Green, J.: Microeconomic Theory. Oxford University Press, Oxford (1995) ISBN 0-19-507340-1

26. Pollak, R.: Conditional Demand Functions and Consumption Theory. Quarterly Journal of Economics 83, 60–78 (1969)

27. Varian, H.R.: Microeconomic Analysis, 3rd edn., ch. 7, 8 and 9. W.W. Norton & Company, New York (1992)
28. Foundation for Intelligent Physical Agents (FIPA): FIPA Contract Net Interaction Protocol. In: Specification [online]. FIPA (2002),
    http://www.fipa.org/specs/fipa00029/SC00029H.pdf
    (cit. June 13, 2011)
29. Vymětal, D., Spišák, M., Šperka, R.: An Influence of Random Number Generation Function to Multiagent Systems. In: Jezic, G., Kusek, M., Nguyen, N.-T., Howlett, R.J., Jain, L.C. (eds.) KES-AMSTA 2012. LNCS, vol. 7327, pp. 340–349. Springer, Heidelberg (2012)

The Alliance between Optimization and Multi-Agent System for the Management of the Dynamic Carpooling

Sondes Ben Cheikh and Slim Hammadi

Ecole Centrale de Lille,
Cit Scientifique, Villeneuve d'Ascq, France
{sondes.ben-cheikh,slim.hammadi}@ec-lille.fr

Abstract. Today, there are several studies that revolve around dynamic carpooling. However, there is a big handicap, due to the problems high complexity, concerning the way to make the process perform efficiently. To address these gaps, we introduce a decomposition process in order to subdivide the global problem into several sub-problems with a reasonable research space. Indeed, we propose to break geographical areas (global problem) into several distinct zones (sub-problem) which each zone is controlled by an agent with an optimized behavior. Therefore, we propose the original alliance between optimization and a multi agent concept to perform parallel Optimized Assignment of Vehicles to users queries. This alliance is characterized by a metaheuristic approach based on a Multi-criterion Tabu Search implemented in the heart of the agent in order to optimize partial requests process which is performed locally in its zone. Moreover, we introduce several agents which are endowed by an evaluator behavior based on the Choquet Integral to evaluate the best solution taking into consideration the interactions among criteria. Finally, to test the validity of the proposed model, some simulation results will be presented.

Keywords: Dynamic Carpooling, Multi-agent System (MAS), Tabu Search, Choquet Integral.

1 Introduction

Urban traffic is responsible for 40% of CO_2 emissions and 70% of emissions of other pollutants arising from road transport [1]. Moreover, the increase in commuting time, traffic congestion and rising fuel prices, lead to a questioning of the massive use of the private car. Hence the emergence of carpooling is one of the planned rationalization of the use of the private car solutions. The carpooling phenomenon has emerged since the mid-1970s. Almost all of them treats static aspect of ridesharing and are rather shown as virtual supports for reservations storage and management (e.g. Covoiturage.fr, carpooling.com). Nevertheless, static systems require booking in advance based on a static reasoning, regardless of instantaneous events. To address this deficit, dynamic carpooling

G. Jezic et al. (eds.), *Agent and Multi-Agent Systems: Technologies and Applications*,
Advances in Intelligent Systems and Computing 296,
DOI: 10.1007/978-3-319-07650-8_20, © Springer International Publishing Switzerland 2014

systems are developed with a real time service management (e.g. greenmon-keys.com, covivo.fr). But most of them no longer exist due to the lack of means or remained in embryonic stage. On the academic side, many researchers have particularly focused their efforts on the problem of carpooling in its dynamic con-text [2]. Moreover, trying to take advantage from new technologies, mobile and lightweight devices, systems integrating the multi-agent concept have emerged ,[3]. For example, Wooldridge, has used the role paradigm to deal with scalabil-ity issues, mainly the response time and the network load balance in large scale systems, in the reference [4], the GAIA methodology uses roles to create a frame-work for abstraction, analysis and design of agent-based systems. Inspired by the existent systems deficit, we propose to address the dynamic carpooling problem searching for the most adequate assignments vehicle/passenger which minimizes travel time, its cost and the CO_2 emission quota. In this context, we introduce our novel approach called Multi-criterion Tabu Search based on the Choquet integral aggregation. Besides, the complexity of the problem addressed and the distributed nature and dynamics of the system in question led us to choose a distributed architect based on the alliance between MAS and optimization.

2   Mathematical Formulation

2.1   Decision Variables

We note $R_p(t)$ the set of instantly received demands where $P = \bigcup_{l=1}^{n}\{P^l\}$ is the set of n passengers. Equation (1) shows passenger P^l requests formulation:

$$R_{P^l}(t) = (P^{l-}, P^{l+}, d^l, a^l, Q^l) \tag{1}$$

Where moving preferences are specified: P^{l-}, P^{l+} are respectively the origin and the destination node asked by P^l, d^l, a^l indicate the preferred earliest departure time and the latest Arrival time. Q^l is the number of passengers including P^l who desire travel together.

Similarly, we note $O_V(t)$ the set of vehicles offers received at time t, where $V = \bigcup_{k=1}^{m}\{V^k\}$ is the set of m vehicles expressing travel deals. Equation (2) shows the formulation of vehicle V^k s offer:

$$O_{V^k}(t) = (V^{k-}, V^{k+}, MD^k, MA^k, L^k, C^k, h^k) \tag{2}$$

V^{k-}, V^{k+} refer respectively the origin and the destination of the vehicle V^k with a capacity L^k. MD^k, MA^k indicate respectively the vector of the departure time denoted D_i^k, and the vector of the arrival time denoted A_i^k of the vehicle V^k on all nodes i between V^{k-} and V^{k+}. These nodes correspond on the Intermediate Destinations specified by the driver. Moreover, each vehicle V^k is characterized by a kilometric cost criterion that we note C^k and a criterion of emission rate of CO_2 per kilometer denoted h^k.

2.2   Criteria

In this paper, we deal with five criteria to evaluate the quality of the generated solution:

The Total Waiting Time: This criterion aims to minimize the P^l s waiting time in origin node ($i = P^{l-}$) and in transfer node denoted j.

$$WT_{lkk'} = Max(0, (D_i^k - d^l) * X_{lk} + (D_j^{k'} - A_j^k) * X_{lk'}) \tag{3}$$

Where X_{lk} is a decision variable which is equal to 1 if P^l is assigned to V^k, 0 otherwise. Then, The total passengers waiting time is calculated as follows:

$$TWT = \sum_{l=1}^n \sum_{k=1}^m \sum_{k'=1, k' \neq k}^m WT_{lkk'} \tag{4}$$

The Total Delay Time: This criterion seeks to minimize the delay time in arrival node ($i = P^{l+}$): $DT_{lk} = Max(0, (A_i^k - a^l))$ $\tag{5}$
Consequently, the total delay time is calculated as:

$$TDT = \sum_{l=1}^n \sum_{k=1}^m DT_{lk} * X_{lk} \tag{6}$$

The Total Route Time: For each passenger P^l assigned to V^k from i to j , the route time is stated as follows: $RT_{lk} = (A_j^k - D_i^k) * X_{lk}$ $\tag{7}$
Then, its global route time is formulated as: $RT_{lkk'} = RT_{lk} + RT_{lk'}$ $\tag{8}$
Indeed, the total route time of all passengers which is determined as:

$$TRT = \sum_{l=1}^n \sum_{k=1}^m \sum_{k'=1, k' \neq k}^m RT_{lkk'} \tag{9}$$

Environmental Criteria: This criterion determines the improvement in terms of CO_2 established in both of the following cases:
-Passenger uses his own vehicle V^{k_l} which is characterized by h^{k_l}.
-Passenger chooses to carpool.

– CO_2 emission quota without carpooling: $IC_{lk_l} = h^{k_l} distance(P^{l+}, P^{l-})$ (10)
– CO_2 emission quota with carpooling:

$$SC_{lk} = (h^k * RT_{lk} * Average\ Speed)/(number\ of\ passengers) \tag{11}$$

where SC_{lk} is the emission quota of CO_2 for each passengers sharing the same vehicle V^k which is characterized by h^k. Hence, the gain of CO_2 per passenger is calculated as follows:

– Without transfer: $Gain_{lk} = Max(0, IC_{lk_l} - SC_{lk} * X_{lk})$ $\tag{12}$
– With transfer: $SC_{lkk'} = SC_{lk} * X_{lk} + SC_{lk'} * X_{lk'}$ $\tag{13}$

Then, $Gain_{lkk'} = Max(0, IC_{lk_l} - SC_{lkk'})$ $\tag{14}$

In that case, the total environmental gain realized by all passengers is equal to:

$$TEG = \sum_{l=1}^n \sum_{k=1}^m \sum_{k'=1, k' \neq k}^m Gain_{lkk'} \tag{15}$$

The Economic Criterion: According to the cost per kilometer C^k of each vehicle V^k, this criterion seeks to minimize the trips cost which is determined as follows:

- Without transfer:$CK_{lk} = C^k * RT_{lk} * Average\ Speeed$ (16)
- With transfer:$CK_{lkk'} = CK_{lk} * X_{lk} + CK_{lk'} * X_{lk'}$ (17)

Then, the total cost for all passengers is calculated by:

$$TC = \sum_{l=1}^n \sum_{k=1}^m \sum_{k'=1,k'\neq k}^m CK_{lkk'} \qquad (18)$$

To achieve these objectives, we propose a metaheuristic approach and a Choquet Integral aggregation which are driven by a MAS.

3   MAS Description

We propose to set up a distributed software environment based on a MAS where highly com-municating entities evolve. Indeed, as shown in Fig. 1, each agent has a specific role and is responsible for one task or more contributing to perform the whole optimizing process.

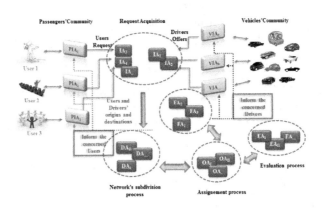

Fig. 1. Several agents to ensure An Optimized Dynamic Carpooling

Hereafter a brief description of each agents functioning:

- Passenger Interface Agent (PIA): transmits users request to the IA and re-covers the corresponding solution.
- Vehicle Interface Agent (VIA): ensures drivers exchange with the involved agents (i.e. IA respectively the FA) transmitting its offer or receiving notifications of pickup and deposit addresses.
- Information Agent (IA): its role mainly consists in receiving users demands and extracting passengers and vehicles trips information and sending them to the DA.

- Decomposing Agent (DA): based on the information provided by the IA, a subdivision principle developed by our team [5] is processed by DA which breaks geographical areas into several distinct zones.Then, the optimized partial requests process is performed locally in each zone.
- Optimizer Agent (OA): Each OA locally processes users requests in the zone it represents performing $TSIDAA$ algorithm which will be described later. The output of this algorithm is the best partial solution.
- Evaluator Agent (EA): the EA agent evaluates the strategies by using an aggregative approach based on the Choquet integral[6]. This approach has proven to be an adequate aggregation operator that takes into consideration the importance and interaction among criteria.
- Fusion Agent (FA): The objective of the FA is to concatenate the partial best solutions sending by the OAs into global best one with respect of the synchronization process.

3.1   Optimizer Agent Behaviour (OA)

Each OA processes a meta-heuristic optimization approach called multi-criterion $TSIDAA$ "Tabu Search Interaction Driven Agent Algorithm", firstly searching for possibilities responding to each global users request. Otherwise, the OA tries to decompose the given request starting from the most approximate node served. This node corresponds to a novel origin node for the next connected area controlled by another agent OA. Once we reach the destination node generated by a current OA, the latest performs valid responses (making the user reach its final destination). In what follows, we will describe the main steps of our $TSIDAA$ for the Dynamic Carpooling Problem (DCP):

Construction of Initial Solution: According to information from the Origin/Destination Matrix (ODM) and regarding vehicles capacities, an initial solution is constructed by randomly assigning passengers to vehicles. Subsequently, the algorithm shall verify whether the passenger is assigned to the same vehicle in the departure and transit point. Otherwise, the passenger needs two vehicles to ensure his journey respecting the constraint "no more than one tolerated transfer". Then the system checks if the two vehicles have a common intermediate destination with a reasonable synchronization time window.

Tabu List: Thanks to Tabu Moves, we can keep the search bias toward point with lower objective function values and escape from local optimum solution. In our case and in order to avoid returning to the local optimum already visited, the Tabu Move is define as the prohibited transfer between vehicles V^k and $V^{k'}$ for the passenger P^l. Then, the triplet $\{P^l, V^k, V^{k'}\}$ is declared forbidden and it is saved in the Tabu List.

Neighbourhood Construction: To obtain the neighborhood solution, our system searches in the Tabu List the unallocated passengers and reassigns them to other vehicles. In case where the Tabu List is empty, the system chooses randomly certain passengers and reallocates them to other vehicles in order to avoid a local optimum solution.

Aspiration Criterion: Since, the Tabu List may forbid certain worthy or interesting assignments vehicle/passenger possibly leading to a better solution than the best one found so far. An aspiration criterion is used to allow tabu assignments to be released if they are judged to be worth or interesting. In our case, the aspiration criterion is defined by the possibility that a vehicle can make a detour to retrieve a passenger in order to provide more flexibility. In other word, the aspiration criterion is to allow "excellent" Tabu Moves to be selected if the aspiration level is attained.

Algorithm 1. TSIDAA for the DCP

Require: $P' = \bigcup_{l=1}^{n'}\{P^l\}$ and $V' = \bigcup_{k=1}^{m'}\{V^k\}$ are respectively the set of n' passengers and m' vehicles sending by DA, where origins nodes belong to the same local zone.

Ensure: MBS Matrix of Best Solution, Tabu_List, Choquet_Integral_Score

    Initialization

1: $Tabu_List = \phi$
2: Initialize Aspiration criterion
3: Solution_Construction(MIS) // MIS: Matrix Initial Solution
4: Send the MIS to the EA to be evaluated// *Communication protocol between OA and EA*
5: Receive the *Choquet_Integral_Score*(MIS) sending by the EA // multi-criterion evaluation TWT, TDT, TRT, TEG, TC
6: $MBS \leftarrow MIS$
7: Define Terminated Conditions **Treatment**
8: **while** $Done = False$ **do**
9:     Neighberhood_Construction(MIS)
10:     Solution_Construction(MCS) // *MCS : Matrix of Current Solution*
11:     Send the MCS to the EA to be evaluated
12:     Receive the Choquet_Integral_Score(MCS) sending by the EA
13:     **if** $Choquet\_Integral\_Evaluation(MBS) \prec Choquet\_Integral\_Evaluation(MCS)$ **then**
14:         $MBS \leftarrow MCS$
15:         Update Choquet_Integral_Score
16:         Update Tabu_List
17:         Update Aspiration Criterion
18:     **end if**
19:     **if** Terminated condition is satisfied **then**
20:         $Done = True$
21:     **end if**
22: **end while**
23: **if** destination node contains in the MBS itineraries **then**
24:     OA sends MBS to FA
25: **end if**

3.2   The Global Behaviour of the System

After describing the individual behavior of different agents, we present a summary of the different interactions between entities to better understand the functionality of the system. We illustrate the behavior of our Multi-agent architecture by a sequence diagram in UML. The sequence diagram expresses the dynamic structure modeling. It shows the communication and interaction between the different agents in order to achieve a common group goal (Fig. 2).

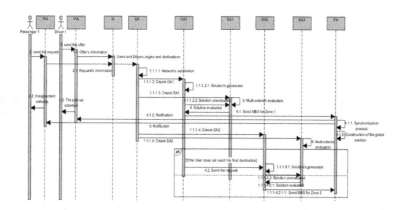

Fig. 2. The global system behavior

4   Simulation Results

Solution is obtained thanks to communication and collaboration between the agents; the programming is developed under the $JADE$ platform (Java Agent De-velopment Framework). In order to improve the effectiveness of our approach and to study the merits of alliance between optimization and the MAS for dynamic carpooling, we propose a ridesharing service in the French city of Lille (Nord department) during a transports disturbance. Lets suppose that between $7a.m.$ and $10a.m.$ a technical problem takes place on a section of the metro line 1 (yellow line) and the bus 42 (green line). We note that there are two transit areas. In this context, we present here a scenario composed of 30 passengers subdivided into 15 requests, and 20 vehicles travelling on the network (Fig.3).

4.1   The Networks Subdivision

Based on the subdivision process performed by DA, our network is subdivided into two zones: zone1 (yellow line) and zone2 (green line). Then, the system assigns the OA_1 to manage the carpooling service in zone1 and similarly the OA_2 for zone2. In addition, the DA sends the requests data including the origins and destinations nodes that belong to the same zone for the corresponding agent in order to optimize their itineraries by applying our $TSIDAA$ algorithm. In case

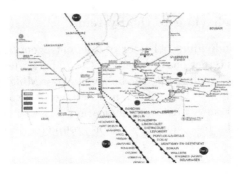

Fig. 3. The transportation network of Lille

origins and destinations nodes for several passengers belong to different zones, the DA sends the data requests containing the origin node to the corresponding OA_1. After the optimization process, the latest sends the data requests involving the transfers and destinations nodes to it connected agent OA_2.

4.2   The Assignment Process

OA_1 and OA_2 have an optimizer behavior by applying the $TSIDAA$ algorithm locally in their corresponding zones. Following the construction of the partial feasible solution, OA_1 and OA_2 begins respectively a communication protocol with EA_1 and EA_2 to evaluate their corresponding solutions. In addition, EA_1 and EA_2 serve for attributing a global Choquet Integral score calculated according to the importance and the interaction between criteria. After the establishment of the partial best solutions provided by OA_1 and OA_2 agents, they are sent to the FA in order to construct the global best solution corresponding to the following criteria scores:

Table 1. Solution evaluation

TWT	TDT	TRT	TEG	TC	Global Score
0.8	0.9	0.6	0.5	0.7	0.93

Therefore, the global best solution, constructed by the FA, which corresponds to the highest score of Choquet Integral is showen in fig.4 and fig.5. To evaluate the efficiency of the best global solution, we present curves below which summarize results for the 15 requests.

The Fig. 6 indicates that majority of passengers arrived at their destinations with an average waiting time of nine minutes, an average route time which may not exceed ten minutes and sometimes without delay. According to Fig. 7, we note that the gain realized by each passenger may exceed 1500 g of CO_2 and the cost of travel can be less than one euro unlike the unified price of public transport tickets which does not take into account the traveled distance.

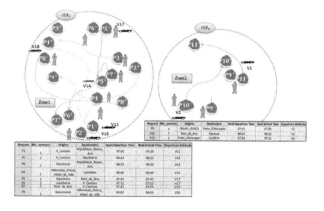

Fig. 4. Zone1 and Zone2 managed respectively by OA_1 and OA_2

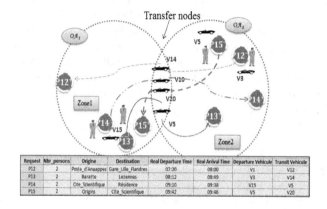

Fig. 5. Coalition between OA_1 and OA_2

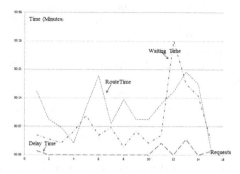

Fig. 6. Values of WT, DT, and RT (minutes) / Request

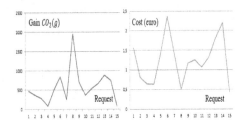

Fig. 7. Gain of CO_2 (g) and Cost (Euros) per request

5   Conclusion and Perspectives

Based on the alliance between MAS and optimization concept, a distributed architecture is proposed to develop a real-time carpooling which optimizes several criteria: travel time, travel cost and environmental gain. In the future work, we will improve the synchronization process between partial solutions generated by agents by developing a coalition algorithm between them. Indeed, we will propose a three-level agent-based system. The main level will contain the MAS modeling the Wooldridges roles concept in each coalition. In this level agents will collaborate and negotiate in order to make an optimizing dynamic carpooling. The decisions made will depend on received real time data from the bottom level corresponding to the carpooling network. The higher level will contain optimization tools, metaheuristics and different mathematical models.

References

1. Bhm, M.: In-Time: Intelligent and Efficient Travel Management for European Cities. In: POLIS Conference (2009)
2. Kleiner, K., Nebel, B., Ziparo, V.: A mechanism for dynamic ride sharing based on parallel auctions. In: Proc. of. the 22th International Joint Conference on Artificial Intelligence (IJCAI), Barcelona, Spain, pp. 266–272 (2006)
3. Kothari, A.B.: Genghis-A multi-agent carpooling system. B.Sc. Dissertation work in Computer Science, San Francisco (1999)
4. Wooldridge, M., Jennings, N.R., Kinny, D.: The Gaia methodology for agent-oriented analysis and design. Journal of Autonomous Agents and Multi-Agent Systems, 266–272 (2001)
5. Sghaier, M., Zgaya, H., Hammadi, S., Tahon, C.: A distributed dijkstras algorithm for the implementation of a real time carpooling service with an optimized aspect on siblings. In: 13th International IEEE Conference on Intelligent Transportation Systems, Maderia Island, Portugal, pp. 795–800 (2010)
6. Ben Cheikh, S., Hammadi, S.: An optimized evolutionary Multi-agent approach for regulation of distributed urban transport. International Journal of Modern Engineering Research, 3841–3851 (2013)

Analyzing the Efficacy of Passive Investment Strategies through Agent-Based Modelling: Overconfident Investors and Investors with Better Predictive Power

Hiroshi Takahashi

Graduate School of Business Administration, Keio University,
4-1-1 Hiyoshi, Kohoku-ku, Yokohama-city, 223-8572, Japan
htaka@kbs.keio.ac.jp

Abstract. This study analyses the efficacy of a passive investment strategy - which is one of the most popular investment strategies in the asset management business - thorough agent-based modelling. As a result of intensive experimentation, the following conclusions were confirmed: (1) overconfident investors could achieve a positive excess return in the market where there are no passive investors. However, (2) even if overconfident investors have better predictive power, they couldn't survive in a market where passive investors exist. These results suggest the effectiveness of a passive investment strategy. The results are of both academic interest and practical use.

Keywords: Finance, Agent-based Modelling, Behavioral Economics, Overconfidence, Asset Management.

1 Introduction

The story of globalization is, to a great extent, the story of the growth and integration of the world's financial systems. In the asset management business, various kinds of investors provide capital to financial markets. In financial markets, institutional investors have more influence on asset prices. Passive investment strategy -which tries to maintain an average return using benchmarks based on market indices- is one of the most popular investment strategies in the asset management business and is consistent with traditional asset pricing theories and is considered to be an effective method in efficient markets [1].

In recent years, there has been rising interest in a field called behavioral finance, which incorporates psychological methods in analyzing investor behavior. There are numerous arguments in behavioral finance that investors' decision-making bias can explain phenomenon in the financial market which had previously gone unexplained. Such arguments often point out the limitations of arbitrage and the existence of systematic biases in decision-making [2][3][4][5]. Behavioral finance has examined a wide range of phenomena in the market and among investors, drawing a number of provocative conclusions. There are, for

G. Jezic et al. (eds.), *Agent and Multi-Agent Systems: Technologies and Applications*,
Advances in Intelligent Systems and Computing 296,
DOI: 10.1007/978-3-319-07650-8_21, © Springer International Publishing Switzerland 2014

example, studies which suggest that overconfident investors could survive in the market.

With this background in mind, the purpose of this research is to analyze the performance of passive investors in a market where overconfident investors exist. To address this problem, we have employed agent-based modeling in this analysis [6][7]. Agent-based modeling has many applications, and none more suitable than for the creation of an artificial market. For example, Arthur et al. [8] analyse the market under conditions where heterogeneous investors trade and concluded that complex conditions emerge. Using agent-based modeling, Takahashi et al. [9] found that irrational traders could survive in the market. A later paper [10] suggests that the combination of behavioral biases and financial constraints causes a significant deviation from fundamental values. Analyses which attempt to replicate realistic market conditions and dynamics present a greater challenge to the researcher than traditional forms of research. Due to the efficacy of this type of advanced analysis, there is a greater demand for research conducted employing these kinds of models. There is, therefore, a need for analyses using this more current approach, in addition to original methods. Agent-based modeling is making an increasingly valuable contribution to financial research.

The next section describes the model used in this analysis. Section 3 shows the results of the analysis. Section 4 summarizes this paper.

2   Model

A computer simulation of the financial market involving 1000 investors was used as the model for this research. Shares and risk-free assets were the two types of assets used, along with the possible transaction methods. Several types of investors exist in the market, each undertaking transactions based on their own stock evaluations [8][9]. This market was composed in three major stages; (1) generation of corporate earnings, (2) formation of investor forecasts, and (3) setting transaction prices. The market advances through repetition of these stages. The following sections describe negotiable transaction assets, modeling of investor behavior, transaction price setting, and rules of natural selection in the market.

2.1   Negotiable Assets in the Market

This market has both risk-free and risk-associated assets. There are risk-associated assets in which all profits gained during each term are distributed to shareholders. Corporate earnings (y_t) are expressed as $y_t = y_{t-1} \cdot (1 + \varepsilon_t)$. However, they are generated according to the process of $\varepsilon_t \sim N(0, \sigma_y^2)$ with share trading being undertaken after the public announcement of profits for the term [11]. Each investor is given common asset holdings at the start of the term with no limit placed on debit and credit transactions (1000 in risk-free assets and 1000 in stocks). Investors adopt the buy-and-hold method for the relevant portfolio as a benchmark to conduct decision-making by using a one-term model.

The buy-and-hold method is an investment method to hold shares for medium to long term.

2.2   Modeling Investor Behavior

Each type of investor handled in this analysis is organized in Table 1. This analysis covers most major types of investor [2]. The investors can be classified into two categories: active investors (Type 1-4, and 6) and a single passive investor type (Type 5). Active investors in this market evaluate transaction prices based on their own forecasts of market movements, taking into consideration both risk and return rates when making decisions. Passive investors employ a buy-and-hold strategy [12]. A passive investment strategy is one of the most popular investment strategies in the asset management business. Each active investor determines the investment ratio (w_t^i) based on the maximum objective function$(f(w^i t))$, as shown below [13].

$$f(w_t^i) = r_{t+1}^{int,i} \cdot w_t^i + r_f \cdot (1 - w_t^i) - \lambda(\sigma_{t-1}^i)^2 \cdot (w_t^i)^2. \tag{1}$$

Here, $r_{t+1}^{int,i}$ and σ_{t-1}^i in the eq. (1) express the expected rate of return and risk for stocks as estimated by each investor i. r_f indicates the risk-free rate. w_t^i represents the stock investment ratio of the investor i for term t. λ shows degree of investor risk aversion. The value of the objective function $f(w_t^i)$ depends on the investment ratio(w_t^i). The investor decision-making model here is based on the Black/Litterman model that is used in the practice of securities investment [14].

The integrated expected rate of return for shares is calculated as follows [14]:

$$r_{t+1}^{int,i} = \frac{c^{-1}(\sigma_{t-1}^i)^{-2}r_{t+1}^{f,i} + (\sigma_{t-1}^i)^{-2}r_t^{im}}{c^{-1}(\sigma_{t-1}^i)^{-2} + (\sigma_{t-1}^i)^{-2}}. \tag{2}$$

Here, $r_{t+1}^{f,i} \mathrm{Cr}_t^{im}$ in the eq. (2) express the expected rate of return, calculated from short-term expected rate of return, and risk and gross current price ratio of stocks respectively. c is a coefficient that adjusts the dispersion level of the expected rate of return calculated from risk and gross current price ratio of stocks [14].

Table 1. List of investor types

No.	Investor types
1	Fundamentalist
2	Forecasting by past average (most recent 10 days)
3	Forecasting by trend (most recent 10 day)
4	Latest Price
5	Passive investor
6	Investor who have perfect forecast (1 day)

The short-term expected rate of return $(r_t^{f,i})$ is obtained where $(P_{t+1}^{f,i}, y_{t+1}^{f,i})$ is the equity price and profit forecast for term $t+1$ is estimated by the investor, as follows: $r_{t+1}^{f,i} = ((P_{t+1}^{f,i} + y_{t+1}^{f,i})/P_t - 1)$.

The price and profit forecast$(P_{t+1}^{f,i}, y_{t+1}^{f,i})$ includes the error term $(P_{t+1}^{f,i} = P_{t+1}^{f,typej} \cdot (1+\eta_t^i), y_{t+1}^{f,i} = y_{t+1}^{f,typej} \cdot (1+\eta_t^i),$ where $\eta_t^i \sim N(0, \sigma_n^2))$ reflecting that even investors using the same forecast model vary slightly in their detailed outlook. The stock price $(P_{t+1}^{f,i})$, profit forecast $(y_{t+1}^{f,i})$, and risk estimation methods are described in the following paragraph.

The expected rate of return obtained from stock risk and so forth is calculated from stock risk (σ_{t-1}^i), benchmark equity stake (W_{t-1}), degree of investor risk aversion (λ), and risk-free rate (r_f), as follows [15]: $r_t^{im} = 2\lambda(\sigma_{t-1}^i)^2 W_{t-1} + r_f$.

Stock Price Forecasting Method. The fundamental value is estimated by using the discounted cash flow model (DCF), which is a well known model in the field of finance. Fundamentalists estimate the forecasted stock price and forecasted profit from profit for the term (y_t) and the discount rate (δ) as $P_{t+1}^{f,typej} = y_t/\delta, y_{t+1}^{f,typej} = y_t$ (Table 1, Type 1). This study also considers investors who have better predictive power. In this analysis, investors who have information about the next step's profit are incorporated (Table 1, Type 6).

Forecasting based on trends involves forecasting the next term's stock prices and profit through extrapolation of the most recent stock value fluctuation trends. Stock price and profit of the next term are estimated from the most recent trends of stock price fluctuation (a_{t-1}) from time point $t-1$ as $P_{t+1}^{f,typej} = P_{t-1} \cdot (1 + a_{t-1})^2, y_{t+1}^{f,typej} = y_t \cdot (1 + a_{t-1})$ (Table 1, Type 3).

Forecasting based on past averages involves estimating the next term stock prices and profit based on the most recent average stock value (Table 1, Type 2).

2.3   Risk Estimation Method

Stock risk is measured as $\sigma_{t-1}^{s,i} = s_i \cdot \sigma_{t-1}^h$. In this case, σ_{t-1}^h is an index that represents stock volatility calculated from price fluctuation of the most recent 100 steps, and s_i is the degree of overconfidence. The presence of a strong degree of overconfidence can be concluded when the value of s_i is less than 1, as estimated forecast error is shown as lower than its actual value. The investors whose value of s_i is less than 1 tend to invest more actively. For example, when such investors predict that stock prices will increase, they invest more in stock than ones whose value of s_i is 1.

2.4   Determination of Transaction Prices

Transaction prices are determined as the price where stock supply and demand converge $(\sum_{i=1}^M (F_t^i w_t^i)/P_t = N)$. In this case, the total asset (F_t^i) of investor i is calculated from transaction price (P_t) for term t, profit (y_t) and total assets from the term $t-1$, stock investment ratio (w_{t-1}^i), and risk-free rate (r_f), as $F_t^i = F_{t-1}^i (w_{t-1}^i (P_t + y_t)/P_{t-1} + (1 - w_{t-1}^i)(1 + r_f))$.

2.5   Natural Selection in the Market

Investors who are able to adapt to and, hence, profit from the market as it fluctuates will remain in the market and their position will grow stronger. Conversely, investors who are unable to do this will drop out of the market. Such a pattern is very suggestive of what might be termed 'Natural Selection' in the market. The driving force behind this 'Natural Selection' is the desire for cumulative excess profit [16]. Two aspects of this pattern are of particular interest: (1) the identification of investors who alter their investment strategy, and (2) the actual alteration of investment strategy [9].

Each investor must decide whether he should change investment strategies based on the most recent performance of each 5 term period (after 25 terms have passed since the beginning of market transactions). The higher the profit rate obtained most recently is, the lesser the possibility of strategy alteration becomes. The lower the profit, the higher the possibility becomes. (In the actual market, evaluation tends to be conducted according to baseline profit and loss.) Specifically, when an investor could not obtain a positive excess profit for the benchmark portfolio profitability, they are likely to alter their investment strategy with the probability below:

$$p_i = \min(1, \max(-100 \cdot r_i^{cum}, 0)). \tag{3}$$

Here, however, r_i^{cum} in the eq. (3) is the cumulative excess profit for the most recent benchmark of investor i. Measurements were conducted for 5 terms, and the cumulative excess profit was calculated as a one-term conversion. For example, if excess profit over a 5 term period is 5 %, a one term conversion would show this as a 1 % excess for each term period.

When it comes to deciding on a new investment strategy, an investment strategy that has a high cumulative excess profit for the most recent five terms (forecasting type) is 'naturally' more likely to be selected. Where the strategy of the investor i is z_i and the cumulative excess profit for the most recent five terms is r_i^{cum}, the probability p_i that z_i is selected as a new investment strategy is given as $p_i = e^{(a \cdot r_i^{cum})} / \sum_{j=1}^{M} e^{(a \cdot r_j^{cum})}$. Selection pressures on an investment strategy become higher as the coefficients' value increases. Those investors who altered their strategies make investments based on the new strategies after the next step.

2.6   Parameter List

Table 2 lists the major parameters of the financial market designed for this paper.

3   Results

The first set of results is from a model in which investors who have strong predictive power are analyzed. The second set presents a situation in which passive investors are present.

Table 2. Major parameters of the financial market

M	Number of investors (1000)
N	Number of shares (1000)
F_t^i	Total asset value of investor i for term t $(F_0^i = 2000$: common)
W_t	Ratio of stock in benchmark for term t $(W_0 = 0.5)$
w_t^i	Stock investment rate of investor i for term t $(w_0^i = 0.5$: common)
y_t	Profits generated during term t $(y_0 = 0.5)$
σ_y	Standard deviation of profit fluctuation $(0.2/\sqrt{200})$
δ	Discount rate for stock$(0.1/200)$
λ	Degree of investor risk aversion (1.25)
σ_n	Standard deviation of dispersion from short-term expected rate of return on shares (0.05)
c	Adjustment coefficient (0.01)

3.1   Case 1: Investors Who Have Better Predictive Power

This section analyses the case where investors who have better predictive power (perfect forecast) are present in the market. In this case, there is the same number of five types of investors in the market (Table 1, Type 1-4, and 5). Fig. 1 shows the transitions of transaction prices. The horizontal axis in the graph shows time steps and the vertical axis shows stock prices. Two transitions are shown: Fundamental values and transaction prices, and it can be seen that transaction prices are almost consistent with fundamental values throughout the entire transaction period. Fig. 2 shows the transition of the number of investors. As time steps pass, the number of investors with perfect forecast increases. Looking at transitions in the degree of overconfidence, a strengthening degree of overconfidence can be seen in the behavior of the remaining investors as market transactions move forward (see Fig. 3). These results suggest that there is something going on in the market which allows overconfident investors - with their biases in investment decision-making - to survive. This would be in clear contradiction of traditional financial theory.

3.2   Case 2: Passive Investors

This section analyses a situation where passive investors invest in the market. There is the same number of six types of investors (Table 1, Type 1-6). The fig. 4, 5, and 6 present the transition of share prices, the number of investors, and the degree of overconfidence, respectively. From the transition of stock prices (see Fig. 5), it is confirmed that, as time steps go, passive investors survive in the market. These results suggest the effectiveness of passive investment strategies, even if there are investors who have better predictive power. However it is also the case that market prices reach a point where they begin to deviate from fundamental values (see Fig. 4). These latter results indicate possible drawbacks of passive investment strategies. From the transitions of degree of overconfidence (see Fig. 6), it can be seen that investors who survive in the market do not display overconfidence. These results suggest that, in this model, overconfident investors do not survive, and a passive investment strategy is superior in its efficacy.

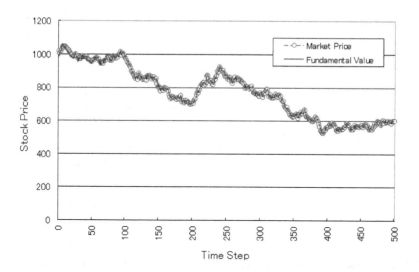

Fig. 1. Price Transitions (Fundamentals, Latest, Trend, Average, Perfect Forecast (1Day))

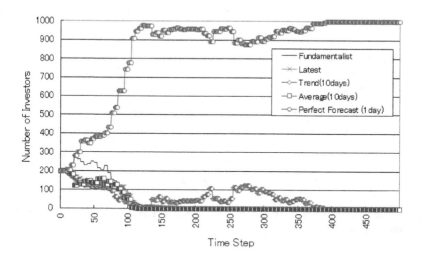

Fig. 2. Transitions of Number of Investors (Fundamentals, Latest, Trend, Average, Perfect Forecast (1Day))

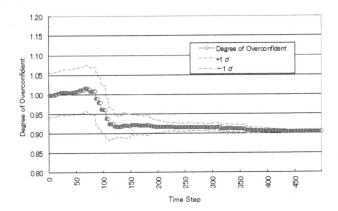

Fig. 3. Transition of degree of overconfidence (Fundamentals, Latest, Trend, Average, Perfect Forecast(1Day))

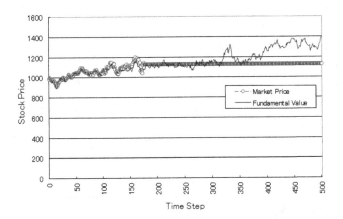

Fig. 4. Price Transitions (Fundamentals, Latest, Trend, Average, Perfect Forecast(1Day), Passive)

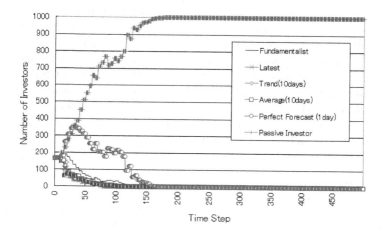

Fig. 5. Transitions of Number of Investors (Fundamentals, Latest, Trend, Average, Perfect Forecast (1Day), Passive)

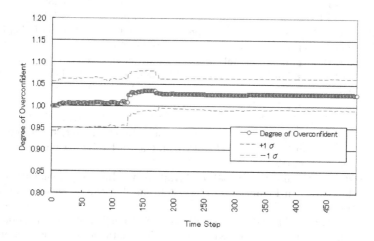

Fig. 6. Transition of degree of overconfidence (Fundamentals, Latest, Trend, Average, Perfect Forecast (1Day), Passive)

4   Summary

Using agent-based modeling, this research looked at the efficacy of a passive investment strategy (consistent with traditional financial theory and a popular investment strategy in the asset management business). As a result of this computer-based market analysis, we found that, when passive investors are absent, overconfident investors could achieve a positive excess return in the market.

However, we also found that when passive investors are introduced, overconfident investors couldn't survive in a market even if overconfident investors have better predictive power. This confirms the conventional view and is in line with traditional financial theory. These results contribute to clarifying the mechanism of financial markets and are of interest in themselves and merit further study. This research mainly analyzed the market where fundamentalist have significant influence on the market. A more detailed analysis that considers both the actual investment environment and extreme market conditions, such as a "bubble" or "crash", should be included in future research.

References

1. Sharpe, W.F.: Capital Asset Prices: A Theory of Market Equilibrium under condition of Risk. The Journal of Finance 19, 425–442 (1964)
2. Shleifer, A.: Inefficient Markets. Oxford University Press (2000)
3. Kahneman, D., Tversky, A.: Prospect Theory of Decisions under Risk. Econometrica 47, 263–291 (1979)
4. Tversky, A., Kahneman, D.: Advances in Prospect Theory: Cumulative representation of Uncertainty. Journal of Risk and Uncertainty 5, 297–323 (1992)
5. Shiller, R.J.: Irrational Exuberance. Princeton University Press (2000)
6. Axelrod, R.: The Complexity of Cooperation -Agent-Based Model of Competition and Collaboration. Princeton University Press (1997)
7. Takahashi, H., Terano, T.: Analyzing the Influence of Overconfident Investors on Financial Markets Through Agent-Based Model. In: Yin, H., Tino, P., Corchado, E., Byrne, W., Yao, X. (eds.) IDEAL 2007. LNCS, vol. 4881, pp. 1042–1052. Springer, Heidelberg (2007)
8. Arthur, B.W., Holland, J.H., LeBaron, B., Palmer, R.G., Taylor, P.: Asset Pricing under Endogenous Expectations in an Artificial Stock Market. In: Arthur, W.B., Durlauf, S.N., Lane, D.A. (eds.) The Economy as an Evolving Complex System II, pp. 15–44. Addison-Wesley (1997)
9. Takahashi, H., Terano, T.: Agent-Based Approach to Investors' Behavior and Asset Price Fluctuation in Financial Markets. Journal of Artificial Societies and Social Simulation 6 (2003)
10. Takahashi, H.: An Analysis of the Influence of dispersion of valuations on Financial Markets through agent-based modeling. International Journal of Information Technology & Decision Making 11, 143–166 (2012)
11. O'Brien, P.: Analysts' Forecasts as Earnings Expectations. Journal of Accounting and Economics, 53–83 (January 1988)
12. Malkiel, B.G.: Passive investment strategies and Efficient Markets. European Financial Management 9, 1–10 (2003)
13. Ingersoll, J.E.: Theory of Financial Decision Making. Rowman & Littlefield (1987)
14. Black, F., Litterman, R.: Global Portfolio Optimization. Financial Analysts Journal, 28–43 (September-October 1992)
15. Sharpe, W.F.: Integrated Asset Allocation. Financial Analysts Journal, 25–32 (September-October 1987)
16. Goldberg, D.: Genetic Algorithms in Search, Optimization, and Machine Learning. Addison-Wesley (1989)

Intelligent and Collaborative Multi-Agent System to Generate Automated Negotiation for Sustainable Enterprise Interoperability

Manuella Kadar and Maria Muntean

Computer Science Department, "1 Decembrie 1918" University, Alba Iulia, Romania
{mkadar,mmuntean}@uab.ro

Abstract. The survival of traditional enterprises within the global economy relies on their ability to embrace new ideas and new organizational forms and to imagine new approaches to collaborating in dynamic networked environments. This paper proposes a system for promoting sustainable interoperability between enterprises involved in complex networked environments through multi-level negotiation, communication and information sharing. The proposed solution is based on a multi-agent system architecture that applies rule-based negotiation at various organizational levels such as: business, ICT, workflows, data systems and people. This architecture has been tested in the case of a collaborative networked environment in which several entities, namely funding authority, beneficiary of funds, contractors and sub-contractors have to be contractually engaged through multi-level negotiations.

Keywords: multi-agent systems, sustainable interoperability, rule-based negotiation.

1 Introduction

The recent global financial and economic crisis has demonstrated that new ways of doing business and new inspirations are required. The survival of traditional enterprises within the global economy relies on their ability to embrace new ideas and new organizational forms and to imagine new ways of delivering value to customers, new approaches to collaborating in a dynamic networked environment.

Organizations can only reach the full collaboration potential if partners develop enhanced capabilities to seamless communicate, coordinate, cooperate, collaborate, and most importantly, interoperate in spite of different organizational structures, technologies or processes [1]. A broad definition of interoperability is referring to the ability of two or more systems to exchange information and use it accurately. Consequently, the lack of interoperability disturbs the creation of new markets, networks, and diminish innovation and competitiveness of business groups [2]. Apart from being only a technical issue, interoperability challenges the enterprise at organizational and semantic level, underlying the need for patterns and solutions that support the seamless cooperation among ICT systems, information and knowledge, organizational structures and people [3].

G. Jezic et al. (eds.), *Agent and Multi-Agent Systems: Technologies and Applications*,
Advances in Intelligent Systems and Computing 296,
DOI: 10.1007/978-3-319-07650-8_22, © Springer International Publishing Switzerland 2014

The lack of interoperability as identified in several industrial sectors and in complex collaborative environments has a major cost, blocking the achievement of the time-to-market demanded by today's competitive environment [4].

This paper proposes a knowledge based adaptive dynamic system, namely k-MAPE (knowledge based cycle of monitoring, analyzing, planning and executing). The final goal of such system is the achievement of sustainable interoperability between enterprises involved in complex networked environments through multi-level negotiation, communication and information sharing.

The reminder of the paper is structured as follows: Section 2 describes the background for sustainability of interoperable solutions. Section 3 discloses the Multi-Agent based Negotiation System within the collaborative networked environment, the multi-layered negotiation process and communication in case of the braking down of interoperability and the requirement of re-negotiation. Section 4 provides a case study of a rule based negotiation process with detailed explanations on: i) agentification of the negotiation process and offer request design workflow; ii) communication protocols and agent behaviors; iii) rule based negotiation of an online auction and iv) implementation of the rule-based offer request design by intelligent and collaborative agents. Full implementation in the Rule Engine for the Java Platform (JESS) and Java Agent Development Framework (JADE) is provided and discussed. In Section 5 follows discussions and conclusions by pointing to future works.

2   Background for Sustainability of Interoperable Solutions

Competitive markets are becoming increasingly complex and dynamic, with companies not surviving and prospering solely through their own individual efforts [5]. Each one's success depends on the activities and performance of others to whom they do business with, and hence on the nature and quality of the direct and indirect relations [6]. Such relationships involve a mix of cooperative and competitive elements, that to cope with them, organizations need to focus on their core competencies by improving their relationships with customers, streamlining their supply chains, and by collaborating with partners to create valued networks between buyers, vendors and suppliers [7]. The collaborative process may be described as sum of coordinated and synchronous activities characterized by reciprocal interactions at high frequency that normally require the transfer of information among several organizations, i.e. knowledge sharing [8]. An emergent research challenge in seamless interoperability is rising. It focus on the sustainability within collaborative business networks, addressed by a wide complexity of interactions and a high probability of changing requirements, in the view that enterprises are complex, and adaptive systems (CAS), with factors that are making interoperability difficult to sustain over time [9]. The research challenges in the area of sustainable interoperability are several and include the study of: enterprise interoperability itself; system monitoring, behavior and adaptability; the "system" aspects of interoperability, from software component design to organizational structure to the communication, collaboration and coordination facilities; decision support to minimize the impact of changing requirements and information models; interoperability of digital ecosystems as complex systems of systems.

This approach considers that one important means to produce sustainable interoperability is designing of adapted and intelligent software components that may support effective communication, coordination and collaboration at all levels within the enterprise and the networked environment, as well. Complex tasks such as the evaluation of the maturity level of an enterprise or the design of any offer request can be successfully achieved by intelligent agents embedded with elements of artificial intelligence, e.g. expert systems and data mining classifiers. Two main aspects concerning the lack of enterprise interoperability in the digital networked environment are hereby discussed, namely: i). weak decision support at business level and ii). behavior monitoring and collaboration interruptions within the various structures of the enterprise. As concerns the decision support we proposed an enterprise interoperability maturity level assessment system carried out by intelligent agents . For instance, it is possible to create agents with Expert Systems as presented in [10]. In this paper the discussion is focused on Artificial Intelligence (AI) and algorithms that mirror the supervised learning process, that is achieved primarily through Restricted Bolzmann Machines (RBM) [11], [12] employed as a latent factor analysis application [13]. Maturity level of enterpise interoperability can be modeled by a RBM that uses a layer of binary hidden units (latent feature detectors) to model the higher-order correlations between various features of enterprises. There are no direct interactions between the hidden units that represent the latent features and no direct interactions between the visible units that represent the selected enterpresises. A simple and efficient method called „contrastive divergence" is employed to learn a good set of feature detectors from a set of training enterprises [10]. An agent endowed with a RBM is able to assess the maturity level of any enterprise before involving into a negotiation process and to inform the decision body, e.g. the Chief Negotiator on the credentials of the potential enterprise willing to get into a certain negotiation process. However, such technique provides flexible ways of making inferences, it has not been widely exploited in Multi-Agent Systems to make Negotiation Process more flexible. Consequently, modelling and implementing a MAS that employs co-ordinately learning and decision making capabilities, becomes a great challenge to provide flexible mechanisms to cope with changing market conditions, and accordingly accommodate a negotiation process with multiple offers. Behavior monitoring at various structural levels can be successfully achieved by collaborative agents and also each phase of the negotiation process and building up of the offer requests are tasks to be successfully achieved by collaborative agents endowed with suitable expert systems. The main contribution we present in this paper is the design and implementation of a collaborative Multi-Agent System to improve the Automated Negotiation Process in a dynamic networked environment.

This approach do not promote the entire substitution of human decision, it rather propose the agent technology acting in the *backstage*, helping to communicate results among different software systems. Agent technology with the inclusion of algorithms emulates intelligence and provides flexible software systems. In this sense, rule-based systems are the main AI technique that has been added to software agents. This paper pioneers on how to incorporate supervised learning in software agents to deal with complex decisions in a negotiation environment. This paper also illustrates a realistic

MAS, in which two different AI techniques are employed coordinately to accomplish a negotiation process. The design and implementation of the Multi-Agent System serve to break boundaries within a negotiation environment represented by contracting authority, contractors and subcontractors. The proposed solution is presented in the following chapters.

3   Multi-agent Based Negotiation System

Agents can be regarded as computer systems situated in some environment, being capable of autonomous action in this environment, in order to meet their design objectives [11], i.e., an agent federation that is built on some of the main attributes of software agents, namely: autonomy, social ability, reactivity and pro-activeness. Agent federations share the common characteristic of a group of agents which have ceded some amount of autonomy to a single delegate which represents the group [12]. The delegate is a distinguished agent member of the group (Chief Negotiator). Simple group members interact with an intermediary agent (BN1, CN2, SC1, etc.), which acts as an intermediary between the group and the Chief Negotiator, who further interact with the outside world, as shown in Figure 1.

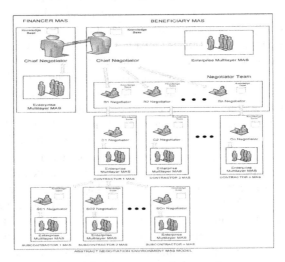

Fig. 1. MAS for abstract negotiation environment

Typically, the intermediary agent, namely, B1 Negotiator, C1 Negotiator, SC1 Negotiator, etc. accepts skill and requirements descriptions from the local agents of the enterprise multilayered MAS. The information is transmitted to the Negotiator Team who uses such information to match with requests from intermediaries representing other groups. In this way the group is provided with a single, consistent interface. The intermediary agent receives messages from its group members. These may include skill descriptions, task requirements, status information and application-level data. Such information will be communicated using some general, declarative communication language which the Negotiation Team of agents and further the Chief Negotiator understand. Outside of the group, the intermediary agent sends and receives information with the intermediaries of other groups. This could include task requests, capability notifications and application-level data routed as part of a previously created commitment. Implicit in this arrangement is that, while the intermediary must be able to interact with both its local federation members (enterprise multi-layered MAS) and with other intermediaries, individual normal agents do not require a common language as they never directly interact. This makes this arrangement particularly useful for integrating legacy or an otherwise heterogeneous group of agents.

However, a fundamental mechanism for managing inter-agent dependencies at runtime is negotiation, that is the process by which a group of agents come to a mutually acceptable agreement on some matter. Negotiation underpins attempts to cooperate and coordinate (both between artificial and human agents) and is required both when the agents are self interested and when they are cooperative. The need for a multi-agent negotiation approach is illustrated in Figure 1 in which negotiator agents represent the interests of different organizational entities (e.g., financer, beneficiary, contractors, sub-contractors).

4   Implementation of the Rule-Based Automate Negotiation with MAS

4.1   Agentification of the Negotiation Process

The environment is a contracting authority that has to accomplish a complex task, the restoration of an ancient Fortress. Besides several works, there is a road construction for which he has to negotiate with several contractors and subcontractors. The type of negotiation is an online auction. The contracting authority set forth the technical and financial characteristic of the offer request and also the potential contractor's profile requirements. The maximum admitted total price of the road construction and the technical features such as: time of achievement, length of the road, raw materials, equipment employed, archaeological rescue plans, work warranty are designed by the Beneficiary's departments. The potential contractor's profile requirements are set up by the Beneficiary's Accountant and Juridical Departments and refers to the turnover, profit, number of employees, previous similar works achieved, flexibility and capacity of working in special settings, IT infrastructure, and equipment. The offer request is launched by the Beneficiary's Chief Negotiator to one of his Negotiators responsible for negotiating Construction works, namely the B1 Negotiator Agent. B1Negotiator Agent further launches the offer request to Constructors. The Constructor's Negotiator Agent is responsible for providing data to it's own agents in order to elaborate the offer request and to further send the offer to B1Negotiator Agent. The Designer Agent as well as the Raw Material Agent as part of the Constructor's MAS face the complex problem of elaborating the best suitable offer according to the requirements. To handle such complexity a rule based expert system has been attached to the MAS.

The design of the negotiation process as a workflow is presented in Figure 2.

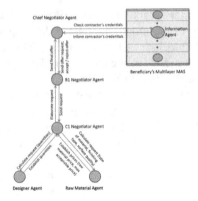

Fig. 2. Design of the negotiation process as a workflow between agents

4.2   Communication Protocols and Agents Behaviors

The MAS is globally coordinated by means of a communication protocol. In Figure 3 we present a simplified version of the proposed model. The left side shows the protocol between the Chief Negotiator and the B1Negotiator agents. On the right side, a similar model is used between the C1Negotiator and the Designer and Raw Material agents. According to the AUML notation , the solid arrows at the end of each message represent synchronous exchanges. The communication between the Chief Negotiator and the B1Negotiator is not initiated until the C1Negotiator agent sends the offer request. The entire communication process ends as soon as the Chief Negotiator updates the value of the offer in the EIS. When this occurs, the negotiation process is ended. The actual message content that is uttered by the ChiefNegotiator, B1Negotiator, C1Negotiator, Designer, RawMaterial agents comes from the emulation of intelligent behaviour.

Fig. 3. Communication protocol

JADE uses the Behavior abstraction to model the tasks that an agent is able to perform and the agents instantiate their behaviours according to the needs and capabilities.

In order to develop an agent, the Agent class should be extended and the agent-specific tasks should be implemented through one or more Behaviour classes. Finally, these classes will be instantiated and added them to the agent. The Agent class, which is a common superclass, allows to inherit a basic hidden behaviour (that deals with all agent platform tasks, such as registration, configuration, remote management, and so on), and a basic set of methods that can be called to implement the application tasks of the agent (e.g. send/receive messages, use interaction protocols, query a knowledge base). The behaviors are described in Table 1.

Table 1. Behaviours

Crt. no.	Behaviour
1.	StartNegotiation = "100 StartNegotiation|"
2.	AcceptNegotiation = "101 AcceptNegotiation|"
3.	DenyNegotiation = "102 DenyNegotiation|"
4.	SendRequest = "200 SendRequest|"
5.	AckRequest = "201 AckRequest|"
6.	WaitingForResponse = "203 WaitingForResponse|"
7.	SendingResponse = "300 SendingResponse|"
8.	AckResponse = "301 AckResponse|"
9.	ConfirmedReceivedResponse = "302 ConfirmedReceivedResponse|"
10.	EndNegotiation = "500 EndNegotiation|"
11.	Error = "9999 Error|"
12.	ConversationIdB1NegotiatorC1Negotiator = "MessageB1NegotiatorC1Negotiator"
13.	ConversationIdC1NegotiatorDesigner = "MesssageC1NegotiatorDesigner"
14.	ConversationIdC1NegotiatorRawMaterial = "MesssageC1NegotiatorRawMaterial"
15.	ConversationIdB1NegotiatorChiefNegotiator = "MesssageB1NegotiatorChiefNegotiator"
16.	JaRBM = "jarbm"

4.3   Rule Based Negotiation for an Online Auction

The experiment describes an online auction for a contract of road construction and uses FIPA (Foundation for Intelligent Physical Agents) standards organization. Jess (the Rule Engine for the Java TM Platform) [14] is a rule-based system shell entirely written in Java. For representing the "accept" or "reject" proposals JESS templates have been used. The considered facts were the negotiation price, the construction time (in months) and the surface of the construction (in square meters).

The next rule fires if the negotiation price between the Contractor and one of its Sub-contractors is greater than 100000, and if the construction time is greater than five months and if the construction surface is less or equal to 500 square meters.

```
(defrule negotiation-protocol-reject
    "This is the negotiation protocol between contractor and subcontractor1, the
case: reject proposal"
    (negotiation {price > 100000})
    (negotiation {construction_time_months > 5})
    (negotiation {surface_square_meters <= 500})
    =>
    (printout t "Reject proposal" crlf))
```

If the price is less than 100000 and the construction time is less or equal to 5 months and the construction surface is greater than 500 square meters, than the "accept proposal" rule will be fired.

```
(defrule negotiation-protocol-accept
    "This is the negotiation protocol between contractor and subcontractor1, the
case: accept proposal"
    (negotiation {price < 100000})
    (negotiation {construction_time_months <= 5})
    (negotiation {surface_square_meters > 500})
    =>
    (printout t "Accept proposal" crlf))
```

4.4   Experimental Results and Discussions

The negotiation environment was tested with agents that were designed with specific rule templates, where rules assert information in their private fact base. The agents respond to this information by sending messages according to the Send - Receive Protocol defined in the system. Once the system agent receives the confirmation of starting a negotiation, an agent will negotiate with his counterpart and automatically make decisions based on his negotiation parameters. The system also provides a functionality of viewing the communications between agents and the systems via a sniffer agent. Various types of rules and queries were used for representing valid, posted and active proposals, depending on the phase of the negotiation process.

For instance, in the C1Negotiator-Designer negotiation, the Designer agent has to evaluate if the offer proposed by C1Negotiator agent exists in his fact base. In order to do this, he runs the following query:

```
(defquery search-by-name
    "Finds road design"
    (declare (variables ?name))
    (design
(name ?name)
(type_of_construction_material ?type_of_construction_material)
    (quantity ?quantity) ) )
```

In the C1Negotiator-RawMaterial negotiation, the RawMaterial agent fires two rules in order to get the total price of the offer:

```
(defrule getTotalPrice
    "Get the total price of the proposed offer"
    ?p <- (item)
    =>
    (add  item_quantity
((?p.name str-cat _item_quantity)
(?quantity1.quantity) (?quantity1.quantity * ?p.price) (?quantity1.maximum_price) ) )
    (printout t (str-cat ?p.name _item_quantity)
", " ?quantity1.quantity  ","
(* ?quantity1.quantity  ?p.price) "," ?quantity1.maximum_price crlf) )
```

and to check if the total price against the maximum price defined the our system:

```
(defrule checkTotalPrice
    "Check price against maxPrice"
    ?p <- (item_quantity
(total_price > maximum_price} )
    =>
    (retract ?p )
(printout t "Retract: " ?p.name ", "    ?quantity1.quantity  ","
        (* ?quantity1.quantity  ?p.price) crlf))
```

If the evaluated offer has a greater price than the maximum price considered, it will be retracted from the fact base and a message will be displayed.

The negotiation protocol was parameterized with specific rules taking into account the sustainable interoperability between agents. The negotiation process has been achieved according to the designed workflow, the entire communication between the agents involved is presented in the viewer. Negotiation time is assessed and calculated for the entire process in various negotiation conditions for further optimisations.

```
msg = 101 AcceptNegotiation|, Time = 2014-03-10 11:03:05 duration: 0 seconds
msg = 200 SendRequest|(request (operation "construction")(structure "paved_road")), Time = 2014-03-10 11:03:05 duration: 0
msg = 201 AckRequest|, Time = 2014-03-10 11:03:05 duration: 0 seconds
msg = 203 WaitingForResponse|, Time = 2014-03-10 11:03:05 duration: 0 seconds
msg = 300 SendingResponse|, Time = 2014-03-10 11:03:05 duration: 0 seconds
msg = 301 AckResponse|, Time = 2014-03-10 11:03:05 duration: 0 seconds
msg = 302 ConfirmedReceivedResponse|, Time = 2014-03-10 11:03:05 duration: 0 seconds
msg = 500 EndNegotiation|, Time = 2014-03-10 11:03:06 duration: 1 seconds

msg = 101 AcceptNegotiation|, Time = 2014-03-10 11:03:05 duration: 0 seconds
msg = 201 AckRequest|, Time = 2014-03-10 11:03:05 duration: 0 seconds
msg = 300 SendingResponse|, Time = 2014-03-10 11:03:05 duration: 0 seconds
msg = 302 ConfirmedReceivedResponse|, Time = 2014-03-10 11:03:05 duration: 0 seconds
1
erial, msg = 100 StartNegotiation|, Time = 2014-03-10 11:03:05 duration: 0 seconds
erial, msg = 101 AcceptNegotiation|, Time = 2014-03-10 11:03:06 duration: 1 seconds
erial, msg = 200 SendRequest|, Time = 2014-03-10 11:03:06 duration: 1 seconds
erial, msg = 201 AckRequest|, Time = 2014-03-10 11:03:07 duration: 2 seconds
erial, msg = 203 WaitingForResponse|, Time = 2014-03-10 11:03:07 duration: 2 seconds
erial, msg = 300 SendingResponse|, Time = 2014-03-10 11:03:08 duration: 3 seconds
erial, msg = 301 AckResponse|, Time = 2014-03-10 11:03:08 duration: 3 seconds
erial, msg = 302 ConfirmedReceivedResponse|, Time = 2014-03-10 11:03:09 duration: 4 seconds
:erial, msg = 300 SendingResponse|, Time = 2014-03-10 11:03:09 duration: 4 seconds
erial, msg = 500 EndNegotiation|, Time = 2014-03-10 11:03:09 duration: 4 seconds

erial, msg = 101 AcceptNegotiation|, Time = 2014-03-10 11:03:06 duration: 0 seconds
erial, msg = 201 AckRequest|, Time = 2014-03-10 11:03:07 duration: 1 seconds
erial, msg = 300 SendingResponse|, Time = 2014-03-10 11:03:08 duration: 2 seconds
erial, msg = 302 ConfirmedReceivedResponse|, Time = 2014-03-10 11:03:09 duration: 3 seconds
gotiator
iefNegotiator, msg = 101 AcceptNegotiation|, Time = 2014-03-10 11:03:12 duration: 0 seconds
iefNegotiator, msg = 201 AckRequest|, Time = 2014-03-10 11:03:14 duration: 2 seconds
iefNegotiator, msg = 300 SendingResponse|Price OK 100000:  road_construction = 100000, Time = 2014-03-10 11:03:15 duration:
iefNegotiator, msg = 302 ConfirmedReceivedResponse|, Time = 2014-03-10 11:03:16 duration: 4 seconds
```

Fig. 4. A sample of results

5   Conclusions and Future Work

This paper proposes a novel multi-agent system that adopts a rule-based approach to implement automated negotiations, to support the sustainability of interoperability and adaptation of the networked organizations. The developed intelligent system integrates responses from the collaborative networked environment and actively contributes to the harmonization of breakings through re-negotiation of various jobs and tasks in order to enhance the sustainable interoperability within the network.

The design and implementation of the proposed MAS serve to break boundaries within an automated negotiation environment.

Future work outlooks: i) the completion of the integration of the rule-based framework into our abstract negotiation environment; ii) the assessment of the generality of this implementation by extending it with several types of negotiations occurred in case of breaking downs of interoperability; iii) the allowance of the logical specification of the rules in order to asses their correctness; iv) the investigation of the e-effectiveness by system validation in several collaborative networked environments.

References

1. Pine, B.J., Gilmore, J.H.: The experience economy. Harvard Business Press (1999)
2. Ray, S.R., Jones, A.T.: Manufacturing interoperability. Journal of Intelligent Manufacturing 17(6), 681–688 (2006)
3. Jardim-Goncalves, R., Grilo, A., Steiger-Garcao, A.: Challenging the Interoperability in the Construction Industry with MDA and SoA. Computers in Industry 57(8-9), 679–689 (2006) ISSN 0166-3615
4. White, W.J., O'Connor, A.C., Rowe, B.R.: Economic Impact of Inadequate Infrastructure for Supply Chain Integration. NIST Planning Report 04-2. National Institute of Standards and Technology, Gaithersburg (2004)
5. Friedman, T.: The World is Flat. Farrar, Straus & Giroux (2005)
6. Wilkinson, I., Young, L.: On cooperating: firms, relations and networks. Journal of Business Research 55(2), 123–132 (2002)
7. Amin, A., Cohendet, P.: Architectures of knowledge: firms, capabilities, and communities. Oxford University Press (2004)
8. Beck, P.: Collaboration vs. Integration: Implications of a Knowledge-Based Future for the AEC Industry (2005), http://www.di.net/articles/archive/2437/ (accessed September 22, 2009)
9. CompTIA: European Industry Association. European Interoperability Framework. white paper - ICT Industry Recommendations (2004)
10. Kadar, M.: Assessement of Enterprise Interoperability Maturity Level through Generative and Recognition Models. In: Proceedings of the 2013 International Conference on Economics and Business Administration (EBA 2013), Rhodes Island, Greece, July 16-19 (2013) ISBN 978-1-61804-200-2
11. Wooldridge, M., Jennings, N.R.: Intelligent Agents: Theory and practice. The Knowledge Engineering Review 10(2), 115–152 (1995)
12. Kadar, M., Cretan, A., Muntean, M., Goncalves, R.: A Multi-Agent based Negotiation System for Re-establishing Enterprise Interoperability in Collaborative Networked Environments. In: Proceedings of 2013 UKSim 15th International Conference on Computer Modelling and Simulation, Cambridge University (Emmanuel College), April 10-12, pp. 190–195 (2013) ISBN 978-0-7695-4994-1
13. Sycara, K., Dai, T.: Agent Reasoning in Negotiation. In: Handbook of Group Decision and Negotiation, Part 4. Advances in Group Decision and Negotiation, vol. 4, pp. 437–451 (2010)
14. Java Agent DEvelopment Framework, http://jade.tilab.com/

Multi-robot Hunting Using Mobile Agents

Naoya Ishiwatari[1], Yasunobu Sumikawa[1],
Munehiro Takimoto[1], and Yasushi Kambayashi[2]

[1] Department of Information Sciences, Tokyo University of Science,
2641 Yamazaki, Noda 278-8510, Japan
[2] Department of Computer and Information Engineering,
Nippon Institute of Technology,
4-1 Gakuendai, Miyashiro-machi, Minamisaitama-gun 345-8501, Japan
{n-ishiwatari,yas,mune}@cs.is.noda.tus.ac.jp, yasushi@nit.ac.jp

Abstract. Multi-robot hunting problem is one of the popular issues treated with multi-robot systems. The purpose of the problem is to search and capture a target using an invisible signal, through which the multi-robots can sense the distance to the target. This paper proposes a new method in which the multiple robots cooperatively search for a target using mobile agents. In the method, we employ multiple mobile software agents. The mobile agents traverse mobile robots through migrations while collecting the information of the target. Since each robot just needs to establish a connection with another robot for migration of a mobile agent, our method reduces the total communication cost of the system. Also, the mobile agents' migration manner is restricted within the view range of the camera of a robot, and thus mainly makes robots around the target active, which contributes to suppressing moving cost. We have implemented a simulator for the mobile agents based hunting system. We show the effectiveness of our method through numerical experiments on the simulator.

1 Introduction

In the last decade, robot systems have made rapid progress not only in their behaviors but also in the way they are controlled. In particular, a control system based on multiple software agents can control robots efficiently [1,2]. Multi-agent systems introduced modularity, reconfigurability and extensibility to control systems, which had been traditionally monolithic. It has made easier the development of control systems on distributed environments such as multi-robot systems.

On the other hand, the excessive interactions among agents in the multi-agent system may cause problems in the multiple robot environments. In order to mitigate the problems of excessive communication, mobile agent methodologies have been developed for distributed environments [3]. In a mobile agent system, each agent can actively migrate from one site to another site. Since a mobile agent can bring the necessary functionalities with it and perform its tasks autonomously, it can reduce the necessity for interaction with other sites. Mobile agent systems

G. Jezic et al. (eds.), *Agent and Multi-Agent Systems: Technologies and Applications*,
Advances in Intelligent Systems and Computing 296,
DOI: 10.1007/978-3-319-07650-8_23, © Springer International Publishing Switzerland 2014

are especially useful in an intermittently connected ad hoc network environment. In the minimal case, a mobile agent requires that the connection is established only when it performs migration [4].

We propose a new method that reduces the communication cost among multiple robots for the *multi-robot hunting* problem [5] using the property of the mobile agents. The objective of the multi-robot hunting is to search and capture a target, but the robots are assumed not to have any device by which they directly recognize the target such as vision sensor. In other words, it cannot know the correct coordinates of the target. What the robots know is just relative intensity of the signals emitted from the target. We assume that the signal radiates uniformly in all directions such as audio signal or electromagnetic radiation. Though this paper is focusing on simple hunting a target by multiple robots, the concept can be applied to various applications. Such applications include urban search and rescue (USAR) operations such as searching for survivors, and locating sources of hazards such as chemical or gas spills, toxic pollution, pipe-leaks, radioactivity. In the problem, it is effective to integrate the information obtained at several locations in order to discern more correct coordinates of the target. Such an approach was previously proposed as the manner in which each robot communicates with other robots to share the information, but in that manner, communication cost tends to increase as the number of robots increases. Excessive communication is not only disadvantage in terms of scalability but also serious problem for robots working by batteries.

In our system, on the other hand, each robot does not directly communicate with other robots, but share information given by mobile agents. Mobile agents traverse robots and collect the information the robots have. As mentioned above, mobile agents need communication just for its migration, and therefore, reduce the total cost of communications. Therefore collecting information by way of agent migrations also contributes to saving energy consumption. Furthermore, the mobile agent based approach makes only the robots around the target active, because the range where mobile agents migrate is restricted within the view ranges of the cameras of the robot. Therefore the information about the target is propagated to only the robots within that range. This property of the system also suppresses total moving cost of robots.

In summary, the contributions of this paper are:

1. We propose a model where mobile agents are integrated into the traditional multi-robot hunting system.
2. We have implemented our model through constructing a simulator to show that our model is practically feasible.
3. We demonstrate the effectiveness of our model for suppressing energy consumption through numerical experiments.

The structure of the balance of this paper is as follows. In the second section, we describe related works. In the third section, we describe the traditional approach that is the multi-robot hunting with swarm intelligence. In the fourth section, we describe the details of the model that extends the traditional approach using mobile agents. In the fifth section, we describe a simulator that implements our model, and demonstrate the effectiveness through numerical experiments. Finally, we conclude our discussions in the sixth section.

2   Related Works

Wang et al proposed a robot controller based on spiking neural network (SNN) for the coordinated hunting of multi-robot system [6]. The controller utilizes twelve direction-sensitive modules to encode and process the inputs including the environment and location information of the target by a biologically inspired coding technique based on spike timing called time-to-first-spike coding and then, the motor neurons generate the control signals for the motors according to the winner-take-all strategy. Spikes (pulses) are used to deliver the information between neurons. Therefore SNN processes the information in the form of spikes that brings temporal structure and extends the functionality of SNN [7]. They implemented their strategies in a simulator, and demonstrated that the controller can make the multiple robots coordinate with each other to finish the hunting task by using the time-to-first-spike coding and winner-take-all strategy.

Ant Colony Optimization (ACO) algorithms is one of the swarm techniques [8]. ACO is an algorithm that mimics the behavior of ants in real world. When Real ants find foods, they carry them to their nest with putting *pheromone* on the ground as paths from the foods to their nests. The pheromone trails lead another ants to the foods, and then the ants will carry the foods walking on the same paths. As all ants put pheromone whenever they bring foods to the nest, the pheromone trails are enhanced. In addition, the pheromone trail of shorter paths is more attractive than longer ones. It is possible to find the shortest path by this mechanism.

ACO can be applied to many kinds of NP-hard and NP-complete problems like traveling salesman problem, finding optimal routes, and scheduling. These problems are represented as graph, and then ants walk on the graph. The routes made by the ants walking are the candidate of the optimal solution.

In various problem, ACO produces good results, but it is possible to converge to a local optimum in some case. To prevent the problem, some new techniques are proposed. Stutzle et al. proposed Max-Min Ant System (MMAS) that restricts the value of pheromone to be in interval $[\tau_{min}, \tau_{max}]$ [9]. Because this restriction gives ants some choice to go to next node, MMAS prevents getting local optimum.

3   PSO Based Approach

For multi-robot hunting systems without guidance from any external agency, Kennedy and Eberhart proposed a technique based on Particle Swarm Optimization (called PSO)[10]. The technique is one of the swarm based algorithms [11].

PSO is performed as follows: 1) the particles are located randomly as the initial state, 2) the location in the step i of each particle is computed along the following equation, where X_i and V_i are the location and the speed of the particle respectively:

$$X_i = X_{i-1} + V_{i-1} \tag{1}$$

$$V_i = \omega V_{i-1} + C_1 Rand_1(P_i - X_i) + C_2 Rand_2(P_g - X_i) \tag{2}$$

Equation 1 means the current location is determined by the location X_{i-1} and the speed V_{i-1} at the previous step. Equation 2 means that the current speed is determined by the speed V_{i-1} at the previous step and the distance to the target. In equation 2, P_i is the particle's own most fitted location gotten through the fitness function. P_g is the location of the particle that is the closest to the target, and it is globally shared by all robots as the best fitness value. The distances from the current location to them adjusts the base speed ωV_{i-1} with the random weight $Rand_{1/2}$, where the ω is empirically determined value more than or equal to 0 and less than 1, and the $Rand_{1/2}$ is a random number over the range $[0, 1]$. Also, C_1 and C_2 are empirically determined values. Notice here that terms $C_1 Rand_1(P_i - X_i)$ and $C_2 Rand_2(P_g - X_i)$ mean that the speed of a particle decreases as the robot is close to the target.

In applying PSO to the multi-robot hunting problem, each robot is regarded as a particle. All robots and a target are initialized to random locations in a work field, where the target has tendency to be located around the center, and the robots are located in such a way they surround the target. Each robot captures the target through communications with other robots without knowledge about the location of the target.

Each robot shares the following information with other robots.

– The location of the robot closest to the target in a group of the robots, and
– Strength of the signal received at the location.

The values of signal the strength at different locations are accumulated in each memory, and can be used to determine whether the robot approaches to the target or not. Also, each robot has a GPS device that gives its absolute location and an omni-directional camera for identifying other robots around it. Notice that the camera cannot identify the target.

Each robot behaves along the equations 1 and 2. They lead the robot to the target. As shown by the equation 2, P_g has to be updated as the globally best fitness value to be shared by all robots. Every time each robot gets better fitness than P_g at the current location, it sends the location to the other robots to update P_g. The updating process may cause too much communication traffic.

4   Mobile Agent Based Approach

As mentioned in the introduction, mobile agents can convey a lot of information as states and program code in a short time. We take advantage of this property of the mobile agents so that robots can share global information with minimum cost. We call the mobile agents that convey information *communication mobile agents* (CMAs). CMA has the information of the location P_g that is closest to a target and the strength at P_g as stored data. We especially call the P_g in a CMA P_a distinguishing from P_g information on each robot. The P_g is selected from the location information P_gs on the robots that CMA has traversed.

In order to traverse the herd of the multiple robots, CMA has to identify the destination robot and migrate to it, and hence, the range in which CMA can migrate is restricted within the view range of the camera equipped each robot. If there are several robots in range, CMA migrate to the robot in the forward direction. The traversal is a sequence of such short hops; the order of the hops depends on how to select the destination. Once CMA migrates to a robot, it compares the signal strength at P_a with one at P_g on the robot, and if the strength at P_g is more than at P_g, P_g information on the robot is replaced with P_a; otherwise, P_a information on the CMA is replaced with P_g. Notice that both of P_g and P_i are initialized to the current location X_i, and therefore the initial speed of a robot depends on the base speed, which initially corresponds to zero in our system.

Since CMA traverses robots to propagate the information to be shared, the traversal has to be efficient. CMA records the trace of migrations to avoid revisits to the same robot, but the trace information is flushed once P_a is updated, which always gives the fresh P_a to robots. Also, employing of several CMAs contributes to the efficient propagation of the P_g. In such propagation, the case where several CMAs simultaneously migrate to the same robot has to be considered because each CMA behaves independently.

In our system, once several CMAs migrate to the same robot, they compare their own signal strength information with the other ones, and replace the strength information with the maximal one. At this time, their P_as are also replaced with the P_a corresponding to the maximal strength. The way of the cooperative propagation enables rapidly propagating P_g information without losing the consistency of shared information.

Furthermore, we suppress ineffective motion of robots taking advantage of the signal strength at the P_a. Once CMA migrates to a robot, it observes a signal there. If the strength of the signal is less than at the previous robot, the CMA decreases the constant ω in the base speed term. The decrease of the base speed suppresses the motion of robots far from the target, which contributes to suppressing energy consumption without sacrificing the efficiency. In our system, ω is alternatively switched between the two values ω_1 and ω_2, which satisfy that ω_1 is more than ω_2. In decreasing ω ω_2 is used as the ω, while ω_1 is used once the CMA migrates to the robot with more signal strength.

In order to demonstrate the effectiveness of our CMA based approach, we compare our approach with the traditional PSO in terms of the followings:

1. The number of total steps.
2. The volume of communications.
3. The average of moving distance for a robot.

From the observations of the experiments, we show the optimal number of CMA in our approach.

5   Experimental Results

We have implemented a simulator to demonstrate the effectiveness of our CMA based approach in comparison with the traditional PSO approach. Fig. 1 shows a snapshot of the output of our simulator. As shown by the figure, it is assumed that there is no obstacle in the field. The red circles represent the target. The blue and gray circles represent robots with constant ω_1 and ω_2 as ωrespectively. The white line on each blue circle represents the forward direction of each robot. In our simulator, the number of robots can be set in any scale, though twenty robots are cooperatively hunting a target in the snapshot. The number of agents can be set in any scale. The order of traversal of CMAs can be determined by how to select a target to migrate of the robots within the visual range.

5.1   Simulator

In the initial field, a target and robots are randomly located, where the target tends to lie around the center, and the robots lie surrounding the targets as previously mentioned.

Each robot does not have the location information of the target, even though it can receive the signal from the target. The robot can just sense the strength of the signal, which is inversely proportional to the distance from the target. In the traditional PSO, the robots share the information of the location closest to the target and the signal strength at the location through a WiFi network. CMA also migrates through the WiFi network. The robot can move to any locations without collision among the robots. The distance that each robot moves in the step is set to 1, and the state of the field is updated in each step. The details of other parameters of the simulator are as follows:

- The size of a robot: 1×1
- The size of the field: 1000×1000
- The maximal speed of a robot: 10
- The orientation a robot can face is 360°
- The number of robots: 100

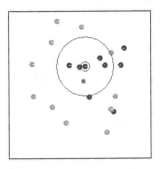

Fig. 1. Simulation performed by 20 robots

– The radius of visual range of a robot: 200

Furthermore, we determine the constants on Equation 1 and Equation 2 as follows:

– Inertial constant (ω_1): 0.9
– Inertial constant (ω_2): 0.3
– PI learning constants (C_1): 2.0
– SI learning constants (C_2): 2.0

5.2 Comparison with the PSO Approach

To investigate the property of our approach, we have implemented CMA based PSO approaches.

For these approaches, we have conducted experiments till some robots capture the target, and we measured the number of total steps, the volume of communications, and the average of moving distance for one robot. We have obtained results as shown in Fig. 2a (The number of steps), Fig. 2b (The volume of communication), and Fig. 3a (The average travel distance). The horizontal axes show the number of robots.

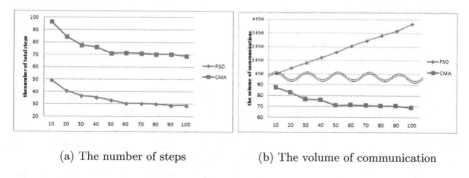

(a) The number of steps   (b) The volume of communication

Fig. 2. The number of steps and the volume of communication in terms of the number of robots

As shown in Fig. 2a (The number of steps), the traditional PSO based system captures the target quicker than CMA based system. This is because PSO can share global information more quickly through direct communications with the other robots instead of the propagation of CMA through migrations.

On the other hand, as shown in Fig. 2b (The volume of communication), the volume of communications of CMA based PSO is remarkably less than traditional PSO. The result shows CMA based PSO approaches have the effectiveness of suppressing energy consumption for the communication, which contributes to prolonging the time-to-live of each robot.

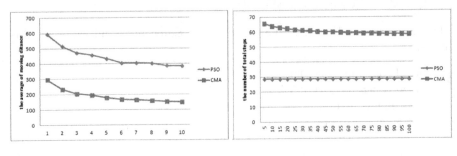

(a) The average travel distance   (b) The number of steps

Fig. 3. The average travel distance and the number of steps in terms of the number of agents

Fig. 3a (The average travel distance) shows the CMA based PSO approach makes the travel distance of entire robots shorter than the traditional PSO. This effect also contributes to suppressing energy consumption in addition to the communication cost saving.

The latter two results show that CMA based PSO approach has remarkable long life time in comparison with traditional PSO approach.

5.3   Multiple Agents

We have shown the result for CMA based PSO approach with a single mobile agent. In this subsection we show the results of the number of the total steps, the volume of communications, and the average of total moving distance for CMA based PSO with multiple CMAs in Fig. 3b (The number of steps), Fig. 4a (The volume of communication), and Fig. 4b (The average travel distance), respectively. Notice that in the figures, the number of robots is fixed to one hundred, and the horizontal axes show the number of CMAs.

As shown in Fig. 3b (The number of steps), the increase of the number of CMAs hardly affects the number of total steps, though we expected that it decreased. On the other hand, as shown in Fig. 4a (The volume of communication) and Fig. 4b (The average travel distance), the number of communications and the moving distance for each robot increase as the number of mobile agents increases.

From the observations of these results, if the efficiency of the solution is the most important requirement, the traditional PSO approach should be chosen; otherwise, CMA based PSO approaches are effective alternatives because they can suppress energy consumption till a target is captured. Most robots work with batteries, and therefore, the energy consumption restricts the total amount of achievements and the size of search area.

Therefore, it is worth investigating on the reason why CMA based PSO approach results in such the energy saving to obtain for the further improvements.

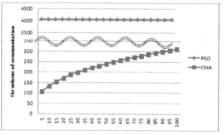

(a) The volume of communication

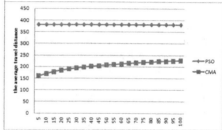

(b) The average travel distance

Fig. 4. The volume of communication and the average travel distance in terms of the number of agents

In the approaches based on PSO, the robots far from the target do not contribute to updating P_g or capturing the target.

Also, in the case of implementing multi-robot hunting based on the PSO, most robots tend to face with the direction to the target, which means that in front of the robot, there are likely other robots closer to the target. If the robot actually faces with the target, CMA migrate to the robot closer to the target, so that the location information of the target stored in CMA is quickly updated to fresh information, which contributes to suppressing the volume of communications. On the other hand, the robot far from the target will not be migrated by the CMA, so that P_g is not updated, and kept far from the target, which contributes to decreasing speed and suppressing moving distance.

6   Conclusions

We have proposed a new method for the multi-robot hunting system based on swarm intelligence. Our method uses mobile agents in older to reduce the volume of communication traffic and to eliminate ineffective motion. We have implemented a simulator to demonstrate the effectiveness of our method. We have conducted experiments of our approach, and have compared the results with the traditional approach. As a result, we have shown that the CMA based PSO approach of ours remarkably suppressed the volume of communication and moving distance, even though our mobile agent based approaches took more time than the traditional approach. The property of our agent base approach contributes to suppressing energy consumption.

Considering the time-to-live of each robot, our approach would be a practical alternative.

References

1. Mizuno, M., Kurio, M., Takimoto, M., Kambayashi, Y.: Flexible and efficient use of robot resources using higher-order mobile agents. In: Proceedings of the 2006 Conference on Knowledge-Based Software Engineering: Proceedings of the Seventh Joint Conference on Knowledge-Based Software Engineering, pp. 253–262 (2006)
2. Takimoto, M., Mizuno, M., Kurio, M., Kambayashi, Y.: Saving energy consumption of multi-robots using higher-order mobile agents. In: Nguyen, N.T., Grzech, A., Howlett, R.J., Jain, L.C. (eds.) KES-AMSTA 2007. LNCS (LNAI), vol. 4496, pp. 549–558. Springer, Heidelberg (2007)
3. Satoh, I.: Mobilespaces: A framework for building adaptive distributed applications using a hierarchical mobile agent system. In: Proceedings of the The 20th International Conference on Distributed Computing Systems, ICDCS 2000, pp. 161–168. IEEE Computer Society (2000)
4. Binder, W., Hulaas, J.G., Villaz, A.: Portable resource control in the j-seal2 mobile agent system. In: Proceedings of the fifth international conference on Autonomous agents, AGENTS 2001, pp. 222–223. ACM (2001)
5. Nighot, M., Patil, V., Mani, G.: Multi-robot hunting based on swarm intelligence. In: 2012 12th International Conference on Hybrid Intelligent Systems (HIS), pp. 203–206 (2012)
6. Wang, X., Cao, Z., Zhou, C., Hou, Z., Tan, M.: Coordinated hunting based on spiking neural network for multi-robot system. In: Multi-Robot Systems, Trends and Development, pp. 327–338 (2011)
7. Kasabov, N.: To spike or not to spike: A probabilistic spiking neuron model. Neural Netw. 23(1), 16–19 (2010)
8. Dorigo, M., Maniezzo, V., Colorni, A.: Ant system: Optimization by a colony of cooperating agents. Trans. Sys. Man Cyber. Part B 26(1), 29–41 (1996)
9. Stutzle, T., Hoos, H.H.: Max-min ant system. Future Gener. Comput. Syst. 16(9), 889–914 (2000)
10. Kennedy, J., Eberhart, R.: Particle swarm optimization. In: Proceedings of IEEE Intermational Conference on Neural Networks, vol. 4, pp. 1942–1948 (1995)
11. Dorigo, M., Gambardella, L.M.: Ant colony system: A cooperative learning approach to the traveling salesman problem. Trans. Evol. Comp. 1(1), 53–66 (1997)

Norms Disappearance in Open Normative Multi-agent Communities: A Conceptual Framework

Muhsen Hammoud[1], Azhana Ahmad[2], Moamin A. Mahmoud[2],
Alicia Y.C. Tang[1], and Mohd. Sharifuddin Ahmad[2]

[1] College of Graduate Studies, Universiti Tenaga Nasional, Selangor, Malaysia
mouhsen21@hotmail.com, aliciat@uniten.edu.my
[2] College of Information Technology, Universiti Tenaga Nasional, Selangor, Malaysia
{azhana,sharif}@uniten.edu.my,
moamin84@gmail.com

Abstract. The dynamisms in open normative multi-agent communities produce changes to enacted norms where some becomes obsolete while others appear and emerge. Such changes result in obsolete norm derogation and eventually disappearance. Norm disappearance happens when the majority removes it from their belief base to avoid conflicts in the enactment of old and new norms. This paper presents a conceptual framework for norms disappearance in open normative multi-agent communities. Norms disappearance is influenced by five elements which are norms conflict; norm's trust decay; high assimilation cost of low norm's yields; norm's yields decay, and violation cost decay. In addition, four parameters that affect on the values of these elements and thus on the decision of an agent to remove a norm from its cognitive structure are assimilation cost; yields value; trust value; and violation rate. Negative changes on these parameters prompt agents to think about the possibility of removing the norm.

Keywords: Social Norm, Norms, Normative Systems, Norms Disappearance, Normative Multi-agent Systems, Software Agents.

1 Introduction

Numerous definitions of norms have been defined by researchers in social and computer sciences. The core idea of these definitions is that norms are practices and constraints that are obligated upon the incumbent. According to Elster [1], "The simplest social norms are of the type: Do X, or, Don't do X". From this definition we can say that social norms generally impose or prohibit an action or behavior. Thus, norms are not used to govern outcomes, it is used to govern actions. Another definition is presented by Ostrom [2], who proposed that "Norms are shared understandings about actions that are obligatory, permitted, or forbidden".

Recently, numerous models on multi-agent systems have been investigated and this includes norms in agent architectures [3], [4]. Researchers have discussed two issues; the first issue is about the effect of norms and normative systems on agents'

G. Jezic et al. (eds.), *Agent and Multi-Agent Systems: Technologies and Applications*,
Advances in Intelligent Systems and Computing 296,
DOI: 10.1007/978-3-319-07650-8_24, © Springer International Publishing Switzerland 2014

performance. The second issue is about the norm's life cycle in normative multi-agent systems. However, the first issue has been discussed extensively [5], [6], [4]. For the second issue, there are four models of norm's lifecycle discussed in the literature [7], [8], [9], [10]. In general, these models emphasize on the processes of norms Creation, Emergence, Enforcement, Assimilation, Internalization, and Evolution.

Evolution process consists of four possibilities which are: Norms legislation, Norms removal, Norm disappearance, and Norms modification. The usage of norms removal and norms disappearance terms is ambiguous in the literature. In this paper, we will clarify the meaning of each term in Section 3. The term disappearance is used in this paper to refer to the process of norm disappearance from society. While the term removal is used to refer to the process of removing a norm from agent's cognitive structure (belief base), not from society. After that, norm disappearance concept will be introduced. A search on empirical research on norms disappearance returns almost non-existent result. The existing research only mentioned the case of disappearance without building any concrete concept for it [11], [12], [9]. Consequently, this paper addresses this issue for investigation by introducing a conceptual approach of norms disappearance.

Norms removal from agents cognitive structure is the process of eliminating obsolete norms when conflicts occur between the domain's current norms [9]. The removing process is theoretically important when the system has been updated and becomes more complicated or it is limited in resources [13]. Norm disappearance happens when the majority of certain society stops adopting it. After that, agents will decide whether to remove the disappeared norm from their cognitive structure or not. The literature provide other terms for norms disappearance such as norms derogation [14], norms decay [15], and norm deletion [16].

This paper, which reports the work-in-progress of our research in norms, presents a conceptual approach for norms disappearance. In this framework, we introduce the issue of norm disappearance from a community over a period of time. However, we discuss first the possible cases which lead to norms disappearance and subsequently present the proposed framework. The cases are norms conflict [9], norm's trust decay [11], high assimilation cost of low norm's yields [9], norm's yields decay [15], and violation cost decay [15]. When agents observe one or more of these cases occur on a norm, they begin to violate or abandon the norm. Over a period of time, the norm is considered insignificant for the agents' cognitive structure (belief base). Agents, consequently, change their behaviors after their beliefs changed.

The objectives of this study are, (i) to highlight the issues of norms disappearance in open agent communities where norms are not offline-designed, and (ii) to develop a norms' disappearance framework. However, this paper considers the first attempt in analyzing and formulating the phenomenon of norms disappearance in a community. Our contribution in this paper is three-fold. Firstly, we define removal and disappearance terms. Secondly, we analyze the possible cases that lead to norms disappearance. Thirdly, we develop a framework for norms disappearance.

The next section dwells upon the related work on norms, normative systems, normative multi-agent systems and norms change. Section 3 contains terms definitions. Section 4 details out the theory of norms disappearance. Section 5 proposes the norms disappearance framework along with formalization. Section 6 will contain a discussion, and Section 7 concludes the paper.

2   Related Work

2.1   Norms, Normative Systems, and Normative Multi-Agent Systems

Norms generally are used to regulate the individuals interactions within a society [11]. Hollander and Wu [8] described norms as any behavioral rule that is valid by the majority of a population. According to them, norms can be socially enforced until the agents internalize them. after internalization, norms will spread either by active or passive transmission mechanisms.

Normative systems are the type of systems which is regulated by norms. The main goal of having normative systems is to minimize deviance by using social enforcement such as sanctions or rewards. Sanctions and rewards are also considered as norms about norms, which is termed as meta-norms[8].

Combining norms and multi-agent systems produce normative multi-agent systems, or so-called NorMAS. There are varieties of definitions for NorMAS presented by many researchers. All of these definitions are common in core idea, which is: "Norms are used to coordinate the interaction among agents" [17].

One of the definitions describing the mechanism of NorMAS: "A normative multi-agent system is a multi-agent system organized by means of mechanisms to represent, communicate, distribute, detect, create, modify, and enforce norms, and mechanisms to deliberate about norms and detect norm violation and fulfillment"[18].

Another definition of normative multi-agent system is presented by Hollander and Wu [8]. They defined NorMAS as a system that combines the concept of norms together with normative information scheme represented explicitly to provide a solution to problems related to open multi-agent systems. This is achieved by building a normative agents with the ability to create, modify, detect, transmit, and reason about norms.

2.2   Norms Change

The literature discusses norms change which is considered a part of norm evolution phase. However, the literature do not provide any formal study on norms removal or disappearance [8], [9].

Numerous researchers presented the norm change concepts in their work. For example, Bou et al. [19] and Campos et al. [20] investigated on norm change at run time in electronic institutions. Another example is the infrastructure presented by Artikis [21], which allows agents to modify a set of laws at runtime. Another work is presented by Oren et al. [22], who considered how norms are created, deleted, and modified by normative powers, which describes agents' ability to act in some ways.

The most recent work on norms change is presented by Dastani et al. [23]. They presented a generic programming structure to facilitate norm modification at runtime.

Norms might be changed either by external agent, or by the normative framework. This is done by representing the norm in the form of conditional, obligations, and prohibitions.

3   Terms Definition

Disappearance: According to Oxford dictionary, disappearance is defined as "an act of someone or something ceasing to be visible" or "an act or the fact of someone or something going missing". Also according to Merriam-Webster dictionary, disappearance is defined as "to stop being visible" or "to stop existing" or "to become lost"

Removal: According to Oxford dictionary, removal is defined as "the action of taking away or abolishing something unwanted". Also according to Merriam-Webster dictionary, removal is defined as "the act of moving or taking something away from a place" or "the act of making something go away so that it no longer exists".

The aim of defining disappearance and removal is to differentiate between two cases: Norm Disappearance and Norm Removal.

Norm disappearance means that the norm is not being practiced by any of the society members. They stopped practicing this norm because it is not beneficial for them anymore. Or they are suffering from negative consequences on practicing. The most important issue here is that the negative consequences of practicing a norm is related to an individual himself only. Negative consequences are like the high cost of norm practicing. While norms removal means that the members of certain community are affected by the negative consequences of practicing specific norm, consequently they agree to remove it permanently from the society. In this case, society members should be convinced that negative consequences of practicing this norm is affecting them all.

An example of norm disappearance is using landlines in communication. This example is limited to communication between normal people only (individual usage), not communication between people and organizations. Using landline to communicate with other people was the norm for a long time before inventing mobile phones. When mobile phones became common among people, they started to use it instead of landlines. The new norm of using mobile phones has emerged due to its high benefits, which is clearly much more than landlines. Accordingly, the norm of using landlines started to derogate and eventually disappeared. Norm derogation and eventually disappearance happened here because a new more efficient norm emerged, although the old norm was not of negative effects on society. Nowadays people ask for mobile phone number by default, they rarely or never ask for landline number.

4   A Proposed Theory of Norms Disappearance

The possible cases which might lead to norms disappearance from a community are reviewed and we observed five cases as follows:

Norms Conflict: Norms conflict happens when a new norm emerges while agents are adopting and practicing a similar effect norm (as shown in Fig. 1).

In such case, agents must decide whether to adopt the new norm and remove the old one or just ignore the new one. However, their decisions depend on the norm's enforcement degree as well as its cost compared with the old one [9]. When the enforcement degree is high, e.g., from father to son or from leader to follower, it results in the tendency to adopt the new norm and ignore the old norm [12].

Another case is when the associated cost of the new norm is less than the cost of the old norm [9]. In this case, agents may adopt the new norm and remove the old one. However, when the enforcement is low or the cost of the new norm is high, agents would ignore the new norm and keep practicing the old one. For example, reading newspaper was the norm of many communities. But when a new norm emerged, causing people to read news online, most of them adopted the new norm and ignored the old norm. The reason is the assimilation cost of the old norm is high (costly and inconvenient) compared with the new norm (free and easy).

Norm's Trust Decay: In this case, as shown in Fig. 2, a norm disappears or is ignored because the trust between agents has decayed. Andrighetto et al. [11] provided an example of a bus stop scenario of a particular community, in which when people arrive at the bus stop, the norm is, they do not form a queue but instead they sit on a bench and memorize who came earlier than them. In such situation, because people trust the norm of others, they adopt the norm. If it happens that the norm is violated by some or more people, then the norm's trust gradually decay, ignored, forgotten and eventually disappeared.

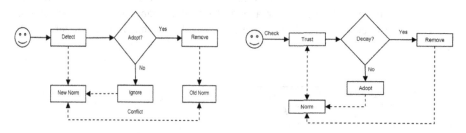

Fig. 1. A Norms Conflict Case   **Fig. 2.** Norm's trust decay

High Assimilation Cost of Low Norm's Yields: Another case for norms disappearance is when the associated cost of a norm is high and assimilating it requires great effort although it has low yields. Such norms often belong to traditions. For example, in a community's tradition of celebrating one's success in some undertaking by having a party and inviting relatives and friends, such norm may disappear through time due to high assimilation cost that does not commensurate with its outcomes or yields. Fig. 3 presents the case.

Norm's Yields Decay: When the yields of a particular norm decay in a community, it may subject the norm for disappearance over a period of time. For example, the norm of selling newspapers by peddlers decayed because the associated yields have decayed (very few people read newspapers). Fig. 4 explains the norm's yields decay case.

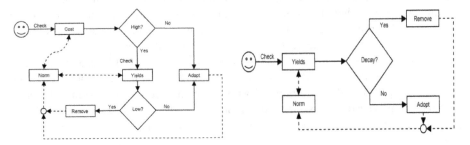

Fig. 3. The assimilation cost of low norm's yields

Fig. 4. Norm's yields decay

Violation Cost Decay: The violation cost of a particular norm is considered high when penalties are applied strictly on violators. Any reduction in this cost affects negatively on the norm's assimilation which may lead to the total disappearance of the norm. For example, sanction is applied strictly to any member who does not wear formal dress during working hours. However, any slack or decay in applying sanctions relieves the agents from strictly complying with and violate the norm by wearing informal dresses. Over a period of time, many other members violate the norm with consequential disappearance of the norm. Fig. 5 illustrates this case.

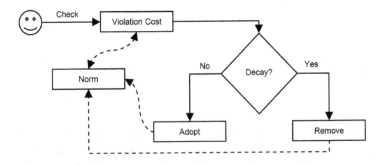

Fig. 5. Violation Cost Case

5   A Conceptual Framework for Norms Disappearance

In the first section we will present the conceptual design of this framework. The second section will be about framework formalization.

5.1   Conceptual Design

In this framework, we introduce the concept of norms disappearance from a normative multi-agent community, in which agents begin violating some norms and subsequently removing the norms from their cognitive structures. Consequently, norms disappearance in a community undergoes two stages which are, (i) agents start violating some particular norms, and subsequently (ii) removing the norms from their cognitive structures. However, for the first stage to occur, one or more of the possible cases for norms disappearance occur. For example, norm's yields decayed or a new emerged norm conflicts with an old one producing better yields.

As shown in Fig. 6, an agent first observes the possible cases for norms disappearance (1). Any case occurs on a specific norm (2), the agent begins to violate the norm (3a), or, the agent maintains the norm if none of the cases occurred (3b). After the violation stage, norm removal stage starts (4). The norm is then removed from agent's cognitive structure (5), which updates the agent behavior (6).

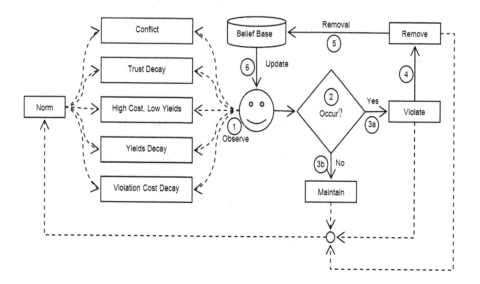

Fig. 6. A Framework for Norms Disappearance

5.2   Formalization

Having presented the framework, we introduce here the proposed approach for norms disappearance. As we mentioned earlier, there are five cases to norms disappearance, and from these cases, we conceive four parameters that affects agents' decision to remove the norms. The parameters are:(1) Assimilation cost, δ; which represents the cost of assimilating the norm, (2) Yields value, γ; which represents the value of the benefit gained from adopting the norm, (3) Trust value, σ; which represents the norm trust level, and (4) Violation rate, λ; which indicates to enforcement level applied on the norm. Consequently, for each norm there exist these four parameters. Any negative change that occurs in any of these parameters or a new emerged norm conflicts with the existing one and possesses better parameters' values subjects the norm to unstable situation due to violation by member agents and finally disappears from the community.

- **Case 1:** If N_C is the norms conflict; α is an agent; n_1 an old norm; and n_2 is a new norm, then,

$$N_C: remove\ (\alpha, n_1) \wedge adopt\ (\alpha, n_2) \Leftrightarrow (\delta n_1 > \delta n_2) \vee (\gamma n_2 > \gamma n_1) \vee (\sigma n_2 > \sigma n_1) \quad (1)$$

This means that the agent,α removes an old norm and adopts the new one if and only if its assimilation cost is higher than the new one; its trust value is less than the new one; or its yields is less than the new one. However, there is another situation, i.e., when the new one has higher trust but lower yields. We leave this issue for our future work.

- **Case 2:** If N_T is the norm's trust decay, and n is an existing norm, then,

$$N_T: decay\ (\sigma(n)) \rightarrow remove\ (\alpha, n) \quad (2)$$

Which means that the norm, n, is removed because its trust value, σ, decayed.

- **Case 3:** If N_S is the assimilation cost of low norm's yields, then,

$$N_S: high\big(\delta(n)\big) \wedge low(\gamma(n)) \rightarrow remove\ (\alpha, n) \quad (3)$$

This means that the norm is removed by the agent because its cost, δ, is high and its yields, γ, is low.

- **Case 4:** If N_Y is the norm's yields decay, then,

$$N_Y: decay\ (\gamma(n)) \rightarrow remove\ (\alpha, n) \quad (4)$$

Which means that the agent removes the norm, n, because its yields, γ, decayed.

- **Case 5:** If N_V is the norm's violation cost decay, then,

$$N_V: decay\ (\lambda(n)) \rightarrow remove\ (\alpha, n) \quad (5)$$

Which means that the agent removes the norm, n, because its violation cost, λ, decayed.

In the above cases, the decision to remove the norms is a consequence of its output based on the input. We shall explore the decision issue further in our next publication.

6   Discussion

This paper is considered to be the first attempt in clarifying norms removal and disappearance terms in normative multi-agent system. Besides it contains a conceptual framework of norms disappearance. Norms removal and disappearance are considered to be an important processes in norms lifecycle according to numerous researchers. In this paper we started with clarifying the terms removal and disappearance since there is an ambiguity in using these terms in the literature. For most researchers both terms refer to the same process. They define removal as the process of removing a norm from agent cognitive structure. But our review in social science returned a different aspect for norms removal. Norms removal is the process of eliminating a negative norm from a society using some kind of beliefs change and removal enforcement. Based on that, the process of removal is different from the process of disappearance. Norms removal is related to a shared agreement between individuals to stop adopting specific norm and remove it from the society. While norms disappearance happens based on individual decision to violate or abandon the norm, and when society majority is not adopting the norm it will disappear. Our next publication will contain more clarification on this matter.

7   Conclusion and Future Work

In this paper, we present our research on norms disappearance in an open multi-agent community where norms appear and disappear. We observe that there are two stages for norms disappearance within a community, which begins when agents violate a norm and subsequently remove it from their cognitive structures. The occurrence of norms disappearance is influenced by five cases, which are norms conflict; norm's trust decay; high assimilation cost of low norm's yields; norm's yields decay, and violation cost decay.

From these cases, we discovered four parameters that affect the decision of an agent to remove a norm from its cognitive structure which are, assimilation cost; yields value; trust value; and violation rate. However, any negative changes occur on these parameters prompt the agent to think about the norm and the possibility of removing it. In our future work, we shall formulate a theory for the proposed framework and develop a method to calculate the values of the parameters.

Acknowledgment. This work is supported by the Exploratory Research Grant Scheme (ERGS) by the Ministry of Education Malaysia under the Grant Ref. No. ERGS/1/2013/ICT01/UNITEN/02/02.

References

1. Elster, J. (ed.): The cement of society: A survey of social order. Cambridge University Press (1989)
2. Ostrom, E.: Collective Action and the Evolution of Social Norms. The Journal of Economic Perspectives 14(3), 137–158 (2000)
3. Ahmad, A., Ahmed, M., Yusoff, M.Z.M., Ahmad, M.S., Mustapha, A.: Resolving Conflicts Between Personal and Normative Goals in Normative Agent Systems. In: The Seventh International Conference on IT in Asia 2011 (CITA 2011), Kuching, Sarawak (2011)
4. Sadri, F., Stathis, K., Toni, F.: Normative KGP agents. Computational and Mathematical Organization Theory 12(2), 101–126 (2006)
5. Ahmad, A.: An Agent-Based Framework Incorporating Rules, Norms and Emotions (OP-RND-E). PhD Thesis, Universiti Tenaga Nasional (2012)
6. Castelfranchi, C., Conte, R., Paolucci, M.: Normative Reputation and the Cost of Compliance. Journal of Artificial Societies and Social Simulation 1(3), 3 (1998)
7. Finnemore, M., Sikkink, K.: International Norm Dynamics and Political Change. International Organization 52(4), 887–917 (1998)
8. Hollander, C., Wu, A.: The Current State of Normative Agent-Based Systems. Journal of Artificial Societies and Social Simulation 14(2), 6 (2011a)
9. Mahmoud, M.: Norms Detection and Assimilation in Open Normative Multi-agent Communities. PhD Thesis, Universiti Tenaga Nasional (2013)
10. Savarimuthu, B.T.R.: Mechanisms for norm emergence and norm identification in multi-agent societies (Thesis, Doctor of Philosophy). University of Otago (2011)
11. Andrighetto, G., Governatori, G., Noriega, P., van der Torre, L.W.: Normative Multi-Agent Systems (2013)
12. Hollander, C.D., Wu, A.S.: Using the process of norm emergence to model consensus formation. In: Fifth IEEE International Conference on Self-Adaptive and Self-Organizing Systems (SASO), pp. 148–157 (October 2011b)
13. Grossi, D., Gabbay, D., van der Torre, L.: The Norm Implementation Problem in Normative Multi-Agent Systems. In: Specification and Verification of Multi-Agent Systems, pp. 195–224. Springer US (2010)
14. Noriega, P., Chopra, A.K., Fornara, N., Lopes Cardoso, H., Singh, M.: Regulated MAS: Social Perspective. In: Normative Multi-Agent Systems, pp. 93–134 (2013)
15. Balke, T., da Costa Pereira, C., Dignum, F., Lorini, E., Rotolo, A., Vasconcelos, W., Villata, S.: Norms in MAS: Definitions and Related Concepts. In: Normative Multi-Agent Systems, pp. 1–31 (2013)
16. Governatori, G., Rotolo, A.: Norm compliance in business process modeling. In: Dean, M., Hall, J., Rotolo, A., Tabet, S. (eds.) RuleML 2010. LNCS, vol. 6403, pp. 194–209. Springer, Heidelberg (2010)
17. Boella, G., Hulstijn, J., van der Torre, L.: Interaction in normative multi-agent systems. Electronic Notes in Theoretical Computer Science 141(5), 135–162 (2005)
18. Boella, G., Van Der Torre, L., Verhagen, H.: Introduction to the special issue on normative multiagent systems. Autonomous Agents and Multi-Agent Systems 17(1), 1–10 (2008)
19. Bou, E., López-Sánchez, M., Rodríguez-Aguilar, J.A.: Adaptation of autonomic electronic institutions through norms and institutional agents. In: O'Hare, G.M.P., Ricci, A., O'Grady, M.J., Dikenelli, O. (eds.) ESAW 2006. LNCS (LNAI), vol. 4457, pp. 300–319. Springer, Heidelberg (2007)

20. Campos, J., López-Sánchez, M., Rodríguez-Aguilar, J.A., Esteva, M.: Formalising situatedness and adaptation in electronic institutions. In: Hübner, J.F., Matson, E., Boissier, O., Dignum, V. (eds.) COIN@AAMAS 2008. LNCS, vol. 5428, pp. 126–139. Springer, Heidelberg (2009)
21. Artikis, A.: Dynamic protocols for open agent systems. In: Proceedings of the 8th International Conference on Autonomous Agents and Multiagent Systems (AAMAS), pp. 97–104 (2009)
22. Oren, N., Luck, M., Miles, S.: A model of normative power. In: Proceedings of the 9th International Conference on Autonomous Agents and Multiagent Systems (AAMAS), pp. 815–822 (2010)
23. Dastani, M., Meyer, J.-J., Tinnemeier, N.: Programming norm change. Journal of Applied Non-classical Logics (2012) (published online)

Teaching Simulation on Collaborative Learning by the Complex Doubly Structural Network

Setsuya Kurahashi[1], Keisuke Kuniyoshi[1], and Takao Terano[2]

[1] University of Tsukuba, Graduate School of Business Sciences,
3-29-1 Otsuka, Bunkyo, Tokyo, Japan
kurahashi.setsuya.gf@u.tsukuba.ac.jp, kuniyoshi@mail.benesse.co.jp
[2] Tokyo Institute of Technology, Computational Intelligence and Systems Science,
4259-J2-52 Nagatsuda-Cho, Midori-ku, Yokohama, Japan
terano@dis.titech.ac.jp

Abstract. In this research, a teaching simulation model is built where the understanding status, knowledge structure, and collaborative effect of each learner are integrated by using a doubly structural network model. The purpose of the model is to analyse the actual conditions of understanding of learners regarding instructions given in classrooms. The influence of teaching strategies on learning effects is analysed in the model. Moreover, the influence of the seating arrangement of learners on collaborative learning effects is discussed. As a result of the simulation, the following points were found: (1) the learning effects depend on the difference in teaching strategies, and (2) a teaching strategy where learning skills, material structure, and collaborative learning are integrated on a doubly structural network model is the most effective.

Keywords: teaching strategy, collaborative learning, classroom, item response theory (IRT), social simulation, a doubly structural network model.

1 Introduction

In education, it is important to understand the status of the understanding of each learner and design instruction content according to their understanding status. Digitalization of learning environment, called e-learning, has enabled the accumulation of records containing a vast amount of information concerning the learning history of students. Many technologies to gain an understanding status of each learner sequentially have been produced [1]. Additionally, there exist relationships between knowledge and the content to be instructed, and it is important to consider the structural dependency relationship when teaching is done. The effectiveness of the collaborative effect among learners has also been clarified.

In the research field of network models, recently, a new model building method, referred to as a complex doubly structured network model, has been proposed [5–7]. This method utilizes two different network models, the internal network

G. Jezic et al. (eds.), *Agent and Multi-Agent Systems: Technologies and Applications*,
Advances in Intelligent Systems and Computing 296,
DOI: 10.1007/978-3-319-07650-8_25, © Springer International Publishing Switzerland 2014

model and the social network. Through this research, we designed a new teaching simulation method (Fig. 1) built by using the complex doubly structured network model consisting of an internal network model, that considers the status of understanding, knowledge structure of the learner, and the social network, that considers the learning space. With this method, we considered a system that supports teaching activities of teachers. By using this complex doubly structured network model in this research, we tried to integrate the understanding status, knowledge structure, and collaborative effect of each learner in order to simulate the actual conditions of the learners' understanding for instructions given in a classroom. Moreover, we set and examined the issues described below by applying the simulation method. (1) What kind of influence could teaching strategies have on learning effects? (Experiment 1) (2) What kind of influence could the seating arrangement of learners have on collaborative learning effects? (Experiment 2).

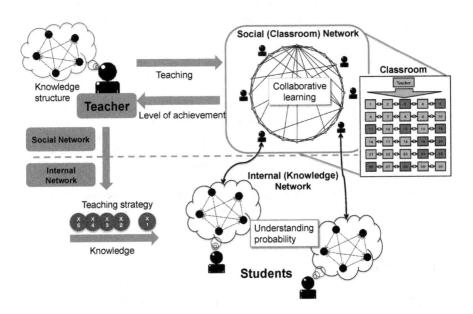

Fig. 1. Overview of a teaching simulation method built by using the complex doubly structured network model

2 Related Work

Learning frameworks have been discussed in a form of a study about learning theory[2][3][9]. Among these studies, learning theories have been transformed as follows:

- Behaviourism: A learning theory based on the concept that behaviour can be explained with a combination of stimulus, which can be observed objectively, and reaction.
- Cognitivism: A learning theory that likens humankind's cognitive processes as an information processing system in order to express the cognitive processes in the form of computer programs.
- Constructivism: A learning theory that considers learning development of a cognitive framework (schema) where each individual learner builds up knowledge.
- Social constructivism: A learning theory that focuses on interactions with others by considering that learning is cultivated within a community amid social activities.

The purpose of this research is to simulate the actual conditions of learners' understanding of instructions given in a classroom, while we attempted to quantify interactions among teachers and learners in the classroom. In this research, we had the concept that individual learners have network relationships, where knowledge nodes are linked according to relevancy, and each of the learners understands what they learn while influenced by the relevancy of their own knowledge networks every time they learn in a classroom. Based on this concept, we tried to quantify the understanding levels. As mentioned above, behaviourism and cognitivism underlies the sense of knowledge where knowledge is considered as universally true, and academic capability is evaluated based on objective evaluation, which is the status of correct answers to questions. Based on this concept and while being aware of the context of social constructivism for interactions and being aware of the context of constructivism for knowledge networks, we made an attempt to build a simulation model.

By gaining the status of understanding of learners, namely academic evaluation has been studied in a form of test theory [8]. In test theory, evaluation methods that use Item Response Theory (IRT) have been proposed, which are moving closer to actual practical use. Based on a response to an item, IRT expresses the learner's ability, the difficulty level of each item, and discrimination as functions. According to the number of parameters, there are calculation models such as the one-parameter logistic (1PL) model, two-parameter logistic (2PL) model, and three-parameter logistic (3PL) model. The 2PL model can be expressed as formula 1. Each term in formula 1 indicates the following points; θ = ability, a_j = discrimination, b_j = difficulty, P_j = probability of a correct response to an item j.

$$P_j(\theta|a_j, b_j) = \frac{1}{1 + exp(-a_j(\theta - b_j))} \tag{1}$$

Discrimination and difficulty are referred to as the item parameter, while the learner's ability is referred to as the ability parameter. Methods for estimating the item parameter and the ability parameter have been studied [9–12].

Meanwhile, methods for quantifying the dependency relationships between events have been studied as probabilistic reasoning, where a logical framework called Bayesian approach was designed [13–16]. The Bayesian network applies to this logical framework, and this network is a calculation model that handles uncertain events. As a study where the Bayesian-approached framework is applied, there is graphical test theory[17, 18]. This theory expresses a learning material structure graph by considering learning items as nodes and the dependency relationship between learning items as directed links.

Regarding a network model, there is agent-based modelling [19, 20] as a framework to integrate multi-layered components, and studies incorporating network models in modelling have advanced. In these studies, a complex doubly structured model [5, 6] that conducts simulation based on a society network and an internal network has been proposed. As an internal network, the complex doubly structured network model expresses knowledge and concept structured using a network structure.

Therefore, the attempt of this research is to clarify the actual conditions of understanding of learning in a classroom. As we have seen, related studies are very effective; however, applying only the methods proposed by these previous studies are insufficient and still has issues to be embodied in our attempt. In this research, we tried to bring about a solution for these issues by using the complex doubly structured network model and incorporating multi-layered IRT and Bayesian networks.

Specifically, the understanding probability of knowledge according to the academic capability of each learner was calculated with IRT, and the relationship of knowledge was estimated as a learning material structure by the Bayesian network. In addition, the relationship between knowledge nodes was linked with conditional probability, and then both relationships were modelled in a form of an internal network. Moreover, the relationship between teachers and learners in the classroom was also modelled as a society network. We also proposed a new teaching simulation method that interfaces these network models as a complex doubly structured network model.

3   In-class Learning Process Simulation

In this research, we tried to build a simulation with a class consisting of 30 learners, where it was assumed that five instructions, from X1 to X5, are used when teaching them. This simulation was to estimate what material should be taught, in what order and how many times, until all learners in the classroom could give the correct answer. In this simulation, we used two criteria, the attainment degree and the average time of teaching. The attainment degree indicates the proportion of correct answer given so that the status where all learners give the correct answer reached 1. The average time for teaching indicates the time until the attainment degree has reached 1, which averages 10 simulation sessions. To build the teaching simulation, we used correct answer history data for model estimations, correct answer data in the class, and seating data. Correct answer history data for model estimation has two values, correct/incorrect answers, of all 300 learners for five questions that correspond to the instructions taught from

X1 to X5. The history data was gathered from an online learning system where primary school children study arithmetic in Japan.

3.1   Definition of the Internal Network

The internal network is composed with multi-layers combined the understanding probability model of knowledge according to the academic capability of each learner and the learning material structure model. When certain knowledge is taught, based on the understanding probability model, the understanding probability according to the academic capability of each learner is calculated. As for knowledge items, the understanding probability propagates along with the material structure model. In this way, the internal network is defined.

Understanding Probability Model. When it comes to the understanding probability model of all knowledge that corresponds to the academic capability of each learner, the item parameter is estimated by conducting the marginal maximum likelihood estimate based on the quasi-Newton's method and the EM algorithm. The ability parameter is estimated by using the experience Bayesian method. By using these estimated values, the understanding probability model is built. Specifically, this estimation is done by using the correct answer history data for model estimation and the ltm package on software R [21]. The result of this estimation is quantified as the form of Ability. The estimated Ability parameter (item characteristic curve), shown in Fig. 2, is set according to the knowledge, and the understanding status for all knowledge of each learner at the point in time before teaching, in order to estimate the understanding probability of knowledge of each learner.

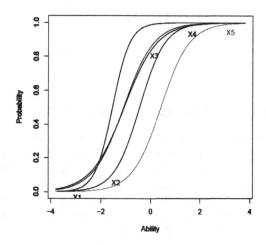

Fig. 2. The estimated item characteristic curves of X1-X5 (Ability)

The Course Material Structure Model. The material structure model was built by utilizing the structure estimation on the Bayesian network. As for model estimation, the correct answer history data for model estimation was used. The result was estimated with the greedy method on the package deal of software R. This was estimated as formula 2.

$$P(X1, X2, X3, X4, X5)$$
$$= P(X1)P(X3|X1)P(X4|X1, X3)P(X2|X1, X3, X4)P(X5|X2, X3, X4) \quad (2)$$

The conditional probability was calculated with the bnlearn package of software R by using the model of formula 2. Fig. 3 shows formula 2 in a diagram with the conditional probability calculated.

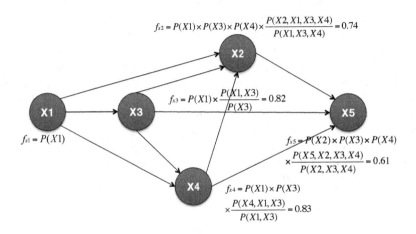

Fig. 3. The bayesian network of the knowledge structure with the conditional probability

3.2   Definition of the Teaching Simulation Model

When it comes to the classroom network, in order to build a model, we assumed an all-together (brick-and-mortar) classroom lecture consisting of one teacher and 30 learners, where collaborative learning would be done between each of learners sitting left to right. Learners were allocated according to seating data. If it was found according to correct answer history data that either those learners on the left or those on the right understood the targeted knowledge taught, he/she should conduct collaborative learning when the teacher teaches that knowledge so that the other learners could also understand the knowledge taught.

Based on the complex doubly structured network model consisting of an internal network and a social network, this simulation estimates the progress of understanding status of the learner when teaching is done by following the procedures described below.

1. Select knowledge to teach, Xn, and move to 2.
2. Select each learner and move to 3. If there is no learner to select, move back to 1.
3. If the learner's Xn is 1, move back to 2. If 0, move to 4.
4. Based on the social network, if a collaborative relationship is generated, it is expressed as $P(Xn) = 1$, and move to 6. If not, move to 5.
5. Based on the internal network, the learner's Ability is estimated and assigned to formula 1. The probability is calculated, expressed as $P(Xn) = 1 * Probability$, and then move to 6.
6. Based on the internal network, select subsequent knowledge connected Xn. According to Xn, the probability is calculated as multiplication of the probability of all precedent knowledge connected and the conditional probability.
7. Based on the probability calculated, the value is conversed into two values 1 and 0.

4   Experimental Results and Discussion

4.1   Experiment 1: Evaluation of Teaching Strategies

In the experiment, we tried to discuss the issue of what kind of influence teaching strategies could have on learning effects. In this experiment, we move our discussion forward by applying the following five teaching strategies for the class pattern 1 in which students are seated randomly, and then by comparing the average time of teaching sessions and the attainment degrees.

TS 1. Teaching along with the complex doubly structured network method
TS 2. Teaching by selecting items to teach in a random manner
TS 3. Teaching strategy 3: Teaching an item where many learners gave wrong answers
TS 4. Teaching by considering the addition of average correct answers per model question to strategy 3
TS 5. Teaching by moving to next item when all learners understood an item by order of the highest correct answer rate according to each model question

As the result of conducting 10 simulation sessions, the average teaching time is shown in Table 1. This result confirmed that learning effects depend on teaching strategies. Fig. 4 shows the transition of attainment degrees for teaching times. When observed from the viewpoint of the average teaching time, in both teaching strategy 1 and 5, the teaching time was less than 10 times. We can consider that these strategies had higher learning effects. From the viewpoint of the attainment degree, teaching strategy 1 has a tendency where the initial growth was higher than the other strategies. For example, when the attainment degrees after the

Table 1. The average teaching time

Teaching Strategy(TS)	Non-collaborative Learning	Collaborative Learning
TS 1	22.5	8.20
TS 2	42.9	17.7
TS 3	32.3	11.8
TS 4	24.0	10.5
TS 5	23.4	9.30

fifth teaching session are compared with the degree of each strategy, teaching strategy 1 was 0.88, 2 was 0.68, 3 was 0.64, 4 was 0.77, and 5 was 0.70. Therefore, this shows that the teaching strategy 1 had the highest attainment degree.

In any of teaching strategies of 1, 2, 3, 4, or 5, the attainment degree did not reach 1. For this reason, in this class, we can see that any of these teaching strategies should be insufficient to have all learners acquire all knowledge items in five teaching times. In other words, if teaching is done one time for every knowledge item, some learners could fall behind in the learning progress. This shows that review or makeup classes are necessary. Next we considered how the average teaching times of teaching strategies of 1, 2, 3, 4, and 5 and the attainment degrees in the teaching time transitioned where there was no collaborative effect. The results are shown in Table 1 and Fig. 4. Where there was no collaborative effect, teaching strategy 1 has the lowest average teaching time. The teaching times increased in the order of teaching strategy 5, 4, 3, and 2. Similar tendencies were shown in the case where collaborative effects were observed. However, some of teaching strategies had more than 20 teaching times, which shows that necessary teaching times significantly increased. This result shows that reviews should be conducted repeatedly in order to facilitate the anchoring of the knowledge in the class where teaching is done without collaborative effects between learners.

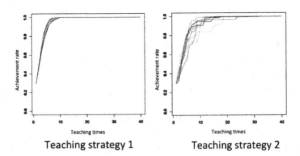

Teaching strategy 1   Teaching strategy 2

Fig. 4. The transition of attainment degrees for teaching times of teaching strategy 1 and 2 (y-axis:achievement rate, x-axis:teaching times)

4.2 Experiment 2: Evaluation of Seat Configuration

The second experiment considered what kind of influence the seating arrangement of learners could have on learning effects. In this second experiment, preparing two different environments for the social network, concentrated arrangement and dispersed arrangement, we conducted 10 simulation sessions by using teaching strategy 1. Afterward, we compared the results. The concentrated arrangement is a model where learners with high academic capability are gathered in one place. The dispersed arrangement is a model where learners with high academic capability next to those learners with low academic capability. About class pattern 1 which was used in the experiment 1, we created particular situations with the concentrated and dispersed arrangements by changing the seating arrangement of learners as shown in Fig. 5. We compared both situations for discussion. In this experiment, we estimated the academic capability of each learner by using IRT based on the correct answer history of the learner. According to the estimated value, we determined those learners with high academic capability. Determining those learners that have above a certain estimated value to be excellent learners, we structured the concentrated arrangement and the dispersed arrangement by changing the seating arrangement of those excellent learners. As for the average teaching times, the concentrated arrangement was 9.5 times, while the dispersed arrangement was 7.7 times. The average teaching time in the first experiment, before the arrangement was changed, was 8.2 times. While the average teaching times increased in the concentrated arrangement, it decreased in the dispersed arrangement. Through this result, we were able to confirm that learning effects vary by making changes in the seating arrangement for learners and the dispersed arrangement could enhance teaching effects.

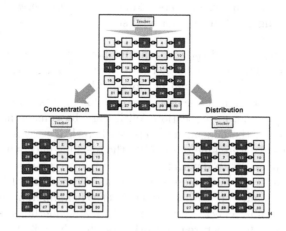

Fig. 5. Experiment 2: Evaluation of seat configuration

4.3   Discussion

We utilised the simulation for in-class learning processes considering academic capability, learning material structure, and collaborative relationship. In the first experiment, using five teaching strategies, we quantified the teaching procedure selected by each teaching strategy and the learning effect status, visualised them in chronological order, and compared the influence of each teaching strategy on learning effects. By so doing, we were able to evaluate the educational effects of each teaching strategy.

In the second experiment, we arranged learners using a concentration arrangement and a dispersed arrangement, quantified the learning effect status, and compared both arrangements. By so doing, we were able to evaluate the influence of seating arrangements on learning effects. Through these evaluations, in the first experiment, we confirmed that learning effects depend actually on teaching strategies, and teaching methods along the complex doubly structured network has a high learning effect. Additionally, we also clarified that if only one teaching session is done for one target knowledge item, some learners could fall behind the learning progress. In the second experiment, we confirmed that teaching would work more effectively where there is a dispersed seating arrangement, not a concentrated seating arrangement.

5   Conclusion

The purpose of this research was to clarify the actual conditions of understanding of teaching done in a classroom. As a means to do so, we proposed a simulation for in-class learning processes with consideration given to academic capability, learning material structure, and collaborative relationships. We built an internal network by estimating the understanding probability network by the use of IRT and estimating the learning material structure model with the use of the Bayesian network.

Furthermore, we modeled the relationships between the teacher and learners in a classroom as a form of a social network. By interfacing these network models as a complex doubly structured network model, we were able to model the actual condition of teaching done in a classroom for the first time, and produced a new simulation model for education. With that, by considering the learner's academic capability, the dependency between knowledge items, and collaborative relationships, we were able to quantify the teaching effects in the classroom and conduct simulations to determine effects. In this research, we were able to acquire three observations through experiments. What kind of influence could teaching strategies have on learning effects? (1) When different teaching strategies are used, learning effects vary, and (2) teaching methods along the lines of the complex doubly structured network procedure have high learning effects, (3) whereas, if teaching is done one time for one target knowledge item, some learners could fall behind in the learning progress.

References

1. Durlach, P.J., Lesgold, A.M. (eds.): Adaptive Technologies for Training and Education. Cambridge University Press, New York (2012)
2. Sawyer, K.: Introduction, The New Science of Learning. In: Sawyer, K. (ed.) The Cambridge Handbook of the Learning Sciences, pp. 1–18. Cambridge University Press (2006)
3. Baker, R.S., Yacef, K.: The State of Educational Data Mining in 2009, A Review and Future Visions. Journal of Educational Data Mining 1(1), 3–17 (2009)
4. Johnson, D.W., Johnson, R.T., Holubec, E.J.: Circles of Learning: Cooperation in the classroom, 5th edn. Interaction Book Company (2002)
5. Terano, T.: A Doubly Structural Network Model. The Operations Research Society of Japan 2(1), 57–69 (2008)
6. Kunnigami, M., Kobayashi, M., Yamadera, S., Terano, T.: A Doubly Structural Network Model and Analysis on Emergence of Money. Information Processing Society of Japan, Mathematical Modeling and Problem Solving 53(12), 661–666 (2009)
7. Kunnigami, M., Kobayashi, M., Yamadera, S., Yamada, T., Terano, T.: A Doubly Structural Network Model and Analysis on Emergence of Money. In: Takadama, K., Cioffi-Revilla, C., Defuant, G. (eds.) Simulating Interacting Agents and Social Phenomena, Agent-Based Social Systems, vol. 7, pp. 137–149. Springer, Tokyo (2010)
8. Birnbaum, A.: Some Latent Trait Models. In: Load, F.M., Novick, M.R. (eds.) Statistical Theories of Mental Test Sores, pp. 97–424. Addison-Wesley, Reading (1968)
9. Ueno, M., Shojima, K.: The New Trend of Learning Evaluation. Asakura Shoten, Tokyo (2010)
10. Toyoda, H.: Item Latent Theory (Beginners' Course). Asakura Shoten, Tokyo (2012)
11. Ueno, M.: An Extension of the IRT to a Network Model. Behaviormetrika 29(1), 59–79 (2002)
12. Wainer, H., Bradlow, E.T., Wang, X.: Testlet Response Theory and Its Appliations. Cambridge University Press, New York (2007)
13. Pearl, J.: Probabilistic Reasoning in Intelligent Systems. Morgan Kaufmann, San Francisco (1988)
14. Shigemasu, K., Motomura, Y., Ueno, M.: General Information of Bayesian Network. Baifukan, Tokyo (2006)
15. Jensen, F.V., Nielsen, T.D.: Bayesian Networks and Decision Graphs, 2nd edn. Springer, Berlin (2007)
16. Koller, D., Friedman, N.: Probabilistic Graphical Models. The MIT Press, MA (2009)
17. Ueno, M., Onishi, M., Shigemasu, H.: An Extension of the IRT to a Network Model. Information Processing Society of Japan (A), J77-A(10), 1398–1408 (1994)
18. Ueno, M.: The Graphical Test Theory from Bayesian Approach. Japan Society for Educational Technology 24(1), 35–52 (2000)
19. Epstein, J., Axtell, R.: Growing Artificial Societies. Brookings Institution Press, Washington, D.C. (1996)
20. Axelrod, R.: The Complexity of Cooperation. Princeton University Press, New Jersey (1999)
21. R Package, http://ran.r-projet.org/web/pakages/

An Ontology-Based Approach to Competency Modeling and Management in Learning Networks

Kalthoum Rezgui, Hédia Mhiri, and Khaled Ghédira

SOIE, University of Tunis
41 rue de la Liberté, Cité Bouchoucha 2000 Le Bardo, Tunis, Tunisie
{kalthoum.rezgui,hedia.mhiri,khaled.ghedira}@isg.rnu.tn

Abstract. With the emergence of the paradigm of Lifelong Learning and the proliferation of the terms "knowledge society", "citizen mobility", or "globalization", competency-based learning and training has known a growing interest in technology-enhanced learning as it provides important benefits for both individuals and organizations by supporting the transformation of learning outcomes into permanent and valuable knowledge assets. In this context, to promote the acquisition and continuous development of new competencies, learning networks have emerged that enables to support the provision of various lifelong learning opportunities. However, lifelong competency development still faces numerous challenges and research issues that need to be addressed, primarily the lack of consensus about a common representation of competencies and competency profiles. This paper analyzes different approaches reported in literature for competency modeling and proposes a competency ontology to formally describe competency-related characteristics of actors and learning resources in learning networks. The proposed ontology also aims to model aspects related to competency information management and tracking in order to support lifelong competency development in learning networks.

Keywords: competency, competency profile, lifelong competency development, learning network, ontology, technology-enhanced learning.

1 Introduction

In the field of education, the concept of competency-based learning was gradually diffused as an alternative approach to teaching and learning that explicitly articulate what students must be able to know and do upon graduation. Initially, it was adopted in the United States in the 1970s, then it was spread in Quebec in the late 1990s, in Belgium in 1993, Australia in 1995, and also in South America and Africa in the 2000s. In France, competencies have been introduced in national education policies in the late 1980s, with a report commissioned by Lionel Jospin, Minister of National Education. Their position was reaffirmed in 2005 with the publication of "*The common core of skills and knowledge*" being integrated into the European requirements. The most important advantages of the competency-based approach can be synthesized in: a better preparation of learners to life, a more active and more sustainable learning,

G. Jezic et al. (eds.), *Agent and Multi-Agent Systems: Technologies and Applications*,
Advances in Intelligent Systems and Computing 296,
DOI: 10.1007/978-3-319-07650-8_26, © Springer International Publishing Switzerland 2014

a less loaded curriculum, clearly defined objectives, a better equity of opportunities, a more moderate and more positive assessment, and a more suitable feedback mechanisms to the learner and the teacher. To concretize such benefits, education and training are expected to tune their curricula toward meeting certain competency requirements in order to appropriately equip their graduates for the labor market.

In the last ten years, several research projects in the field of technology-enhanced learning, such as PROLIX (http://www.prolixproject.org/) and TENCompetence (http://www.tencompetence.org/), have been launched that address new approaches to workplace-based learning and competency building. In these approaches, competencies were considered as the key bridge between the needs of the individual and those of the organization. At the same time, lifelong learning (Beer, 2007) has become an important topic of academic and policy debates on promoting new partnerships and curricula aiming at promoting acquisition and continuous development of new competencies. In short, lifelong learning can be considered as a continuous education process which covers any form of learning (i.e. formal, non-formal and informal) undertaken throughout the life and leading to fostering of continuous development and improvement of knowledge, skills and competencies. In this context, learning networks (Koper, 2005; Sloep, 2008) emerged as alternative and feasible integrated models that merge pedagogical, organizational, and technological perspectives to support the provision of a broad variety of learning opportunities. A learning network is defined as a set of actors, learning resources, and competency maps that are mutually interconnected and supported by information and communication technologies to facilitate lifelong competency development (Miao et al., 2009). However, lifelong competency development in learning networks still faces numerous challenges and research issues that need to be addressed, primarily the lack of consensus about a common representation of competencies and competency profiles. In fact, a common semantic competency model is required to ensure correct interpretation and to achieve interoperability.

Competency modeling can be considered as the activity of describing information about competency definitions and their relationships (Mansfield, 1998; Coi et al., 2007; Sampson and Fytros, 2008). The management of definitions of knowledge, skills and competencies facilitates their search and reuse in different learning scenarios (Najjar et al., 2010). For example, in the design of units of learning, they enable expressing the prerequisites and expected learning outcomes of joining them, and designing authentic assessment activities based on those expected outcomes. In other scenarios, annotating educational resources with competency-related metadata helps individuals and organizations in searching, retrieving, and sharing appropriate resources for competency development programmes (Ng and Hatala, 2007; Sampson, 2009). In the literature, a number of information models have been proposed to describe competencies. Recently, several approaches have discussed the use of ontologies as the enabling semantic infrastructure for competency modeling and management. This paper aims to analyze the different approaches reported in the literature for competency modeling and to propose a competency ontology that enables formal description of competency-related characteristics of actors and learning resources in learning networks. The remainder of this paper is organized as follows: A brief discussion of competency modeling approaches is presented in related work. Next, in Section 3, we present the developed competency ontology. Finally, conclusion is presented in Section 4.

2   Related Work

In the literature, several definitions of the competency concept have been proposed (e.g., Mansfield, 1998; Paquette, 2007; Sampson and Fytros, 2008; EQF, 2008). However, as is discussed in (Sampson and Fytros, 2008), a generic competency model should include these dimensions: *definition* of the competency; *proficiency level* that represent the actual competency level measured based on the individual's performance during the demonstration of competency by an action; and *context* in which the competency is applied or acquired, which may refer to a specific occupation, a set of a domain topics, or learner-related personal settings.

In order to facilitate the exchange of competency descriptions in a standard way between different system implementations, several open competency metadata standards and specifications have emerged (Rezgui et al., 2012), such as IMS RDCEO (Reusable Definition of Competency or Educational Objective) (IMS RDCEO, 2002), IEEE RCD (Reusable Competency Definition) (IEEE RCD, 2008), and HR-XML Competencies (HR-XML, 2006). The IEEE RCD data model is based on the IMS RDCEO specification, i.e., implementations that conform to the IMS specification also conform to the IEEE RCD standard. Mapping the elements of IEEE RCD to those of the HR-XML shows that both specifications provide the following elements: identification and title of the competency, description of the competency, and complete definition of the competency, while HR-XML adds elements for measurable evidences and weights. Besides, competency definitions provided by both specifications do not include information about context. To extend the scope of official competency scheme, some authors have proposed to add new elements primarily related to context and proficiency level (Coi et al., 2007; Sampson, 2009; Sitthisak et al., 2013a). Nevertheless, the fundamental problem with these competency models is that the semantics of the underlying description elements (i.e. title and description) are comprehensible only by humans. However, in order that the semantics of this information becomes accessible for machines, additional knowledge must be added which are drawn from controlled vocabularies or ontologies. Thus, semantic Web technologies and ontologies should be applied to enable different semantics to be used by Web applications and intelligent agents.

In recent years, a number of ontologies have been proposed for the description of competencies and their relationships (e.g., Sicilia, 2005; Schmidt and Kunzmann, 2006; Paquette, 2007; Ng and Hatala, 2007; Byoungchol et al., 2008; Braun et al., 2010, Rogushina and Gladun, 2012; Malzahn et al., 2013; Sitthisak et al., 2013b). However, the problem of unreliability of competency information gathered in learning networks has not been addressed by these ontologies. Nevertheless, solving this problem is crucial for an automated tracking and management of competency, and accordingly for an efficient lifelong competency development in learning networks

(Miao et al., 2009). The problem of unreliability of competency information has been discussed by (Ostyn, 2005; Miao et al., 2009).

In (Ostyn, 2005), the concept of distillation of competency has been proposed as an attempt for solving this problem. According to this approach, a confidence rating is used to qualify competency records and their associated evidence records. It enables to store information about the percentage of reliability assigned to a record which is fixed according to a predefined policy. The reliability of a competency record depends on two factors: the policy used to select the set of evidence records and the competency framework used to interpret those evidences. Following this approach, the competency record or the evidence record with the higher confidence level according to the policy will be chosen as the appropriate competency estimate of the actor. This method of competency estimation was criticized by (Miao et al., 2009) who argued that a predefined policy is not true for all cases and proposed an information fusion approach to solve the problem of unreliability of competency information in learning networks. According to this approach, fusing relevant competency records which are associated with the same actor on the same competency produce an estimate of the current competency state of the actor that is more credible and trustworthy than each single competency record. In our work, we combined the best features of both of these approaches to model aspects related to competency information management and tracking in learning networks.

3   Presentation of the Proposed Competency Ontology

In previous sections, we presented and analyzed different approaches to competency modeling in technology-enhanced competency-based learning. In this section, we present a competency ontology that combines that contributes to overcoming key issues in competency modeling and management. The proposed competency ontology defines a number of concepts to formalize the terms *competency* and *competency profile*. In real-life applications, competency profiles represent the most visible aspect of competency modeling. Actually, they support different tasks, such as creating personal competency profiles which highlight the abilities of an individual, creating job/project competency profiles for hiring or selecting candidates for a particular project, and creating prerequisite/objective competency profiles of available courses and training programs.

Our approach to competency modeling is based on the definition proposed by (Paquette, 2007). For (Paquette, 2007), a competency is the statement of a relationship among a generic skill applied to knowledge at a certain level of performance. For example, "*Apply the process of construction of a use case diagram, without help for a fairly complex system specification*" is a competency statement in which *apply* denotes the generic skill that a learner has to apply to a knowledge element (*the process of construction of a use case diagram*) according to particular performance indicators described in the rest of the competency statement. We chose the definition of Paquette because it seems richer than the other definitions. Furthermore, it is resulting from the study of a set of works made in education (Hotte et al., 2007).

The proposed competency ontology also reuses some classes and properties from well-developed Semantic Web vocabularies for content annotation (e.g., Dublin Core (DC, 2003)) and taxonomy representation (e.g., SKOS Core ontology (Miles and Bechhofer, 2009)) which makes it semantic web compliant. The following figure (Fig.2.[1]) illustrates a graphical representation of the Competency Ontology:

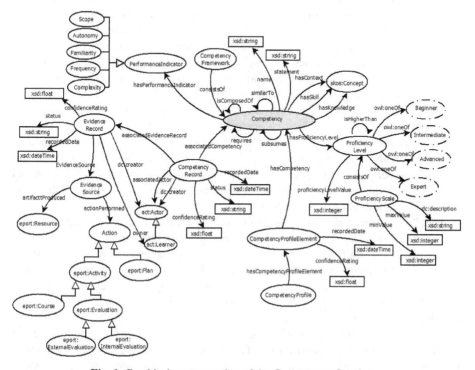

Fig. 1. Graphical representation of the Competency Ontology

In the center of Fig. 2., we find the *Competency* class that formally describes the concept of competency. Each competency belongs to a competency framework formally represented by the *CompetencyFramework* class, and is described by a statement expressed in natural language and by the quadruplet (skill, knowledge, proficiency level, and context). Therefore, the Competency class is related to:

- A skill selected from a learning-domain skills taxonomy/ontology,
- A knowledge (concept) drawn from a learning-domain ontology that specifies a consensual view of a subject matter,

---

[1] The prefix *skos* stands for the namespace of the SKOS Core ontology, the prefix *dc* denotes the namespace of the Dublin Core metadata schema, *act* denotes the namespace of the actor ontology, *eport* denotes the namespace of the e-Portfolio ontology, *owl* stands for the OWL namespace, and *xsd* denotes the XML schema namespaces. Classes and properties without a prefix originate from the Competency Ontology.

- A context selected from a domain ontology, and
- The *ProficiencyLevel* class that enables describing the actual proficiency level.

In particular, elements of taxonomies and domain ontologies (i.e., knowledge entities, skills, contexts) are defined as instances of the **Concept** class from the SKOS Core Vocabulary. Besides, SKOS semantic relations (e.g., broader, related) are used to formally describe hierarchical and associative links between these elements.

The *ProficiencyLevel* class is specified as an enumeration through the *owl:oneOf* element that contains the following instances: *Beginner*, *Intermediate*, *Advanced*, and *Expert*. Furthermore, it includes the *proficiencyLevelValue* property that indicates the rating value for the proficiency level and the *belongsTo* property that refers to the rating scale used for representing proficiency levels.

The *ProficiencyScale* concept subsumes measurement scales used to evaluate the proficiency level of the acquired competency. For example, the scales of Lanz and Friedrich (Lantz and Friedrich, 2003) and Likert (Likert, 1932) can be defined as subclasses of the *ProficiencyScale* class. A *ProficiencyScale* is described by the *description* property that indicates the name of the rating scale or a description of it and the *minValue* and *maxValue* properties that indicate the minimum value and the maximum value of the rating scale, respectively.

The *PerformanceIndicator* class allow precising the skill when it is applied on a knowledge element by specifying particular performance conditions.

The proposed ontology also enables formal representation of diverse kinds of dependency relationships between competencies. For example, the expression stating that a competency *C1* requires the acquisition of a competency *C2* can be modeled using the requires property between *C1* and *C2* otherwise the *isPrerequisiteFor* property in the opposite case. For instance, the competency *"developing dynamic websites"* requires the competency of using a scripting language and the competency of mastering a client server architecture. To represent the fact that one competency is composed of other simpler competencies, the *isComposedOf* property can be used. The *similarTo* property can be employed to state that two competencies *C1* and *C2* are strongly correlated (i.e. they have a strong relationship with each other) or resemble each other. Furthermore, the proposed ontology makes it possible to organize competencies in a hierarchical structure using the subsumption/generalization relationship formally represented by the *subsumes* property. One example for competency generalization could be a competency *"Object-oriented programming"* and a subcompetency *"Programming in Java"*. Similarly, the proposed ontology enables formal representation of subsumption relationships between the elements of the proficiency scale via the *isHigherThan* property.

In addition, for each learner the proposed ontology makes it possible to keep track of competency estimates (*CompetencyRecord*) and their associated *EvidenceRecords*. A *CompetencyRecord* is the result of the distillation process which consists in interpreting a set of evidence records in order to make a judgment stating that a learner has a given proficiency level for a particular competency. In particular, it includes the date and time when the competency record was created (*recordedDate*), the recorded *status* (e.g., valid or expired), and the recorded *ConfidenceRating*. Also, it refers to:

- The acquired competency together with the reached proficiency level,
- The associated learner,
- The associated *EvidenceRecords*, and
- The actor responsible for its creation.

The *EvidenceRecord* class subsumes information objects that serve as kinds of evidence of the competency acquisition by the learner. An *EvidenceRecord* concerns a given learner and includes:

- The date and time when it was created (*recordedDate*),
- The recorded *status* and *ConfidenceRating*,
- A reference to the actor responsible for its creation which may be the learner himself, and
- A list of references to the sources of evidence that are formally modeled by the *EvidenceSource* class. The sources of an evidence record are information collected in a e-Portfolio, such as:
- A description of performance associated with a personal development plan (*eport:Plan*), a unit of learning (*eport:Course*), or a single activity (*eport:Activity*),
- A product (*eport:Resource*), such as reports, developed computer programs, video/audio recordings, and
- A description of assessment results associated with a unit of assessment (*eport:Evaluation*). The *Evaluation* class is broken down into two subclasses: *ExternalEvaluation* (provided by any number of actors of the learning network, such as peer assessment) and *InternalEvaluation* (provided by the learner himself, such as self-assessment and reflection).

The *CompetencyProfile* class enables formal representation of collections of acquired or required competencies known as competency profiles. A competency profile can be defined as a set of competencies together with associated proficiency levels. We differentiate between two types of profiles according to their purpose:

- *RequiredCompetencyProfile* which specifies the requirements in terms of competencies to be fulfilled by a learner, and
- *AcquiredCompetencyProfile* which specifies the accomplishments in terms of competencies of learners.

Each type of profile is composed of a set of *CompetencyProfileElements* that represent requirements or statements of acquired competencies depending on the profile container type. It should be noted that competency statements may or may not be credible, thus the *ConfidenceRating* property was attached to the *CompetencyProfileElement* class. The Competency ontology was constructed using the Protégé Ontology Environment (Protégé, 2007) and was implemented in OWL (Web Ontology Language) (Antoniou and Harmelen, 2004). Fig. 3. visualizes the Competency Ontology using the OntoViz plugin for Protégé. This visualization plugin displays an ontology as a graph. Ontology classes are shown as boxes and the relationships among them are shown as arcs. The label on the arc denotes the name of the relationship.

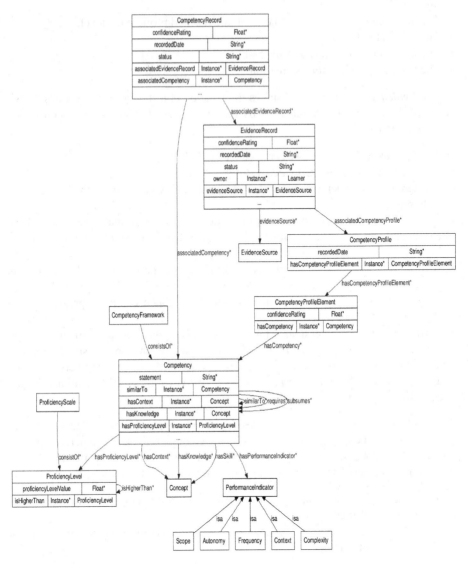

Fig. 2. Visualizing the Competency Ontology using the OntoViz plugin for Protégé

4   Conclusion

In this paper, we presented a review of existing approaches to competency modeling with the aim of identifying the advantages and limits of each approach. Besides, we presented the development of a competency ontology which contributes to overcoming key issues in competency modeling and management. The proposed ontology provides humans with a shared vocabulary that enables capturing competency-relevant characteristics of actors and learning resources in learning networks and serves as the basis for the semantic interoperability for machines. Besides, it supports lifelong competency development in learning networks by modeling aspects related to competency information management and tracking. As a next step, we plan to deal with the problem of competency comparison by proposing an ontology-based matching algorithm of competency profiles. This algorithm could serve to compare competency profiles in different competency assessment scenarios in a learning network.

References

1. Antoniou, G., Harmelen, F.: Web ontology language: OWL. In: Staab, S., Studer, R. (eds.) Handbook on Ontologies, pp. 67–92. Springer, Berlin (2004)
2. Beer, S.: Lifelong learning: debates and discourses. Paper submitted to NIACE's Lifelong Learning Inquiry (2007),
   http://www.niace.org.uk/lifelonglearninginquiry/docs/ConceptualisingLLL.pdf
3. Coi, J.L., Herder, E., Koesling, A., Lofi, C., Olmedilla, D., Papapetrou, O., Siberski, W.: A model for competence gap analysis. In: Proceedings of International Conference on Web Information Systems and Technologies (WEBIST 2007), Barcelona, Spain (2007)
4. DC, Dublin Core Metadata Element Set, Version 1.1: Reference Description (2003),
   http://dublincore.org/documents/2003/02/04/dces/
5. EQF. The European Qualifications Framework for lifelong learning (EQF) (2008),
   http://ec.europa.eu/education/pub/pdf/general/eqf/broch_en.pdf
6. Hotte, R., Basque, J., Page-Lamarche, V., Ruelland, D.: Ingénierie des compétences et scénarisation pédagogique. International Journal of Technologies in Higher Education 4(2), 38–56 (2007)
7. HR-XML. HR-XML Competencies (Measurable Characteristics) (2006),
   http://xml.coverpages.org/HR-XML-Competencies-1_0.pdf
8. IEEE RCD. IEEE std 1484.20.1-2007 IEEE standard for learning technology-data model for reusable competency definitions (2008),
   http://www.doleta.gov/usworkforce/pdf/2007-IEEEcomp.pdf
9. IMS RDCEO. IMS reusable definition of competency or educational objective specification (2002), http://www.imsglobal.org/competencies/
10. Koper, R., Rusman, E., Sloep, P.B.: Effective Learning Networks. Journal of Lifelong Learning in Europe, 18–27 (2005)
11. Lantz, A., Friedrich, P.: Learning in the workplace: An instrument for competence assessment. The Learning Organization 10(3), 185–194 (2003)
12. Likert, R.: A technique for the measurement of attitudes. Archives of Psychology 140, 1–55 (1932)
13. Malzahn, N., Ziebarth, S., Hoppe, H.U.: Semi-automatic creation and exploitation of competence ontologies for trend aware profiling, matching and planning. Knowledge Management & E-Learning 5(1), 84–103 (2013)
14. Mansfield, R.S.: Building competency models: Approaches for HR professionals. Human Resource Management 35, 7–18 (1998)
15. Miao, Y., Sloep, P., Hummel, H., Koper, R.: Improving the unreliability of competence information: an argumentation to apply information fusion in learning networks. International Journal of Continuing Engineering Education and Life-Long Learning (IJCEELL) 19(4/5/6), 366–380 (2009)

16. Miles, A., Bechhofer, S.: SKOS Simple Knowledge Organization System Reference (2009), http://www.w3.org/TR/skos-reference/
17. Najjar, J., Grant, S., Simon, B., Derntl, M., Klobucar, T., Crespo, R.M., Kloos, C.D., Nguyen-Ngoc, A.V., Pawlowski, J., Hoel, T., Oberhuemer, P.: Isure: Model for describing learning needs and learning opportunities taking context ontology modelling into account (2010), http://www.icoper.org/deliverables/ICOPER_D2.2.pdf
18. Ng, A., Hatala, M.: Ontology-based approach to formalization of competencies. In: Competencies in Organizational e-Learning: Concepts and Tools, pp. 185–206. Idea Group, Hershey (2007)
19. Ostyn, C.: Distilling Competency Information. White Paper (2005), http://www.ostyn.com/standardswork/competency/Distilling%20competency%20information.htm
20. Paquette, G.: An ontology and a software framework for competency modeling and management. Educational Technology & Society 10(3), 1–21 (2007)
21. Protégé. The protégé ontology editor and knowledge acquisition system (2007), http://protege.stanford.edu/
22. Rezgui, K., Mhiri, H., Ghédira, K.: Competency Models: A Review of Initiatives. In: Proceedings of the 12th International Conference on Advanced Learning Technologies, Rome, Italy, pp. 141–142 (2012)
23. Rogushina, J., Gladun, A.: Ontology-based Competency Analyses in New Research Domains. Journal of Computing and Information Technology 20(4), 277–291 (2012)
24. Sampson, D.G., Fytros, D.: Competence models in technology-enhanced competence-based learning. In: International Handbook on Information Technologies for Education and Training, pp. 155–177. Springer, Berlin (2008)
25. Sampson, D.G.: Competence-related metadata for educational resources that support lifelong competence development programmes. Educational Technology & Society 12(4), 149–159 (2009)
26. Schmidt, A., Kunzmann, C.: Towards a Human Resource Development Ontology for Combining Competence Management and Technology-Enhanced Workplace Learning. In: Meersman, R., Tari, Z., Herrero, P. (eds.) OTM 2006 Workshops. LNCS, vol. 4278, pp. 1078–1087. Springer, Heidelberg (2006)
27. Sicilia, M.A.: Ontology-based competency management: Infrastructures for the knowledge intensive learning organization. In: Intelligent Learning Infrastructures in Knowledge Intensive Organizations: A Semantic Web Perspective, pp. 302–324. Idea Group, Hershey (2005)
28. Sitthisak, O., Gilbert, L., Soonklang: Integrating Competence Models with Knowledge Space Theory for Assessment (2013), http://caaconference.co.uk/wp-content/uploads/Sitthisak_caa2013_submission.pdf
29. Sitthisak, O., Gilbert, L., Albert, D.: Adaptive Learning Using an Integration of Competence Model with Knowledge Space Theory. In: IIAI International Conference on Advanced Applied Informatics (IIAI AAI), Matsue, Japan, pp. 199–202 (2013)
30. Sloep, P.B.: Building a Learning Network through Ad-Hoc Transient Communities. In: International Conference on Computer Mediated Social Networking, Dunedin, New Zealand (2008)

Role of Agent Middleware in Teaching Distributed Network Application Development

Costin Bădică[1], Sorin Ilie[1], Mirjana Ivanović[2], and Dejan Mitrović[2]

[1] Department of Computers and Information Technology, University of Craiova, Romania
[2] Department of Mathematics and Informatics, Faculty of Sciences, Novi Sad, Serbia

Abstract. In this paper we introduce the structure and educational experiences of our course on Distributed Network Application Development across the last four years. The presentation is focused on the role of agent middleware and multi-agent systems on teaching the various theoretical and practical aspects of the course. In particular, we conclude that the use of agent middleware in general and of JADE platform in particular for teaching Distributed Systems certainly brings many advantages, but also has some limits and poses few difficulties. We provide in this paper a careful discussion of some of these aspects by presenting our own experiences and conclusions.

1 Introduction

Concepts of Distributed Systems, as well as their applications, design and technologies are important components of contemporary Computer Science curricula. We are now in a world of computing where basically everything is distributed in the broader sense, i.e. computing devices are interconnected, they use heterogeneous software and hardware platforms, and they exchange information via heterogeneous network communication channels. Facing this reality, technologies of distributed computing are developed and diversified with the spread of new platforms, architectures and languages for applications that could not have been imagined before, like ubiquitous and pervasive computing, mobile computing, sensor networks, high-performance computing, cloud computing, or the Internet of Things. Therefore, the rigorous design, integration, and harmonization of various topics of Distributed Systems into Computer Science curricula presents a quasi-permanent challenge taking into account the various constraints of time, resources, effort, and expertise of educators and students.

The motivation and discussion of the structure of a core Distributed Systems course, as well as its integration into Computer Science curricula is not an easy task and it would probably require more space than it is available here. Based on our experiences of teaching a one-semester course in Distributed Network Application Development to undergraduate Computer Science curricula during the last 4 years, we focus here on the role played by multi-agent distributed middleware in teaching the various concepts of Distributed Systems.

Our investigation starts with the following core question:

Q1: *Can agent middleware play a relevant role in teaching topics of Distributed Systems in Computer Science curricula?*

G. Jezic et al. (eds.), *Agent and Multi-Agent Systems: Technologies and Applications,* 267
Advances in Intelligent Systems and Computing 296,
DOI: 10.1007/978-3-319-07650-8_27, © Springer International Publishing Switzerland 2014

Then, if the answer to this question is "yes", we follow with our next question:

> **Q2:** *What roles can agent middleware play in teaching topics of Distributed Systems ?*

Based on our experiences, we do believe that agent middleware is relevant for teaching several theoretical and practical aspects of Distributed Systems – i.e. the answer to **Q1** is "yes". We will provide arguments for this in this paper. Moreover, in this context, this paper brings also to the reader our own approach and conclusions on using agent middleware to support the lectures, lab and project activities during the course of Distributed Network Applications Development that was taught during the last four years to Computer Science undergraduates at the University of Craiova, Romania.

The paper is organized as follows. We start in Section 2 with an overview of various sources for developing a Computer Science course focused on topics in Distributed Systems. We follow in Section 3 with a brief overview of our course on Distributed Network Applications Development. In Section 4 we focus on the specific role that agent middleware can play in a Distributed Systems course and actually played in our own course. In Section 5 we discuss our experiences and conclusions, with a special attention to the practical aspects of the course – lab and project.

2   Background

2.1   Computer Science Curricula Recommendations

ACM and IEEE are continuously developing, revising, and refining recommendations to help the academic educators to design and further adapt Computer Science curricula by incorporating the most recent results of Computer Science research and by addressing the new market requirements for Computer Science professionals. The recommendations are available as Computer Science Body Knowledge – CSBK, the last version being issued in 2013 [1].

Moreover, Computer Science scholars are providing textbooks in Distributed Systems, like for example [2]. Those textbooks, as well as CSBK are valuable sources of knowledge and methodology for teaching Distributed Systems topics to Computer Science undergraduates.

CSBK provides a comprehensive structuring of Computer Science knowledge into 18 *knowledge areas* – KA. Each KA is decomposed into a number of *knowledge units* – KU, while each KU is further refined into a number of *topics* with associated *learning outcomes* – LO. Each LO must have associated a certain *level of mastery* from the available set: familiarity, usage, and assessment.

Three CSBK KAs that are relevant for the development of courses including Distributed Systems topics were recently updated to reflect the current developments in Computer Science:

- Networking and Communication (NC) KA was split, because of its growth and divergence. A part of it was included in PBD (see below).

– Platform-Based Development (PBD) is a new KA, with its content mainly grown from the NC KA.
– Parallel and Distributed Computing (PD) is a new KA that acquired topics that were previously spread in other KAs.

2.2 Agent Middleware

Following [3], agent systems are deployed over specialized software infrastructures that provide the basic set of functionalities for the existence of a realistic multi-agent application. Seen from the perspective of Distributed Systems, such infrastructures are placed at the middleware level. They define a software layer that (i) assures platform (here understood as hardware + operating system) independence and (ii) provides a collection of software functionalities and services including: agent lifetime management, agent communication and message transport, agent naming and discovery, mobility, security, etc. An *agent framework* is a software infrastructure that is available as a software library, a language environment, or both and provides the core software artifacts needed for creating the skeleton of a multi-agent system. A software package that provides the core functionalities for deploying and running distributed multi-agent applications is traditionally known as an *agent platform*. Typical *agent middleware* provides both an agent platform and an agent framework.

More than a hundred of agent platforms and toolkits that differ in maturity, quality, standards compliance, and complexity are reported in the literature [3]. One of the most popular, well-documented, FIPA[1] compliant, and easy-to-program agent packages is JADE [4].

3   Overview of *Distributed Network Application Development* Course

In developing our Distributed Network Applications Development (DNAD hereafter) course we took into consideration the definition of course learning objectives and the availability of course prerequisites. The DNAD course is scheduled in the 3rd year, 6th semester of Computer Science curricula.

Firstly, the course learning objectives were formulated in accordance with recommendations provided by CSBK on topics that were considered relevant for Distributed Systems, as well as based on standard textbooks in Distributed Systems:

– **LO1:** To introduce the principles and concepts of distributed software.
– **LO2:** To introduce the basic technologies of distributed software with a focus on core middleware technologies based on Internet.
– **LO3:** To provide an opportunity to obtain practical experience in applying these techniques for programming small-scale distributed software applications.

Secondly, the course prerequisites were established taking into account the current structure of the Computer Science curricula. The students must be familiar with theory

---

[1] http://www.fipa.org/

and practice of Computer Programming (including Object-Oriented Programming using the Java programming language), Operating Systems and Computer Networks. The courses that directly benefit from DNAD are Electronic Commerce and Web Application Design.

Thirdly, a course structure was defined including lectures, labs, and project activities. Finally, we added topics on agent middleware that were aimed to teach several conceptual and practical aspects of Distributed Systems.

The DNAD course is structured in 2 modules with separate grading: (i) Course module comprising lectures and labs with 4 ECTS points; (ii) Project module with 1 ECTS point. Both modules have a duration of 14 weeks with lectures 2h/week, lab 2h/week and project 1h/week. Each module has a duration of 14 weeks.

Our DNAD course uses a technology-centered pragmatic approach to teach Distributed Systems. The course is more focused on applications and programming in the spirit of [5,6], rather than theory and algorithms, as in [7]. Nevertheless, important concepts are introduced and exemplified with the help of technologies and frameworks.

Please note that similar pragmatic approaches for teaching Distributed Systems applications to undergraduates were already used, like for example [8]. However, the special feature of our approach is the significant role that we assigned to distributed agent middleware in teaching the concepts and practical applications of Distributed Systems.

There is no single textbook to cover the course content. Nevertheless, for our course, a useful reference is [2]. The course lectures comprise the following list of topics:

- Introduction to Distributed Systems
  - Definition, classification and characteristics
- Models of Distributed Systems
  - Physical, architectural, and fundamental models
- Inter-process communication in Distributed Systems
  - TCP, UDP, group communication
- Core technologies for Web-based Distributed Systems
  - HTML/CSS, XML, HTTP
  - Web clients and servers
  - Servlets and Apache/Tomcat
- Object-based Distributed Systems and Remote Method Invocation
  - Design of RMI
  - Programming Java RMI
- P2P systems
  - Structured and unstructured overlay networks
- Agent-based Distributed Systems
  - FIPA and JADE
- Web Services
  - Concepts and standards
  - Axis2

Our DNAD course addresses several KUs that are part of three KAs of CSBK, as follows (note that two KUs are on our wish-list, as future work):

- KA: Networking and Communication (NC)
  - KU: NC/Networked Applications
- KU: KA: Platform-Based Development (PBD)
  - KU: PBD/Introduction
  - KU: PBD/Web Platforms
  - KU: PBD/Mobile Platforms (on the wish-list as future work)
- KA: Parallel and Distributed Computing (PD)
  - KU: PD/Distributed Systems
  - KU: PD/Cloud Computing (on the wish-list as future work)

The grading of the DNAD course module is based on final exam and lab assignments. The final exam counts 60% of the final mark and comprises 30% based on a questionnaire of knowledge questions and 30% based on "apply skills" exercise involving the design of a small-scale distributed software application. The lab grading counts 40% of the final mark and it is determined from a set of lab assignments. Their number depends on difficulty, and it is usually chosen between 3 to 5. During the semester, as part of the lab activity, the students are also exposed to a number of tutorials that introduce the software packages required for carrying out their lab and project assignments.

The grading of the DNAD project module is based on a project assignment and associated deliverables: a project report and a software package. The grade is split as 20% for the intermediary report and 80% for the final report and software deliverable.

4   Role of Agent Middleware

Agent middleware can play (and actually played) an important role in teaching our DNAD course. In order to argument this statement we performed a thorough analysis of the JADE agent middleware – JADE [4] with respect to the requirements set by the learning objectives, structure and topics of the DNAD course.

Firstly, JADE agent middleware supports all the course learning objectives. It supports **LO1**, as using JADE examples, we can explain many principles and concepts of Distributed Systems, including: platform heterogeneity management, transport protocols, white and yellow pages (naming and directories), code mobility, fault tolerance (JADE supports a limited form of fault tolerance), and interaction protocols. Conceptually, following the classification of architectural models proposed in [2], JADE uses a software component-based model (i.e. JADE agents are actually software components), it supports asynchronous message passing for agent communication in the P2P style (although JADE itself cannot be characterized as a true P2P system), and it is standards (i.e. FIPA) compliant. Partly, it also supports **LO2**, as JADE itself can be described as a middleware platform. It also supports well **LO3** with the low cost of a smooth learning curve by providing a meaningful and well-documented API that helps students to acquire skills for developing JADE-based small-scale distributed software in due time and with reasonable effort.

Secondly, using JADE as example, we can cover part of the topics included in our DNAD course – both lectures and laboratories. In particular, JADE is a good example of component-based distributed middleware platform. Moreover, the interaction model (one of the fundamental models of a Distributed System [2]) of a Distributed System actually corresponds to a distributed algorithm and it can be described in a disciplined way as a set of communicating state-machines [7]. This model can be naturally implemented using JADE agents with the help of finite-state machine behaviors, i.e. *FSMBehaviour* [4].

JADE can be also used to introduce the service-oriented architecture. JADE agents can expose services registered in a yellow pages directory – the *Directory Facilitator* agent. Services can be named, searched in this directory, and then invoked using interaction protocols, thus supporting one of the basic architectural patterns of service-oriented computing. Services provided by JADE agents can be exposed as Web Services [9], allowing the integration of distributed multi-agent applications into the Web environment, and enabling use of JADE as integration middleware of distributed heterogeneous applications.

Using JADE we can also exemplify an elegant model of object serialization based on Java Beans and Semantic Web technologies. Firstly, JADE provides an API for manipulating ontologies that supports packing and unpacking of complex objects when they are exchanged between agents via FIPA ACL messages. Moreover, the specialized *Ontology Bean Generator* tool is available to facilitate the engineering process of JADE ontologies [10].

One of the weaknesses of JADE, however, is the integration with standard Web technologies. Although possible, the development of a Web-based application that integrates JADE on the server-side is not natural and it requires the use of an additional software glue known as JADE gateway [11]. This facility was nevertheless useful for project activities, where students chose to develop a Web-based interface to a JADE-based distributed multi-agent application.

Finally, is worth mentioning that JADE can also support the development of mobile computing applications on Android-based smartphones [12]. This can be very useful with regard to the extension of our DNAD course with topics covering the *Mobile Platforms* KU. Moreover, JADE can be also useful in the near future to support lab and project activities involving the development of distributed mobile applications with JADE agents running on Android-based smartphones.

Before concluding this section we would like to mention that although JADE (and multi-agent middleware in general) can support the teaching of Distributed Systems in multiple ways, it is by far not the silver bullet. There are many other aspects related to concepts and technologies of Distributed Systems that require additional examples and tools to support a good coverage of the subject. Some example topics that require different tools are Web technologies, RMI, P2P systems, Cloud computing, Ubiquitous and pervasive computing. Taking also into account that it is probably impossible to find a single tool to cover all the topics, we can conclude that the correct approach that we also followed is to carefully analyze the course curricula (lectures, labs, and project) and decide where precisely agent middleware can serve as relevant example, as well as practical implementation tool.

5   Educational Experiences

In this section we present some of our experiences gained with teaching the DNAD course during the academic years 2009-2010, 2010-2011, 2011-2012, and 2012-2013. During the course lectures we followed the course topics outlined in section 3. They included a chapter on agent middleware covering FIPA and JADE. Moreover, agent middleware examples were used in many places to discuss concepts of Distributed Systems. On the other hand, we experimented with different approaches and assignments in each year for the lab and project work, so the following presentation is more focused on those aspects of the course.

Agent middleware was firstly added to the DNAD course during 2009-2010 academic year. The lectures included a chapter on FIPA and JADE. Also, the students were exposed to the design and implementation of a simple multi-agent systems during the laboratory. The process started with the presentation of a scenario, following with the identification of agent types, the design of interaction protocols and agent behaviors. However, while we have noticed during the exam that the students received well the discipline of multi-agent system (MAS hereafter) design, they had difficulties with the implementation of the agent system. Actually very few of them were able to produce a working JADE-based MAS at the end of the lab activity. We learnt that one weakness of our approach was the schedule of JADE introduction too late, towards the end of the course. The students needed more time to gain more familiarity with the technology in order to successfully finalize some concrete programming work.

Therefore, during the next year 2010-2011, we decided to direct more the lab assignments towards multi-agent system design and implementation, as well as on the understanding of the underlying technologies in connection with agent middleware. So, apart from the course lectures dedicated to FIPA and JADE, students received a lab task to implement distributed entities called "agents" that can interact using a simple "ping" protocol. For the implementation the students had to use several Java middleware technologies, including: sockets, RMI, servlets, Web Services and Jade. The students were introduced to the MAS design methodology during the first lab, while the rest of the laboratory work was concentrated on the various implementations.

Learning from past experiences, we helped students by creating a JADE bootstrap class that instantiates the JADE platform as well as the *Remote Management Agent –* RMA that provides a Graphical User Interface. Then we taught students how to easily add agents to their system using sample Java code. This proved to be a good approach since all students were able to create more easily agent running examples .

During 2010-2011 we also directed the project activity towards using agent middleware. Therefore students received an "agent stress" experiment as project assignment. Their first task was to define an agent organization with a fixed communication topology of their choice (i.e. ring, mesh, linear etc.) and to run an experiment to determine how many "ping" protocols they can instantiate on a single machine before their system started to exhibit any kind of failures like lost messages or agent crashes. Their second task was to rerun their experiment on a computer network using more machines and then compare the results. The implementation technology of the "agents" was left at the students' choice. Out of the 28 students that presented the project in the first exam session, 25 students decided to use JADE. However, this result is only partly positive as 29 out of 57 students did not present the project.

We generally concluded that students liked working with JADE. However, as a "side-effect" we noticed that they prefer to use a simple development environment that automatically takes care of general repetitive tasks that are part of the system setup. The conclusion was independent of the technology taught, as we noticed the same problems with teaching Java servlets. Although it can be argued that doing such actions manually can be more educative as students are faced with "more realistic" problems, those activities are also tedious and discouraging. One negative effect is that most students just stop doing the lab work altogether, accepting lower grades. So we decided to focus them more on programming and experimentation activities, rather than forcing them to do themselves the configuration and setup activities.

Doing experiments on multiple computers also captured well the students' attention, in spite of the obvious complexities of setting up a distributed application on a computer network. The JADE RMA agent was very well received by the students for monitoring the setup of their distributed system before the actual starting of the agents interactions.

Most of the students needed personal assistance from the lab professor until they were able to produce a working lab assignment. This resulted in students asking so many questions that the professor ended up by handing out pieces of code to them. The students that prepared their lab assignments in advance, finished their work more quickly and consequently the professor took advantage to assign them with helping slower students that ran out of time. However this approach turned out to be not so "competitive", as slower students received actually more attention. The negative effect was that the number of these students was increasing by the end of the semester. Nevertheless, this was compensated during the grading process.

During 2011-2012 we decided to continue with the same lab assignments, while applying an "elitist" type of assistance. If the student did not present a reasonable personal attempt then he/she would not get any help. But in order to make sure that all students could make that first attempt, we provided them with a review of Java programming in the first lab session. For diversity, we slightly modified the project assignment by adding the requirement to measure the MAS setup time for instantiating an increasing number of ping protocols on a single machine, as well as on 2 machines and then compare the results. As consequence, only 15 out of 45 students were able to present their project work during the first exam session. Also, only one student chose to use sockets instead of JADE agents.

For the following year 2012-2013 we chose to make a radical change for the lab and project activities. We devised 2 large assignments: (i) a distributed master-slave password cracker implemented using a safe socket-based communication protocol over UDP and (ii) a Twitter-like system implemented using JADE agents and equipped with a Web based GUI. We also continued with the presentation of a Java programming tutorial during the first lab. For the project task, the students were asked to perform a "stress" experiment on one of the lab assignments of their choice.

During their assignments, we noticed two most frequent difficulties encountered by the students: (i) dealing with the potential failures of the UDP protocol (ii) interfacing

JADE code with the Tomcat Web server. They determined many students to actually refuse to even try to achieve these particular requirements of the assignments. Out of 90 students, only 21 students were able to implement the safe communication protocol over UDP and only 17 created some kind of Web interface to their agent-based Twitter-like system. Just 25 out of 90 students presented the project in the first exam session, which was in fact an overall negative result. We concluded that the main cause was the significantly higher complexity of the lab assignments, as compared with the previous years.

Based on our experiences we can conclude that agent middleware in general and JADE in particular can offer some advantages as compared with other frameworks for teaching several aspects of Distributed Systems technologies and applications. JADE has a smooth learning curve and requires considerable less effort than other enterprise technologies – like Enterprise Java Beans, for example. Students enjoyed programming simple JADE-based distributed applications. Moreover, agent-based design offered a disciplined approach for design and development of distributed applications.

However, there are aspects where the use of JADE presents difficulties. Students needed help with the setup and creation of simple applications. This was achieved by the creation of a JADE bootstrap class, as well as a special application configuration of Eclipse platform to facilitate the development and running of agent applications. Moreover, implementing a Web-based GUI to a JADE-based MAS requires the tedious and discouraging work of interfacing two different distributed technologies: FIPA agents and Web Servers. This fact requires further investigation before deciding to consider this as a limit of our lab or to find a better way of interfacing JADE with Tomcat Web server.

Finally, we also noted that students do not perceive JADE as an actual enterprise technology. We have reason to believe that this is one of the main causes of their low turn up with the project presentations in the first exam session. It is quite hard to capture the full attention of all the students when the only foreseeable result of using JADE is as a basis for a prototype or concept.

6   Conclusion and Future Works

In this paper we presented the structure and educational experiences of our course on Distributed Network Application Development during the last four academic years. We highlighted the role of agent middleware and multi-agent systems on teaching the various theoretical and practical aspects of the course. Our conclusion is that the use of agent middleware and of JADE platform for teaching topics of Distributed Systems certainly brings many advantages, but also has some limits and poses few difficulties. As future work we plan to adapt our course curricula and methodology to address some of these issues. We also plan to expand our course curricula with adding new topics in Mobile Computing and Cloud Computing, while maintaining the significant role of agent middleware.

References

1. The Joint Task Force on Computing Curricula. ACM and IEEE CS: Computer science curricula 2013. ironman draft (version 1.0) (February 2013)
2. Coulouris, G., Dollimore, J., Kindberg, T., Blair, G.: Distributed Systems. Concepts and Design, 5th edn. Addison Wesley (2011)
3. Bădică, C., Budimac, Z., Burkhard, H.D., Ivanović, M.: Software agents: Languages, tools, platforms. Computer Science and Information Systems 8, 255–298 (2011)
4. Bellifemine, F.L., Caire, G., Greenwood, D.: Developing Multi-Agent Systems with JADE. John Wiley & Sons (2007)
5. Graba, J.: An Introduction to Network Programming with Java. Springer (2007)
6. Ince, D.: Developing Distributed and E-Commerce Applications, 2nd edn. Addison-Wesley (2003)
7. Santoro, N.: Design and Analysis of Distributed Algorithms. John Wiley & Sons (2007)
8. Albrecht, J.R.: Bringing big systems to small schools: Distributed systems for undergraduates. SIGCSE Bull. 41, 101–105 (2009)
9. Bădică, C., Ilie, S., Bassiliades, N., Kravari, K.: Enabling agent reasoning over the web. In: Diamantaras, K.I., Evangelidis, G., Manolopoulos, Y., Georgiadis, C.K., Kefalas, P., Stamatis, D. (eds.) Balkan Conference in Informatics, BCI 2013, pp. 259–266. ACM (2013)
10. van Aart, C.: Organizational Principles for Multi-Agent Architectures. Whitestein Series in Software Agent Technologies. Birkhäuser Verlag (2005)
11. Ilie, S., Bădică, C., Bădică, A., Sandu, L., Sbora, R., Ganzha, M., Paprzycki, M.: Information flow in a distributed agent-based online auction system. In: Burdescu, D.D., Akerkar, R., Bădică, C. (eds.) 2nd International Conference on Web Intelligence, Mining and Semantics, WIMS 2012, p. 42. ACM (2012)
12. Mocanu, A., Ilie, S., Bădică, C.: Ubiquitous multi-agent environmental hazard management. In: 14th International Symposium on Symbolic and Numeric Algorithms for Scientific Computing, SYNASC 2012, pp. 513–521. IEEE Computer Society (2012)

Cloud e-Learning: A New Challenge for Multi-Agent Systems

Krenare Pireva[1], Petros Kefalas[2], Dimitris Dranidis[2],
Thanos Hatziapostolou[2], and Anthony Cowling[3]

[1] South-East European Research Center, 24 P. Koromila, 54622, Thessaloniki, Greece
[2] The University of Sheffield International Faculty, City College,
3 L. Sofou, 54624, Thessaloniki, Greece
[3] The University of Sheffield, Department of Computer Science,
211 Portobello Street, Sheffield, UK

Abstract. The developments of pedagogical models in e-learning together with the advances of learning technologies and cloud computing
give us confidence to believe that the traditional e-learning will evolve
into a process which will put the learner in the center of educational
provision. This paper proposes that Cloud e-Learning, a new approach
to e-learning, will open opportunities for learners, by allowing personalisation, enhancing self-motivation and collaboration. The learners should
be able to choose what to learn, what sources to use, with and by whom,
how and in what pace, what services and tools to use, how to be assessed,
whether to get credits towards a degree etc. In such a dynamic environment, the need for Multi-Agents Systems is necessary. Actors in Cloud
e-Learning would need automated facilitation in all services involved.
We outline few indicative scenaria for Cloud E-Learning in which smart
agents will act on behalf of the learners, teachers and institution in order
to maximise the benefit of the proposed concept.

1 Introduction: From Traditional e-Learning to Cloud e-Learning

Recent advances in Information and Communication Technologies have provided
the opportunity to enhance e-learning with new synchronous and asynchronous
features to both students and instructors. Educational institutions that provide
e-learning can now develop courses and programmes that utilise existing pedagogies and experiment with new ones. It is apparent that these developments
have lately facilitated the accessibility of e-learning through a wide variety of
MOOCs (Massive Open On-line Courses).

A typical e-learning course, whether it is open or private, consists of four
main components. The **pedagogy** should determine a number of characteristics
for this course, such as the way in which the learning outcomes will be met by
delivery and assessment methods as well as the learning path and learning pace
of the group. Pedagogy will in broad terms define the balance between instruction and self-learning, implying also the type and frequency of communication

G. Jezic et al. (eds.), *Agent and Multi-Agent Systems: Technologies and Applications*,
Advances in Intelligent Systems and Computing 296,
DOI: 10.1007/978-3-319-07650-8_28, © Springer International Publishing Switzerland 2014

(synchronous or asynchronous) between teachers and learners. The **content** will include a variety of text and media deemed as appropriate to give opportunities to meet the learning outcomes. The **technological infrastructure** is the set of Learning Technologies tools used by the teachers and learners in order to facilitate knowledge transfer and skill acquisition, such as VLE, teleconferencing tools, wikis, file sharing, social interaction, support and feedback etc. Finally, the **course administration** is a set of regulations and processes as well as their monitoring under which students enrol, attend, progress, etc. Irrespectively of any combination of the above, e-learning inherits some rigidities of traditional face to face learning. The restrictions that characterise both types of learning are:

- teachers apply predefined pedagogies,
- the selection of material is largely done and/or recommended by the teacher,
- the tools of the technological infrastructure are specified by the course provider (teacher or institution),
- regulations and processes are provider/institution specific.

The big contradiction in this situation is that the learner, who is the receiver of the process, should abide by what the course providers have agreed, with no or little involvement in the above. This seems the "rational thing to do" for groups of learners, especially when providers are tied by the general educational framework in which they belong. Thus, for instance, Universities need to follow certain quality assurance requirements in order to award credits for courses and eventually degrees. But even then, course providers have been criticised that they do not apply a learner-centered approach, taking into account the individual types and needs of each learner. In this respect, **Cloud e-Learning** can be considered as an advancement of e-Learning, taking into account that there exist courses in which learners can take the initiative to select:

- the way and pace in which they learn,
- the means through which knowledge and skills are acquired,
- the tools that they are going to use for learning,
- the people (teachers, facilitators, other learners with whom they wish to collaborate etc) and institutions involved in their learning.

An immediate reaction to the above could be rather conservative, given the authors own experience in traditional education. Admittedly, however, reservations that most educators had two decades ago did not prevent the evolution of e-learning courses, e.g. MOOCs, by respectable institutions which are available to masses of learners, even if this is currently done mostly without credits.

It is evident from the above that the proposed Cloud e-Learning is a complex venture that, despite all technical and pedagogical issues, allows for a certain amount of automation. It would be cumbersome if all the tasks are carried out manually on the responsibility of the learner. Facilitation is definitely required. The authors position is that Cloud e-Learning is fertile ground for Multi-Agent Systems which would be responsible for such facilitation.

This is, in principle, a vision paper; its main contribution is to provide a definition of Cloud e-Learning and discuss how Multi-Agent Systems would play a strategic role in the development of this new concept. Section 2 outlines Cloud e-Learning and its benefits to a learner-centered pedagogical approach. In section 3 then discusses Cloud Computing. Intelligent agents and their roles are presented in section 4. Finally, the paper concludes by summarising the next potential research paths.

2   A Vision for Cloud e-Learning

There does not exist a widely accepted definition of Cloud e-Learning. It is meant to be a new term, and so a precise definition will not be attempted here: instead the main characteristics of this concept will be outlined. The aim of **Cloud e-learning (CeL)** is *to provide personalised services that will increase interaction between users (learners, teachers and institutions) by sharing a pool of experiences and knowledge available in cloud and suggest structured courses that match learners preferences.* The important component in CeL is the Cloud and the opportunities it offers together with its existing infrastructure and services. The Cloud has opened up a range of possibilities for:

- enhanced distant collaboration,
- instant availability to web through a variety of devices,
- wide accessibility to information of different type,
- consolidating self-motivation,
- increase personalisation through combination of services,
- a variety of tools and services.

These possibilities can be illustrated by considering some scenarios in CeL. The following may look at first glance like basic pedagogical and technological challenges, but their aim is to show how the full potential of CeL could be unleashed in the medium long term future (Table 1).

Table 1. Fundamental characteristics of CeL

Cloud e-Learning Scenarios - Fundamental Question	
Open syllabus	What to learn?
Open material	What sources to use?
Open group	With and by whom?
Open learning path	How and in what pace?
Open assessment	How to be assessed?
Open VLE	What services and tools to use?
Open accreditation	How to get credits towards a degree?

Collective Creation of Syllabus: Imagine that a collaborative environment could be developed in which learners would be able to determine collectively the learning outcomes of a course. This could be done in accordance to some loose initial template that a teacher sketches. Learners will create a syllabus that emerges through individual preferences. Syllabi emergent learning outcomes will then drive teaching and assessment methods to reflect learners aims.

Collection of Material through a Variety of Sources: Consider the variety of sources and their types (books, notes, libraries, video, audio, etc.) that exist in the web. Given semantic annotation to learning resources and processes, these could form a cloud of knowledge where learners could choose from. Suitability of sources would depend on learners learning style, past experience and popularity among learners and providers.

Selection of Teachers, Learners and Providers: The learners would, in principle, be able to select by whom they are going to be tutored. In a cloud of teachers and providers globally accessible, a matching between learners and tutors would provide better opportunities for better learning experience. The same could apply for the selection of providers as well as fellow learners with common interests and similar personal development plans.

Flexible Learning Paths: Learning paths may be personalised in terms of content, transition between steps, and pace for each step. This would assume existing experience of individual learners as well as other learners on similar course while taking into account individual learning styles, personal commitments, etc.

Personalisation of Assessment: Given the learning outcomes, there would be a variety of assessment methods that meet them. The learners, who would definitely be of different learning types and capacity, would be able to choose in collaboration with their teachers among the most suitable assessment for them, thus having more opportunities to achieve the aim of the course.

A Customisable VLE: Users should be able to choose on a set of tools rather than dealing with the fixed set of tools provided by a specific VLE. Thus, every learner would have a customised environment in which all processes will be accommodated in a way that would not require extra effort or deviation from everyday routine. Similar customisation could apply to teachers also.

Configuration of Course Characteristics That May or May Not Lead to Award of Credits: Learners should be able to configure a course according to their need. Thus, it would be a different course which would satisfy personal interests, another which would be pursued for professional development and another which would lead to award of credits and eventually a degree. That would also need different levels of quality assurance and accreditation that would be specific from case to case.

It is important to note that the above characteristics of the CeL concept suggest that new ways for supporting **learner engagement and motivation** are required. A common assumption is that learners choosing their learning provides intrinsic motivation on its own, but this is a wrong assumption to make. As a number of studies [6,2] suggest, MOOCs currently face this challenge since dropout rates are very high despite the fact that people choose what courses to attend. In fact, all the privileges and flexibility in learning content, pace, methods of delivery and assessment offered by CeL actually bear an increased responsibility for supporting individual learner motivation. A dynamic learning setting that can change from face-to-face to online, within a programme of study or even within a course, that comprises of learners with different learning strengths, needs and backgrounds can be challenging and can easily lead to loss of learner motivation. Therefore, culturally responsive pedagogies that sustain the cognitive, behavioural and emotional engagement of learners must be a priority in CeL.

3   Cloud Technologies and CeL

3.1   Cloud Computing

Cloud computing enables access to a pool of resources which are delivered over the Internet as services on demand. Renting services over the Internet is not something new [15,23], since earlier ISPs provided software as services or applications as services. The main difference is that these earlier services faced long delays, low speed of network connection and lack of resources on demand. In addition, the renting resources needed to be scaled up manually. Cloud computing managed to overcome these problems by relying on a number of existing technologies such as virtualization, grid computing, and web services and by taking advantage of the following characteristics [17]: on-demand self-services, broad network access, resource pooling, rapid elasticity, measured services.

In a widely accepted model definition [17,9,16,25], the cloud model is composed of the following service models:

- **Infrastructure as a Service (IaaS).** The cloud users have access to computing resources, such as processing, storage, and networks that are necessary in order to install operating systems and then deploy their applications.
- **Platform as a Service (PaaS).** The providers provide the necessary programming languages, libraries, services, databases and tools so that cloud users can deploy their own applications onto the cloud in the form of SaaS.
- **Software as a Service (SaaS).** The cloud users have the capability to use applications which are installed in a cloud infrastructure and are being maintained by the service providers. The software services are accessible to users via thin-client interface devices (e.g. via a web-browser) or to other applications via programmatic interfaces.

The above cloud model is fairly standard, although there have been proposals to include additional layers too, such as Business Process as a Service (BPaaS). The vendors are able to assign and split the resources on demand, such as processing and storing capacity. Besides other benefits, data storage space is not limited. This gives great opportunities for learners, teachers and educational institutions to literally access immeasurable amount of information. It is often the case that educational institutions nowadays outsource part of their services to vendors such as Google. As the amount, diversity and quality of information grows, the ability of turning this information into useful knowledge by searching, reviewing, evaluating and synthesising becomes even more difficult, unless automated assistance is provided. This is the primary role of CeL.

3.2   The CeL Layers

As it has been stated above, the aim of CeL is to provide personalized services that will increase interaction between users (learners, teachers and institutions) by sharing a pool of experiences and knowledge available in cloud open courses and suggest structured courses that match learners preferences. Within the CeL platform all social networks as well as other vendors should be involved. CeL is proposed to have three layers where each layer has its own functionality (Fig. 1). The core layer of CeL will have a basic functionality and each outer layer will further increase the functionality of the framework:

- The **Open Course Layer** is proposed for learners who are interested to gain new knowledge or skills on a specific domain, without interest in acquiring accreditation, credits or degrees.
- In principle, CeL should be designed to offer courses with credits, which implies that within these courses the students will be assessed and get credits. Accreditation is required to be sought for Universities which provide such courses. The **Credit Bearing Course Layer** extends the functionality of core layer applies for those learners who are interested in getting credits on the CeL courses that they will attend.
- Finally, CeL should act as a virtual university which will inherit all characteristics of university establishments, such as accreditation, credits, quality assurance and monitoring, regulations etc. The **Degree Award Programme Layer** extends the functionality of second layer further and it offers services for those learners who are interested to acquire degrees from CeL.

The three different layers have different inherent complexities. At the core level, CeL would not so much require collaboration between teachers and institutions. At outer level, where degrees could be awarded would need rather complex arrangements which will mostly deal with accreditation and quality assurance. Thus, for instance, at Open Course layer, learners would need to collaborate for open syllabus as well as emergent collection of appropriate material, which after personalisation, each leaner would follow an individual learning path with no further commitments with regard to assessment. In Credit Bearing level, the teachers of an establishment would need to collaborate in order to establish some requirements, including assessment, under which award of credits from that establishment would be possible. That would need compliance with local quality assurance standards. Finally, at Degree Award level, various institutions need

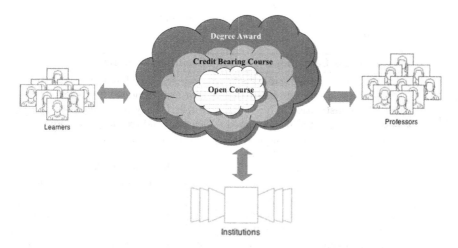

Fig. 1. Layers of Cloud e-learning and its main Actors

Table 2. Requirements for CeL layers and complexities implied

Requirements	Open Course	Credit Bearing	Degree Award
Collaboration between Learners	desirable	desirable	desirable
Collaboration between Teachers	optional	desirable	necessary
Collaboration between Institutions	-	optional	desirable
Quality Assurance at local level	optional	desirable	necessary
Quality Assurance at National level	-	desirable	necessary
Quality Assurance at International level	-	optional	necessary
Accreditation	-	optional	desirable

to collaborate in order to provide courses that meet national and international prerequisites for quality assurance and accreditation. Actually, the outer level of CeL would form the **virtual meta-University** in which the learners should be able to choose among various University providers and available credit bearing courses. In brief, Table 2 summarises the requirements at each layer.

As mentioned already, the CeL main actors would be: (a) Learners, (b) Teachers, and (c) Institutions. Table 3 summarises a comparison between traditional e-learning and CeL for each of these types of actors.

4   The Challenge for Multi-Agent Systems

The key issues of CeL are: (a) learner-centered, (b) openness, (c) personalisation, (d) self-motivation and (e) collaboration. The previous sections have outlined the environment (CeL) and the actors involved (learners, teachers and institutions). It is believed that CeL will be such a dynamic and complex environment that

Table 3. The comparison between traditional e-learning and cloud e-learning from each actor's prospective

	Traditional e-Learning	Cloud e-Learning
Learner	Learners access their university courses. They collaborate internally within their institution. They have access to the material developed by their local teacher. The discussion around the subject of study is mostly constraint within the University.	Learners are offered open materials that are developed by various institutions. They have access to other learners and teachers from other institutions. They use a variety of tools. They are flexible to decide what they want to learn, when to learn, from whom to learn and how to learn.
Teacher	Teachers are restricted to choose among traditional teaching, learning and assessment methodologies for learners, in a kind of one-size-fits-all way. They deal only with students within their institutional class. They are restricted to use the institutional VLE for all activities.	Teachers are open to collaboration and scrutiny from colleagues at other institutions. Competition will act as a driver to achieve better quality and disseminate best practices and inspiration to others. They will use a customisable VLE but some of them will be susceptible to resistance to change.
Institution	Institutions apply their internal monitoring of quality assurance. They define their own programmes and curricula. Learners and teachers abide by the institutional regulations and procedures.	Institutions will be forced to provide better service to learners and better policies for teachers. They will have to negotiate quality assurance and accreditation and as a result upgrade the standards of education provision in global competing market.

the actors could not manage it without help. This is where Multi-Agents Systems (MAS) involvement is needed. In this context, the intelligent (for others just "smart") counterparts of agents are considered. These agents should be able to facilitate the process of learning by acting on behalf of the actors in certain complex tasks. Intelligent agents are distinguished from ordinary programs by the degree of **autonomy, goal-orientation** and ability to reason, **reactiveness** and **interaction, collaboration, negotiation** through communication with other agents or humans. Secondary attributes, such as learning, rationality, mobility could also play an important role in the CeL context.

The Semantic Web is thought to be an incubator where smart agents can grow and unleash their potential. W3C mentions that smart agents can be thought as a semantic web service. However, a semantic web service is an abstract functionality which may be implemented by one or more smart agents. In CeL, semantic annotation and ontologies are by default necessary and therefore the deployment of semantic cloud services through intelligent agents would be the essence of its desired functionality [24].

CeL offers students, teachers and institutions the highest level of learning and teaching flexibility and promotes exploration of new pedagogies. Successful implementation of MAS in such a learning environment mandates the interoperability of e-learning content and systems through standardisation. The need for interoperable e-learning systems has long been recognised and several learning resource specifications have emerged in the last fifteen years. Popular standards include the Learning Resource Metadata Specification and Content Packaging by the IMS Global Consortium [13] and the Learning Object Metadata (LOM) by IEEE [12]. The e-learning standard, however, that has gained the most widespread adoption and is supported by most major vendors of Learning Management Systems [20] is the Sharable Content Object Reference Model (SCORM) by ADL [1]. SCORM actually integrates the e-learning specifications created by IMS and IEEE into a consolidated model that enables the construction of reusable, accessible, durable and interoperable web-based learning content which is assigned descriptive metadata in order to form learning objects. Depending on the nature of a learning object such metadata can be technical (e.g. size, format) or educational (e.g. learning outcome, context, level of difficulty). The CeL characteristics presented in Table 1, however, necessitate the development of new or the extension of existing standards and metadata in order to enable intelligent MAS to support the flexibility that this learning and teaching paradigm offers.

The use of MAS in e-learning is not a new concept. Agents were proposed in many phases of learning, such as formative assessment [19], personal tutors [11,3], skill management [10], learning paths [7], personalised content search [18], communities for group collaboration [26,22], affective facilitation [5], and many more. Various applications in e-learning [14] and frameworks with MAS were also proposed [21,8,4]. With the development of e-learning to CeL, all the above and more automated on line assistants will be needed. As CeL will:

- provide personalised services,
- increase interaction between users,
- allow sharing of pool of experiences and knowledge resources,
- recommend learning structures to users, and
- do matchmaking of resources and users,

the agents employed would have certain characteristics. **Profile Development Agents** will be responsible to build a profile for individual learners, teachers and institutions. Profiles will be based on preset configuration but also learning (gathering and mining data) from individuals behaviours in a period of time. **Matching Agents** will attempt to match profiles of individuals to learning content. Teachers and institutions, taking also into account preferences that other individuals have expressed over a period of time and past experience on successful match making. A type of matching agent, **Collaboration Agents** will create communities of learners and teachers who are most likely to have better collaboration. **Searching Agents** will have the task to perform intelligent search through a cloud of knowledge in order to find the best fit with regards to learning content and actors. A variety of **Course Agents** will use results to compose learning collections, ranging from personal syllabi to learning objects and personalised learning paths. Similarly, **Programme Agents** will perform equivalent tasks for teachers and institutions, taking into account regulatory frameworks for award of credits and degrees. Learners will be assisted in their learning process by **Performance Monitoring Agents** who will be able to recommend alternatives if the progress is different from the desired. **Configuration Agents** will suggest the best tools available to individuals in order to surpass technological difficulties not related directly to the learning process. Finally, **Affective Emotional Agents**, probably presented as anthropomorphic on-line companions, will take into account the emotional state of the learners and adjust the learning process accordingly. Needless to say that interaction between all agent types will be essential to achieve the emergent CeL result.

5   Conclusions

This paper has defined Cloud e-Learning characteristics and justified why CeL goes hand in hand with advances in MAS. This new framework for learning can be used as a gateway to open available courses from different institutions. The proposed layers allow students to learn anything, with or without credits, with or without a degree and be able gain top quality on-line education through choosing the best teachers. Teachers will collectively develop top quality content, and institutions will compete for offering top quality programmes. The big picture of CeL is hard to deploy fully. Every individual aspect of CeL is by itself a research area that can be further developed. The vision for CeL, as described in this paper, provides a number of pedagogical and technological challenges which need to be addressed in the coming decades. The latter are concerned with the deployment of different kinds of agents that are assigned various roles to carry out specific tasks to assist the learners in this complex environment. Future work will include a more in-depth discussion of the role of agents and of the technological challenges of integrating MAS and the cloud.

References

1. ADL Technical Team: SCORM 2004 4th Edition Specification, http://www.adlnet.gov/scorm/scorm-2004-4th/
2. Belanger, Y., Thornton, J.: Bioelectricity: A Quantitative Approach Duke University's First MOOC (2013), http://dukespace.lib.duke.edu
3. Bennane, A.: Tutoring and multi-agent systems: Modeling from experiences. Informatics in Education-An International Journal 9(2), 171–184 (2010)
4. Bhavsar, V.C., Boley, H., Yang, L.: A weighted-tree similarity algorithm for multi-agent systems in e-business environments. In: Computational Intelligence, pp. 53–72 (2003)
5. Chatzara, K., Karagiannidis, C., Stamatis, D.: Students attitude and learning effectiveness of emotional agents. In: 2010 IEEE 10th International Conference on Advanced Learning Technologies (ICALT), pp. 558–559. IEEE (2010)
6. Daniel, J.: Making sense of moocs: Musings in a maze of myth, paradox and possibility. Journal of Interactive Media in Education 3 (2012)

7. Draganidis, F., Chamopoulou, P., Mentzas, G.: An ontology based tool for competency management and learning paths. In: 6th International Conference on Knowledge Management (I-KNOW 2006), pp. 1–10 (2006)
8. Fenton-Kerr, T., Clark, S., Cheney, G., Koppi, T., Chaloupka, M.: Multi-agent design in flexible learning environments. ASCILITE 98, 223 (1998)
9. Galen, G., Eric, K.: What cloud computing really means (April 2008), http://www.infoworld.com/d/cloud-computing/what-cloud-computing-really-means-031
10. Garro, A., Palopoli, L.: An XML multi-agent system for E-learning and skill management. In: Kowalczyk, R., Müller, J.P., Tianfield, H., Unland, R. (eds.) NODe-WS 2002. LNCS (LNAI), vol. 2592, pp. 283–294. Springer, Heidelberg (2003)
11. Gladun, A., Rogushina, J.: An ontology-based approach to student skills in multi-agent e-learning systems. Information Technologies and Knowledge 1 (2007)
12. IEEE Learning Technology Standards Committee: IEEE 1484.12.1-2002 Learning Object Metadata standard, http://ltsc.ieee.org/wg12/index.html
13. IMS Global Learning Consortium Inc.: IMS Learning Resource Meta-data Specification V1.3, http://www.imsproject.org/metadata/index.html
14. Ivanovic, M., Jain, L.C. (eds.): E-Learning Paradigms and Applications - Agent-based Approach. SCI, vol. 528. Springer, Heidelberg (2014)
15. Kaur, M., Kaur, A., et al.: A review article of cloud computing. International Journal of Computers & Technology 4(1), 102–105 (2013)
16. McFedries, P.: The cloud is the computer (April 2008), http://spectrum.ieee.org/computing/hardware/the-cloud-is-the-computer
17. Mell, P., Grance, T.: The nist definition of cloud computing (draft). NIST Special Publication 800(145), 7 (2011)
18. Orzechowski, T.: The use of multi-agents' systems in e-learning platforms. In: Siberian Conference on Control and Communications, SIBCON 2007, pp. 64–71. IEEE (2007)
19. Otsuka, J.L., Bernardes, V.S., Rocha, H.: A multiagent system for formative assessment support in learning management systems. In: Anais do I Workshop Tidia, São Paulo (2004)
20. Shih, W.C., Yang, C.T., Tseng, S.S.: Ontology-based content organization and retrieval for scorm-compliant teaching materials in data grids. Future Generation Computer Systems 25(6), 687–694 (2009)
21. Stamatis, D., Kefalas, P., Kargidis, T.: A multi-agent framework to assist networked learning. Journal of Computer Assisted Learning 15(3), 201–210 (1999)
22. Stamatis, D., Kefalas, P., Tsadiras, A.: Networked academic societies in collaborative development of e-learning courses. In: Fifth International Conference of Networked Learning, Lancaster, UK (2006)
23. Sultan, N.: Cloud computing for education: A new dawn? International Journal of Information Management 30(2), 109–116 (2010)
24. Talia, D.: Cloud computing and software agents: Towards cloud intelligent services. In: WOA, pp. 2–6 (2011)
25. Vaquero, L.M., Rodero-Merino, L., Caceres, J., Lindner, M.: A break in the clouds: towards a cloud definition. ACM SIGCOMM Computer Communication Review 39(1), 50–55 (2008)
26. Yang, F., Han, P., Shen, R., Kraemer, B.J., Fan, X.: Cooperative learning in self-organizing e-learner communities based on a multi-agents mechanism. In: Gedeon, T(T.) D., Fung, L.C.C. (eds.) AI 2003. LNCS (LNAI), vol. 2903, pp. 490–500. Springer, Heidelberg (2003)

Possible Routes on a Highway of eLearning – Promising Architecture for eLearning Systems

Mirjana Ivanović[1], Zoran Putnik[1], Dejan Mitrović[1], and Bela Stantić[2]

[1] Department of Mathematics and Informatics,
Faculty of Science, University of Novi Sad, Serbia
{mira,putnik,dejan.mitrovic}@dmi.uns.ac.rs
[2] School of Information and Communication Technology,
Griffith University, Gold Coast QLD, Australia
b.stantic@griffith.edu.au

Abstract. This paper presents the first and preliminary, original ideas about the possible architecture for the improvement of eLearning use at the university level. We suggest introduction of three-way architecture consisting of ways to: convert traditional teaching resources into the eLearning form of learning objects, methodology to employ some electronic activities as a replacement for traditional classroom activities, and methodology to use software agents for harvesting the necessary additional learning material from open learning repositories. Individual parts of the proposed architecture have been tested in practice, and showed very positive results, so we expect to further improve the application of eLearning by using the architecture as a whole.

Keywords: Learning Objects, Web 2.0, Agents, Learning Repositories.

1   Introduction

Overpowering our teaching life for the last decade or so, eLearning introduced new ways of thinking and organization of educational experiences. More than that, eLearning market is also overloaded with various tools and platforms intended to be used within learning communities. Applications used independently by learners, by combinations of individuals, institutions and systems, or by artificial LMSs' involve a number of separate tools. Consequently, a lot of challenging issues encourage further research [25], which we slightly adjusted to fit better into our purposes and research plans:

- **Convertibility:** What we *do* have already created, either as teaching resource, or teaching activity, should be easily convertible to any other needed form;
- **Reusability:** Learning content has to be organized in a way that it can easily be split up and reused in diverse contexts;
- **Interoperability:** Various learning systems and tools must be able to communicate mutually in order to offer, share, distribute, combine and

G. Jezic et al. (eds.), *Agent and Multi-Agent Systems: Technologies and Applications,*
Advances in Intelligent Systems and Computing 296,
DOI: 10.1007/978-3-319-07650-8_29, © Springer International Publishing Switzerland 2014

suggest parts of learning material as supplementary useful learning resources;

- **Accessibility:** Learning material should be available for access by different learning tools.

Technical part of the most of the problems arising within the field are highly relaxed within the last two decades or so, making possible focusing of the research on more practical details of management, regular exchange or reusability of common electronic teaching material, or on practical application of electronic teaching activities [11]. Reasons are simple and obvious:

- Data storage media became particularly cheap and available;
- Most of the incompatibilities between platforms are solved, those which are not, are overcome by the usage of common mediator – World Wide Web;
- Development of LMS's (and Internet in general) enabled searching, data mining, communication and exchange of teaching media easier than it has ever been,
- Growth of everyday use of elements of Web 2.0 and social networks introduced a lot of communication and collaboration activities into lives of scholars and students, getting them ready for the future life.

Looking into what was left of unsolved issues, we tried to use, explore and search into problems and obstacles surrounding the field of usage of learning management systems in university teaching. Of particular interest were three things:

- Need to convert and use legacy teaching material, prepared once-upon-a-time for different types of usage, yet of high quality, practically tested and employed (for years) in previous courses;
- Need to apply different types of electronic activities, which should replace classic face-to-face classroom activities, and
- Need to research and harvest available learning objects repositories for those resources that we do not have, and that we do not have time or funds to develop.

Lucky for us – considering the third request – there is a lot of electronic educational material in organized repositories (such as Open Educational Resources – OER, http://www.oercommons.org/community) and there is a trend of standardizations and a strong suggestion to make the content, as well as the accompanying metadata, widely available for harvesting by mobile agents. Standards are most often focused on searching and reusability of learning material, usually created in a form of learning objects.

Having these factors in mind, our proposed architecture is designed to facilitate three types of things:

- Software agents that will automatically and dynamically search, find, recognize and propose possible useful learning objects as additional learning material for a student;

- Tools for conversion of teaching resources into the form of learning objects, and
- Methods of transfer of teaching activities into electronic activities.

Considering the first route, the need for a meaningful search, recognition, and acquirement of widely available general learning objects, the most important concept we incorporated in our proposed architecture is *harvesting,* or allowing a software agent to collect different resources from remote repositories. The main problem with this route is the fact that rather frequently this process of "knowledge acquisition" from the distance can be less effective than traditional teaching. The most important reason for this phenomenon is students' inability to select essential information from "informational noise" and the fact that educational material is not always prepared in an appropriate way. Namely, the problem of finding an appropriate OER on the Internet is still widely open and is an important concern.

In order to overcome the open range of problems connected to the preparation, management, modernization, and reusability of learning repositories and courses, a solution is proposed to improve efficiency of knowledge acquisition process. This could be the application of an intelligent multi-agent system, and this paper is trying to help in line with this research. For this branch of our proposed system, we advise an agent-based architecture as a support for suggestion and (semi)automatic collection of additional needed teaching material. This part of the architecture is based on our previous research in the domains of agent languages [2], mobile agents [22] and tutoring systems [18].

With the last two issues, as the main outcomes of our research we thoroughly tested existing tools and techniques and reach the opinion about the best ways to perform conversion. We devised model of conversion, tested it in practice, and tested the use of electronic activities as a replacement of classis classroom activities. We came to a conclusion, and proved it by very positive and favourable student opinions that transfer from traditional classroom teaching activities to eLearning methodology is relaxed and eased for us, when supported by the outcomes of our study.

Ideas researched in this paper considering needed conversion methods were for several years practically tested within the long-lasting educational project, currently joining participants from seventeen universities, belonging to ten countries: Germany, Serbia, FYR of Macedonia and Bulgaria being the core members, and Croatia, Bosnia and Herzegovina, Romania, Albania, Montenegro, and Slovenia as associate members. More about the project, its goals and members can be found in [8], as well as publications [4] and [15], while the experiences gained were described in [5, 6, 7]. Large number of participating universities and interested lecturers ensured thorough testing, analysis and experimentation with all of the open questions, so the results gained are of a practical value.

The rest of the article is organized in the following way: the second section presents parts of the available research in all of the areas we are considering here in this paper. Third section depicts and explains our proposed architecture in more details, looking into each of the suggested routes separately. Besides, this section also presents experiences we gained so far with the use of proposed methodology in practice. Finally the fourth section concludes the paper, giving also insight into possible future application and development of proposed architecture.

2   Related Work

Over the years, the agent technology was suggested as a promising tool for both intelligent tutoring and course recommendation systems. This we can notice in for example, Educ-MAS [12], MathTutor [10], or ABITS [9]. Yet, instead of use of harvesting agents, mentioned systems have different approach, usually offering metadata authoring tools to teachers, who then themselves prepare and organize the course material.

One of the fundamental examples of agents use for metadata harvesting was demonstrated in [24]. This system employs crawler agents to parallelize the process of extracting metadata from web-based content. In addition, there are *Agent Based Search System* (ABSS) [21] and the AgCAT system [3], representing more complete agent-based frameworks used to harvest learning objects. It is interesting to notice that both systems operate by using agents that track changes in remote learning objects repositories.

What we find very important however, is that while metadata harvesting agents in ABSS and AgCAT continuously monitor the changes in distant repositories, we propose that the harvesting agents in our system are dispatched to remote learning objects repositories *in response* to an *automatically detected* decline in student's performance. In other words, we are trying to include intelligent, BDI-style agents in our system that will monitor the student's progress through a course, and obtain new learning materials on a need bases. The newly harvested material is than processed using intelligent data mining techniques and *pushed* to the student, instead of a student having to *pull* it using a search engine.

Considering the other routes of our research – conversion of teaching resources and teaching activities into a digital form – in [14], a suitable model for both creation and conversion of teaching material is presented. If we put it in words, it consists of the following steps:

- Teaching material is created using "favorite" tool – word processing or graphics application, software for design of multimedia resources, and similar;
- File is saved in some standard form;
- Conversion tool is used that transforms created file into a form suitable for eLearning;
- (If possible, file created in the second step is immediately saved in a form suitable for usage within LMS used.)

Problem with this model is lack of time and/or finances for creation of all of the teaching material from scratch. As a conclusion, authors in [14] suggest use of converters for legacy teaching material, and its fast transformation into a form suitable for learning management systems. This we will try to model in the following chapters, and describe the architecture of a system practically usable.

Conversion of teaching activities, from the traditional classroom activities, to digital ones, has not been studied in a form we're doing it here, to the best of our knowledge, but *was* in practice performed as a side-effect of use of Web 2.0 tools and techniques, almost as an accident! More about this line of our research, reader can find in [28, 29]. The most of the results were satisfactory, so this line of study has a definite value.

3   System Architecture

3.1   Part I – Software Agents

With the evolution of agent technology over the last two decades, several types of agents usable for teaching/learning activities emerged [1]. Amongst other things, appropriately developed and used, agents are capable of performing as virtual assistants for human clients, gathering on command specific data and presenting it according to the users' wishes. So, when incorporated in some eLearning system, agents can reduce duties of the human supervisor, helping with the selection of the appropriate learning material [27].

Based on the proven usefulness of various types of software agents, let us propose an agent-based architecture for harvesting learning resources. Our main goal is to suggest an architecture that will cover the most important needs of a user involved with the eLearning management systems, of which software agents are one of the three pillars. Software agents should play the part of the system that will – on-the-need basis – offer appropriate supplementary learning material for students in automatic and intelligent way. The main and the most realistic area where we suggest looking for the needed learning objects are many available OER's. Still, we can also imagine that with some slight changes, the system might be suitable for research on the Web as a whole.

Learning architecture, in more details explained in [16], can be graphically depicted as given in Fig.1. It consists of several major components:

- **SPM** – Student Performance Module – that constantly monitors students' behavior and learning activities, collecting the data about the learning process. Based on the progress through the course, quantity and quality of knowledge acquired, flaws, problems and/or improvements, SPM is capable of deciding on starting of harvesting, when appropriate, in order to obtain additional learning material for the student. When this happens, SPM sends proper information and request for additional learning objects to Central Intelligent Agent;
- **HA** – Harvesting Agent – actually performs the metadata harvesting, selects appropriate learning objects, and delivers the search results back to CIA, and

- **CIA** – Central Intelligent Agent – the central component of the system, assembling the metadata and learning objects received from harvesting agents, and making the selection of the most suitable harvested components, sending those back to the student as "additional teaching material". After receiving the request from SPM, CIA initiates the harvesting procedure.

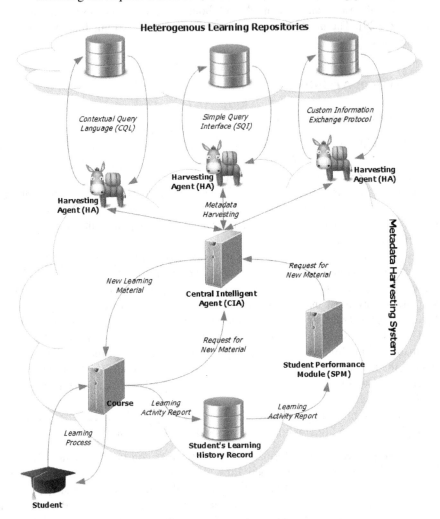

Fig. 1. Agent-based harvesting architecture

3.2   Part II – Conversion of Teaching Resources

The theoretical model for teaching resources conversion presented in [14] is still after a decade of a great value, yet it remains mostly on a philosophical level! Practical, and more down-to-earth model that we suggest here in Fig. 2, can be divided into three phases:

- Import of legacy teaching material, created in random application/form, and its conversion into a working middle-form;
- (Multi-step) Conversion of middle-form into a final, standardized XML form, and
- Export of the final material into (arbitrary) form of standardized learning objects.

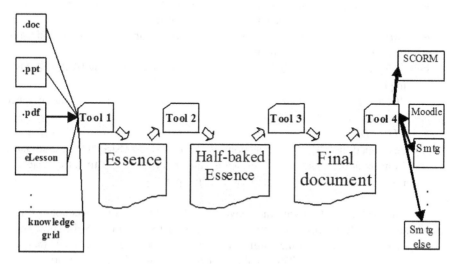

Fig. 2. Detailed model for conversion

On the software market, there is currently a large offer of tools claiming to be able to read those mentioned standard document formats, and convert those into a form of learning objects. This is not as truthful as marketing campaigns suggest, but there exist definitely some excellent tools usable for conversion. Even those highly creative, inventive and above all original parts of the process of teaching resources creation are challenged by the tools developed! More than that, useful results are achieved with the extraction of the essence of the teaching material. Sometime those tools require some additional effort, are not fully suitable for persons whose main expertise is out of the field of computer science, but the results we achieved were worth the effort.

Considering the intermediate representations of teaching material that is achieved by tools while going from traditional format, towards the final format, usable in Learning Management Systems, there is no strict, in advance defined standard format we suggest. Various tools create various formats, that author changes and incorporates into a more complex structures. While these structures are most commonly following some "best practices" (suggested and helped usually by all of the tools), until the final step, there is no need for strictness and hard definition of intermediate results. Yet, for the final format we propose use of XML, since by our experience it is the most suitable form, recognized and supported by all tools and all learning management systems.

In any case, combination of some of the tools and web-services available on software market, and lecturers' effort, can relatively easy and fast guarantee practical conversion of legacy teaching material into a form of learning objects, suitable for use in learning management systems. As possible candidates, out of several tenths of software applications we tested; let us mention those that produced the best conversion results:

- *"Thesis Learning Object Manager"* – conversion tool created by HunterStone, Inc. (available at http://hunterstone-thesis.software.informer.com/);
- *"MyUdutu"* – web service created by Udutu Online Learning Solutions (available at http://www.udutu.com/);
- *"Magic"* – conversion and essence extraction tool created by IBM, described in [19];
- *"Content Transformation Engine"* – again, conversion and essence extraction tool depicted in [26], and
- *"The Darwin Information Typing Architecture – dita"* – originally developed by IBM, later becoming a part of OASIS "Open Standards Consortium, by far the best tool for conversion and adjustment of teaching resources (available at https://www.oasis-open.org).

As a conclusion, situation with the conversion of teaching resources is not as bright as one might conclude from marketing campaigns of today' tools. Still, situation is also not that bad, and after some careful research, it is possible to find tools that will produce satisfactory results. This enables work with not too much effort by user considering the pure technical things, leaving more time and willpower for a profound work on the further development of learning objects.

3.3   Part III – Conversion of Teaching Activities

An actual need for concern in electronic teaching activities arises from the fact that modern students refuse to be only passive users of educational services, and instead want to take active part in creation of those. Collaboration, communication with lecturers and fellow students, and teamwork, are currently a usual and obligatory part of education, particularly on university level. This is above all true for professions that will require collaboration and teamwork later, in real-life situations, such as computer science.

There is also another, practical motive for dealing with electronic activities that developing countries face in the field of computer science, and not only in this field. Namely, majority of students of master studies or higher years of bachelor studies of computer science are already employed, and thus unable to attend classes. Still, even without considering this problem, all of the benefits of eLearning are further enhanced by adding some of the "humane" elements, which by our experience electronic activities based on Web 2.0 and social networking essentials give. While all of the significant learning management systems have the abilities of employment of electronic activities, descriptions we give here are tested within LMS Moodle, learning management system we are using for more than a decade now [20].

Discussion Forums

For a type of assignment where students are required to write a joint seminar paper, within a team of several students, we successfully used discussion forums as additional helpful tool. In the beginning, forums can be used to enable easier communication between team members, being a starting point for discussions about both organizational issues within a team and for discussing research issues. As time progresses, forums can and should be used for exchange of ideas, quotations of things found on the Internet (in connection with their tasks) and similar. In our experience, this influences positively the development of the teamwork, because students adding various individual opinions and discussions diverging from the starting point enabled better insight into the topic covered. What's more, by our experiences, participation of students in this activity was very high: there were no students that didn't participate at all, while on the average, students participated more than ten times a week during the whole semester.

Chat-Rooms

In case that it is possible to organize in advance scheduled chat-rooms, this electronic activity can provide the same functionality as discussion forums, adding to it the element of "live" discussion. The problem with this activity of course can be inability to schedule those events, or even worse, lack of students using it, because of their other obligations or preferences. Our experience unfortunately agrees with this lack of students participating in live chat-rooms, and their explanation that they prefer the "asynchronous version" of it, i.e. discussion forums.

Wikis

Already mentioned joint seminar papers, as well as teamwork assignments, are by far the best performed through the use of wiki technology. By nature of their previous habits and every day understanding and practice, in our experience, even the freshmen students of the first year of studies were successful and without any problems able to use wikis for solving their assignments with no additional preparation classes. On the average, they even showed better understanding of the technical part of the process, while the students of the higher years of studies presented better overall results and gave more meaningful jointly created seminar papers.

An additional value brought by using wikis for assignment solving was introduction of more fair and honest grading, which we explained in more details in [23]. By looking into the history of the development of the joint work – a feature regularly available in all learning management systems – lecturer is able to assess all of the participants for the exact amount and the quality of the work they invested in the final solution, while at the same time retaining team spirit and introduce this valuable part of studies and exercises alive.

Our idea and suggestion is to use wikis in two different ways:

- *Independent wikis* – for analysis of several aspects of a given topics, where students can be required to select one of the topics offered and then research and jointly write about it with the other students who selected the same topic, or
- *Discussion based wikis* – where teams were not only defined in advance, but roles within a team were also given (chief, moderator, researcher, editor, and similar).

Quizzes and Glossaries

As a useful thing for learning, quizzes can be offered to students for self-testing purposes. Of course, in some controlled environment, electronic quizzes can be used for grading also. By our experience again, ability to receive grades immediately, prepared in advance and performed by a computer, was very welcomed by students.

Use of glossaries of important terms and notions in connection with the topic is yet another possibility for improvement of eLearning, another usable electronic activity. Here our experiences are not that bright, since students didn't show too much enthusiasm while using glossaries, but they were not completely neglected, and were not negatively graded.

Assignments

Final electronic activity that can be used and that was already mentioned in combination with several other electronic activities is online assignments and their grading. In general, participation and posts in discussion forums can be immediately graded, but this is not the only possibility. Assignments solved using wiki can also be graded online, where grades are either available to students during the courses, or not, depending on the decision of lecturers. Even for the assignments solved off-line and then submitted to LMS, grades would be available within a system.

While this practice can and is sometimes challenged because of the privacy issues, our research and experience shows differently. Our students preferred this "public" method, considering it as a way to provide transparent and fair conditions, and do not mind if their colleagues can see their grades, which we described further in [17].

4   Conclusions

There is a lot of eLearning environments and applications of such systems in everyday life nowadays. Still we fill that it is necessary to upgrade such systems with additional functionalities. The first things that come to mind are personalization, recommendation, and more productive usage of teaching materials available on the Internet. In line with that, our proposed architecture aims to facilitate ways of better use of learning management systems and software agents to aid students in their efforts to learn.

We propose the improvement of eLearning through use of three-way architecture we described in this paper. We suggest use of software agents to automatically and dynamically recognize, collect and suggest additional learning material for a student, based on the observed behavior and advancements of students. Thus, harvesting agents represent a valuable concept that has been incorporated into the proposed architecture. They should have the task of exploring heterogeneous repositories and suggesting the most adequate learning material to students.

Further, we tested common techniques of Web 2.0 and social networking, trying to develop their place in the field of education. In this area, our experiences are excellent evaluated by both sides – lecturers and students. It is very true that application of these techniques require significant work and a lot of changes in habits of lecturers, but those are very worthwhile. In practice, we were able to distinguish both through students' answers in surveys, and by simple observation of students' behavior, which of the electronic activities have their value and future in education, and which require some additional refinement, so we propose this line as another route within our proposed eLearning architecture.

Considering the third and final branch of our proposed architecture, methodology to convert available, traditional learning resources into the form of learning objects, more suitable for eLearning, we assess that with the proper methodology and model of conversion we propose, it would be relatively easy to convert traditional teaching material into a digital form. Here we talk just about the technical part of the problem, of course. Considering the essential part of this conversion process i.e. the need to extract the essence of the teaching material and adjust it to better suit the new, eLearning facilities, we do not find perfect solutions available. Still, we consider this part of the process to be highly creative, so we did expect that the manual and (naturally) intelligent work should be invested.

Presented tools and techniques have all been individually and initially tested in practice, so the proposed architecture is based on our own preliminary experience. Each of the branches by itself was very usable and helped us in improving the use of eLearning facilities at our institutions, so we expect that the combination of those will have even better effects. Results of conversions of teaching materials have been considered from the viewpoint of a user, author of teaching resources. Results of changes in students' involvement and satisfaction were observed and surveyed – they are major and notable. Technologies used in the design of component which include harvesting agents are based on expertise and know-how previously developed at the Department of Mathematics and Informatics in Novi Sad. This will, in the long run, allow us to perform more serious evaluation and employment of the component in everyday eLearning practice. We expect that integration of new and additional functionalities will be done in a straightforward manner.

After preliminary positive results of usage of proposed architecture we considered possibilities to make it operational within some other learning management systems (LMS). As a possible test-bed for the proposed architecture we are considering a specific educational application for learning Java programming.

Because of the noted situation that learning computer programming can be difficult and can have a high student failure rate, at School of Information and Communication Technology, Griffith University of Australia, lecturers built upon a successful international blended-learning model and introduced Java Programming Laboratory (JPL). JPL is an educational application designed to assist students to learn Java as their first programming language [13]. JPL provides an environment that allows students to develop their programming skills by starting with simple code fragments and slowly transitioning to complete programs.

Technically, JPL is organized as a cloud self-paced learning environment which incorporates a number of features found in other successful programming learning environments. It builds upon them with a range of innovative features, some of which we described in this paper.

Since system enables lecturer or tutor to see how each student is performing at any time during the semester and therefore to identify any potential problem in progress, the idea is to combine this system with our proposed architecture in the future. It has been shown that the JPL positively influences students' retention, learning experience, and poor learning outcomes, so we expect that further development in the direction of application of intelligent eLearning facilities can improve it.

Acknowledgment. This work was supported by Ministry of Education and Science of the Republic of Serbia within the project entitled "Intelligent techniques and their integration into wide-spectrum decision support" (no. OI174023).

References

1. Badica, C., Budimac, Z., Burkhard, H.-D., Ivanović, M.: Software agents: languages, tools, platforms. Computer Science and Information Systems 8(2), 255–296 (2011)
2. Bađonski, M., Ivanović, M., Budimac, Z.: Adaptable Java Agents (AJA): A Tool for Programming of Multi-agent Systems. SIGPLAN Notices 40(2), 17–26 (2005)
3. Barcelos, C.F., Gluz, J.C.: An agent-based federated learning object search service. Interdisciplinary Journal of e-Learning and Learning Objects 7, 37–54 (2011)
4. Bothe, K., Schützler, K., Budimac, Z., Putnik, Z., Ivanović, M., Stoyanov, S., Stoyanova-Doyceva, A., Zdravkova, K., Jakimovski, B., Bojić, D., Jurca, I., Kalpić, D., Cico, B.: Experience with shared teaching materials for software engineering across countries. In: Informatics Education Europe IV, IEE-IV, Freiburg, Germany, pp. 57–62 (2009)
5. Budimac, Z., Putnik, Z., Ivanović, M., Bothe, K., Schuetzler, K.: Conducting a Joint Course on Software Engineering Based on Teamwork of Students, Informatics in Education. International Journal, Institute of Mathematics and Informatics, Lithuanian Academy of Sciences, 17–30 (2007), doi:10.1.1.149.2118
6. Budimac, Z., Putnik, Z., Ivanović, M., Bothe, K.: Common Software Engineering Course: Experiences from Different Countries. In: 1st International Conference on Computer Supported Education, CSEDU, Lisbon, Portugal, vol. 1, pp. 375–378 (2009)
7. Budimac, Z., Putnik, Z., Ivanović, M., Bothe, K., Schuetzler, K.: On the Assessment and Self-assessment in a Students Teamwork Based Course on Software Engineering. Computer Applications in Engineering Education 19(1), 1–9 (2011)

8. Budimac, Z., Putnik, Z., Ivanović, M., Bothe, K.: Transnational Cooperation in Higher Education in Balkan Countries. Novi Sad Journal of Mathematics 43(1), 167–177 (2013)
9. Capuano, N., Marsella, M., Salerno, S.: ABITS: an agent based intelligent tutoring system for distance learning. In: International Workshop in Adaptative and Intelligent Web-Based Educational Systems, pp. 17–28 (2000)
10. Cardoso, J., Guilherme, B., Frigo, L.B., Pozzebon, E.: MathTutor: a multi-agent intelligent tutoring system. In: Bramer, M., Devedzic, V. (eds.) Artificial Intelligence Applications and Innovations. IFIP, vol. 154, pp. 231–242. Springer, Heidelberg (2004)
11. Emin, V.: A Goal-oriented Authoring Approach to Design, Share and Reuse Learning Scenarios. In: 3rd EC-TEL 2008 PROLEARN Doctoral Consortium, European Conference on Technology Enhanced Learning Maastricht, The Netherlands (2008)
12. Gago, I.S.B., Werneck, V.M.B., Costa, R.M.: Modeling an educational multi-agent system in maSE. In: Liu, J., Wu, J., Yao, Y., Nishida, T. (eds.) AMT 2009. LNCS, vol. 5820, pp. 335–346. Springer, Heidelberg (2009)
13. Pullan, W., Drew, S., Tucker, S.: An integrated approach to teaching introductory programming. In: Second International Conference on e-Learning and e-Technologies in Education (ICEEE), Lodz, Poland, pp. 81–86 (2013), doi:10.1109/ICeLeTE.2013.6644352
14. Horton, W., Horton, K.: E-Learning Tools and Technologies. Wiley Publishing Inc. (2003)
15. Ivanović, M., Budimac, Z., Putnik, Z., Bothe, K.: Short Comparison of Tasks and Achievements of Different Groups of Students with the Common Software Engineering Course. In: International Conference on Software Engineering Theory and Practice (SETP 2009), Orlando, USA, pp. 84–91 (2009)
16. Ivanović, M., Mitrović, D., Budimac, Z., Vidaković, M.: Metadata Harvesting Learning Resources – An Agent-oriented Approach. In: ICSTCC: 15th International Conference on System Theory, Control and Computing, pp. 306–311 (2011)
17. Ivanović, M., Putnik, Z., Komlenov, Ž., Welzer, T., Hölbl, M., Schweighofer, T.: Usability and Privacy Aspects of Moodle - Students' and Teachers' Perspective. Informatica, An International Journal of Computing and Informatics 37(3), 221–230 (2013)
18. Klašnja-Milićević, A., Vesin, B., Ivanović, M., Budimac, Z.: Integration of recommendations and adaptive hypermedia into Java tutoring system. Computer Science and Information Systems 8(1), 211–224 (2011)
19. Li, Y., Dorai, C., Farrell, R.: Creating MAGIC: System for Generating Learning Object Metadata for Instructional Content. In: ACM Multimedia 2005, The 13th Annual ACM International Conference on Multimedia, Singapore, pp. 367–370 (2005)
20. Moodle home-page, https://moodle.org/
21. Orzechowski, T.: The use of multi-agents' systems in e-learning platforms. In: Siberian Conference on Control and Communications, SIBCON 2007, pp. 64–71 (2007) ISBN: 1-4244-0346-4
22. Pešović, D.: A high-level language for defining business processes. PhD Thesis, University of Novi Sad, Novi Sad (2007)
23. Putnik, Z., Ivanović, M., Budimac, Z., Samuelis, L.: Wiki – A Useful Tool to Fight Classroom Cheating? In: Popescu, E., Li, Q., Klamma, R., Leung, H., Specht, M. (eds.) ICWL 2012. LNCS, vol. 7558, pp. 31–40. Springer, Heidelberg (2012)
24. Sharma, S., Gupta, J.P.: A novel architecture of agent based crawling for OAI resources. International Journal of Computer Science and Engineering, IJCSE 2(4), 1190–1195 (2010)

25. Soylu, A., Kuru, S., Wild, F., Moedritscher, F.: E-Learning and microformats: a learning object harvesting model and a sample application. In: 1st International Workshop on Mashup Personal Learning Environments (MUPPLE 2008) at The 3rd European Conference on Technology Enhanced Learning (EC-TEL 2008), Maastricht, The Netherlands (2008)
26. Su, J.M., Tseng, S.S., Chen, C.H., Sung, Y.C., Su, T.H., Tsai, W.N.: A Study of Standardization of Traditional Teaching Materials. In: International Conference on Engineering Education ICEE 2003, Barcelona, Spain (2003)
27. Woda, M., Piotr, M.: Distance learning system: multi-agent approach. Journal of Digital Information Management 3(3), 198–201 (2005)
28. Zdravkova, K., Ivanović, M., Putnik, Z.: Evolution of Professional Ethics Courses from Web Supported Learning towards E-Learning 2.0. In: Cress, U., Dimitrova, V., Specht, M. (eds.) EC-TEL 2009. LNCS, vol. 5794, pp. 657–663. Springer, Heidelberg (2009)
29. Zdravkova, K., Ivanović, M., Putnik, Z.: Experience of Integrating Web 2.0 Technologies, Educational Technology Research & Development. ETRD 60(2), 361–381 (2012), doi:10.1007/s11423-011-9228-z

A Collaborative Environment for E-training in Archaeology

Manuella Kadar and Maria Muntean

"1 Decembrie 1918" University of Alba Iulia, Romania
{mkadar,mmuntean}@uab.ro

Abstract. This paper describes CREST (Collaborative Environment for Students and Teachers) a novel integrated environment for collaborative content retrieve and annotation and e-training in the field of Archaeology that is used by teaching stuff and students, as well. CREST is used in a broad range of collaborative applications and enables multi-authoring, using information in educational interactions by indicating information source, maintaining information, structuring information, adding meta-information, and sharing information among participants. An ontology enabled annotation and knowledge management environment was developed and endowed with collaborative information searching agents. Two agents were implemented in order to search and download aggregated metadata and text documents from the Web and to store information into a documents corpus repository. The corpus repository is further accessed by users through the annotator interface. This original approach is based on open standards and integrates open source services that improve discovery and reasoning across domain specific collections. CREST allows users to create, edit, store, and retrieve objects and annotations, promotes development and re-use of meaningful content. Such environment has a great utility for the development of virtual learning and research spaces.

Keywords: Domain knowledge representation, social tagging, information retrieval, 2D/3D digital objects annotation, ontology.

1 Introduction

The wide adoption of technologies that enable users to connect to each other and to contribute to the online community has changed the way that content is organized and shared on the Web. Social tagging to annotate resources represents one of the innovative aspects introduced with Web 2.0 and the new challenges of the semantic Web 3.0. In many online applications, it is possible for users to upload their own content or links to existing content and to organize it by use of tags, i.e., free-form keywords. Such applications, examples of which are *delicious*, *flickr*, and *BibSonomy*, are commonly referred to as collaborative tagging systems and they use the Internet to harness collective intelligence.

This paper presents the design and implementation of a **CollaboRative Environment for Students and Teachers**, named **CREST**. CREST allows a broad range of

G. Jezic et al. (eds.), *Agent and Multi-Agent Systems: Technologies and Applications*,
Advances in Intelligent Systems and Computing 296,
DOI: 10.1007/978-3-319-07650-8_30, © Springer International Publishing Switzerland 2014

collaborative applications and enables multi-authoring by: using information in educational interactions, indicating information source, maintaining information, structuring information, adding meta-information, and sharing information among participants. The participants are required to contribute through annotations that may include features such as comments on multimedia objects: text, 2D/3D objects, audio, video, virtual graphical spaces.

Current ways for Web indexing are not sufficient for learning resources. Indeed, automatic indexing, e.g. Google, can hardly rise above the syntax level of contents while indexing by human experts implies high costs. Recent approaches like semantic web and participative web (Web 2.0) offer promising solutions. This approach is based on semantic web technologies and ontology development used for building educational hypermedia systems. Another challenge addressed in this paper is about the study of functionalities on participative websites and the adding of content and metadata by visitors. In this respect, this paper presents a model of participative web interface adapted to communities of students and teachers.

Formal domain ontologies generally produced by experts are opposed to heterogeneous tags added by numerous users with various profiles. The presented model takes advantage of both semantic and participative approaches by populating formal domain ontologies with automatically extracted information from annotated multimedia objects. The goal of this model is to help the development of applications for sharing resources into communities of practice. It is based on a progressive indexing in which users progressively structure metadata, to finally allow semantic reasoning by computers and a shared vision of the domain by humans. This model integrates through the annotator interface several tools such as: social bookmarking tool, SemanticScuttle [1], that offers original features like tags structured by relations of inclusion and synonymy, or wiki spaces to describe tags. Another integrated facility is the tagging system of 2D objects achieved through Annotation Pilot [2] and a 3D ontology-enabled semantic annotator, ShapeAnnotator [3]. The environment was developed and tested with students and teachers in the field of Archaeology.

The Collaborative Environment for Students and Teachers – CREST aims to identify, implement and evaluate semantic approaches and enable academic institutions to exploit the full potential of community annotation/tagging systems. CREST was developed in order to enable annotation and knowledge management environment and to provide semantic web services. Personalized annotation is used to equip the collaborators with Web based authoring tools for commenting, knowledge articulation and exertion. The environment has enhanced conventional annotation system by extracting metadata from both the annotated content and the annotation itself, and establishing ontological relation between them. The aim was to develop an efficient environment based on open standards and comprising a set of open source services that improves discovery and reasoning services across domain specific collections by meeting the following main objectives:

- to identify a common model for representing tags and annotations on text, 2D and 3D digital objects, virtual graphical spaces and to enhance the interoperability of tags/annotations from distributed sources e.g., different communities using different systems;

- to develop a set of easily deployable tools and services for attaching annotations to multimedia including text, 2D/3D objects, virtual graphical spaces for harvesting annotations, and aggregating distributed annotations with metadata;
- to search and download aggregated metadata and documents from the web with information searching agents and to collect searched results into a corpus repository.

The remainder of this paper is organized as follows: section 2 discusses previous related work and the technical issues that influenced system design and implementation, section 3 describes the common, extensible model employed for representing annotations/tags and CREST architecture, section 4 provides examples of operation within CREST, section 5 presents results and discussions, and finally future work plans and conclusions are pointed out in sections 6.

2   Background and Related Work

2.1   Web-Based Annotation Services for Documents

Annotation of documents can be achieved by specialists according to a model produced by experts such as Learning Object Metadata [4]. In such cases, experts create ontologies, which are models of domains on which computers can perform reasoning. According to [5] and [6], this approach includes at least two weaknesses. Firstly, the creation of a common model is a difficult task — even for experts — requiring an important negotiation phase. Secondly, the annotation process is costly because it must be operated by specialists who understand and are able to apply the pre-defined model. Consequently, this centralized approach can be difficult to manage as difficulties encountered in the application of LOM. Nevertheless, tagging reaches its limits in the fact that it does not allow advanced structuring. First of all, tags leads to flat structures, letting few ways for users to organize their own tags. Moreover, tags' efficiency decreases because of problems of typography or synonymy. On the other hand, the rigidity of approaches based on controlled vocabularies (thesaurus, ontologies) realized by experts seems incompatible with the flexibility of tags built in a very distributive way. One of the big efforts to integrate annotations with ontologies is the Semantic MediaWiki (SMW) open-source project to which many people and organisations have contributed. Semantic MediaWiki (SMW) is an extension of MediaWiki – the wiki application best known for powering Wikipedia – that helps to search, organise, tag, browse, evaluate, and share the wiki's content. While traditional wikis contain only text which computers can neither understand nor evaluate, SMW adds semantic annotations that allow a wiki to function as a collaborative database. Semantic MediaWiki introduces some additional markup into the wiki-text which allows users to add "semantic annotations" to the wiki [7].

2.2   Web-Based Annotation Services for 2D/3D Objects

Learning systems should be provided with sufficient visualization of multidisciplinary contents for educational activities, especially by 3D models, 3D animations and simulations, which facilitate the learner's immersion in a hidden world [8]. The coding of additional knowledge in the form of structured metadata is important for the development of learning environments that take into account not only the geometry of shapes but also their semantics or meaning. At the same time, structural decompositions allow to consider a shape not only as a whole, but also as the collection of its parts. An approach to structural decomposition is augmented reality that can be used as system for annotating real-world objects. In this approach a virtual form (model) of a real-world object is matched to the real object, allowing one to visually annotate the real components with information from the corresponding model. Augmented reality provides a natural method for presenting the "enhancing" computer-based information by merging graphics with a view of the real object. User queries on the real object can be translated into queries on the model, producing feedback that can augment the user's view of the real world [9].

At present, several research projects are focused on the issue of the content-based information attached to 3D data generated by 3D scanning (achieved by scanner devices) photogrammetry (reconstruction of 3D data from 2D images) and procedural/parametric shape design (creation of new shapes from existing similar, parameterized shapes). Due to the nature of the data type and complexities involved in acquisition, production and processing of 3D data, a number of serious issues exist regarding 3D data acquisition, representation, encoding, content mark-up, and data history management. To date, these problems have not been sufficiently solved, and represent a major obstacle to a full integration of the 3D data type into digital archives. Among the 3D annotation tools that have been developed so far, are to be mentioned: SpacePen that allows a team to collaboratively work on a building design by annotating Java3D models using a web browser and a pen-based interface for drawing suggested modifications on the model [10], AnnoCryst [11] that enables users to annotate 3D crystallography models through a JMOL viewer and store the annotations on a shared web server and 3DSEAM designed to enable 3D scenes represented using the Extensible 3D (X3D) standard to be annotated using MPEG-7 [12]. All of these systems enable users to attach annotations to 3D models and to browse annotations added by others, asynchronously.

3   Architectural Model

CREST stores the annotations on a server that is separate to the server hosting the multimedia data. It has been assumed that there may be multiple communities, who may be annotating the same set of objects.

The Web annotation system comprises an annotation creation/authoring and attachment interface, an annotation web browser, information searching agents with retrieval interface, annotation storage and indexing component that stores the information into a documents corpus repository. CREST architecture is presented in Fig. 1.

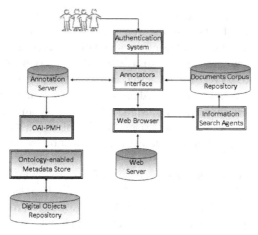

Fig. 1. CREST architecture

In the proposed model, decoupling of the annotations from the content allows more control and flexibility over how the annotations are accessed, processed, presented and re-used. The annotations are stored on a separate server that facilitates easy access, authoring and posting of responses that are controlled and restricted to a particular community of users, in this case, students and teachers from the Archaeology specialization. The separation also allows a single resource to be annotated in many different ways by users on the same annotation server or on different annotation servers, by using different community-specific terminologies or ontologies. By separating the annotations from the resources, the copyright issues that arise when having to store a copy of the digital resources on the social tagging site are also avoided.

The importance of being able to aggregate metadata from a range of sources has been recognized by a number of projects. Annotea [13] has recognized the need for standardized ways of defining annotations and tags so they can be shared between communities. However, the problem of annotation aggregation is still largely unresolved.

CREST proposes retrieving stored annotation data through the Open Archives Initiative Protocol for Metadata Harvesting (OAI-PMH). This approach involves mapping the annotations stored on Annotea server to the collections' metadata schema and periodically harvesting the annotations/tags by having the central agency send OAI-PMH (HTTP) requests to the server (Fig.1).

CREST adopts the "ontology-enabled folksonomy" approach in which users are provided with suggested Dublin Core metatags [14] and popular tags of an ontology (specified at system configuration). Users have also the option to define their own unique tags (Fig. 3). When an information searching agent browses on a parent tag, all items with the parent tag, synonym tags or children tags are retrieved. The class hierarchies are also incorporated within the tag cloud to embed multi-level structuring. In addition, CREST restricts access to the annotation server through an identity management system. This novel approach provides annotation services for closed communities with specific knowledge or expertise – students and teachers in a specific field that reduces the proportion of incorrect, inappropriate or misleading tags.

Within CREST community annotations are stored on an annotation server that is separate from the data collections or the web sources that the users annotate. An OAI-PMH interface has been built on top of the annotation server. This enables the periodic harvesting of new annotations (since the last harvest) by sending OAI-PMH (HTTP) requests to the server. The harvested annotations are then aggregated with the institutional metadata, to enrich the metadata store with community knowledge [15].

The information searching agents download documents from the Internet by using a list of words as thesaurus. The downloaded files are placed in the documents corpus repository and are processed in order to extract information by using specific annotation tools.

The information searching agents consist of the following classes: *Receiver-Agent, ReceiverBehaviorReceivePing, SenderAgent, SenderBehaviorReceiveAnswer, SenderBehaviorSendPing, download, yahoo, MessageManager, RunJade* and *RunMe*. The main classes are presented in the class diagrams in Fig. 2.

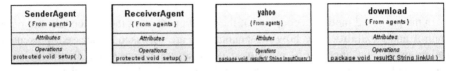

Fig. 2. Class diagrams

SenderAgent class aims to create intelligent agent "a1" which extends the *Agent* class. Method *setup()* initializes the agent, and by calling *addBehavior* method, behavior is added. Thus, agent "a1" has two behaviors: it receives messages from the agent "a2" through *SenderBehaviorReceiveAnswer* and it sends messages through *SenderBehaviorSendPing*.

```
public class SenderAgent extends Agent {
      protected void setup() {
             //Add SenderBehaviourReceiveAnswer behavior
             addBehaviour(new SenderBehaviourReceiveAnswer(this) );
             // Send messages to "a2" agent
             addBehaviour(new SenderBehaviourSendPing(this)); } }
```

ReceiverAgent class aims to create "a2"agent, which also extends the *Agent* class. The *setup()* method initializes the agent, and by calling *addBehavior* methos, it receives a behavior. Thus, "a2" receives messages from the agent "a1" and sends its reply through *ReceiverBehaviorReceivePing*.

```
public class ReceiverAgent extends Agent {
      protected void setup() {
             addBehaviour(new ReceiverBehaviourReceivePing(this)); } }
```

This behavior is used by the agent "a1" to send to the "a2" keywords for searching and downloading *.pdf* or *.html* documents from the www. *SenderBehaviorSendPing* class extends *OneShotBehavior* class, having a one-shot type behavior, i.e. sends all the keywords, and then it stops working.

SenderBehaviorReceiveAnswer class is the second "a1" agent behavior and aims receiving messages from "a2" through conversation with id "message". This behavior stops working when *done()* method is called and then it returns the value "true".

ReceiverBehaviorReceivePing class represents the "a2" agent behavior, which has the following functions that are repeated for each keyword received from the agent "a1":

1. Gets the word through conversation with id "message";
2. Calls the method *results1* from *yahoo* class sending as parameter the keyword;
3. Takes the search results from *yahooresults.html* file and extracts through regular expressions the target links to which documents to be downloaded are related;
4. Sends response to "a1" agent with the links found, one by one;
5. Sends the links found, one by one, to the *download* class by calling *results3* method that will download documents from the www to a local folder.

All this take place within the *action()* method, and when the *done()* method returns true, the agent "a2" stops its activity.

Yahoo class is called inside *ReceiverBehaviorReceivePing* class and through *results1()* method, it performs the following operations:

6. Add keyword received as a parameter to a URL through is perform a search with the www.yahoo.com popular search engine;
   ```
   URL url1 = new URL("http://search.yahoo.com/search?p="+inputQuery);
   ```
7. Create a *yahooresults.html* document that will store bit by bit all page content resulting from the search by given keyword:
   ```
   FileOutputStream fs = new FileOutputStream("D:\\DownloadPDF\\yahooresults.html");
   ```

4   Examples of operation of CREST

This section presents an example of how users annotate historical objects, based on the information previously searched by agents and automatically stored into the documents corpus repository. The information searching agent starts to perform search by keywords such as: *axe.pdf, socket.pdf* and *sharpaxe.pdf*. A piece of the output generated by running application is shown below:

```
Sender: I am a1 and I am sending input queries
Receiver: I am Agent a2 and I have received: axe.pdf from a1
Sender: I am a1 and I received: Search in the browser performed from a2
Sender: I am a1 and I received: I extracted the link: http://www.flickr.com/ from a2
Sender: I am a1 and I received: I downloaded a .pdf or .html file from the address:
http://www.flickr.com/   from a2
Sender: I am a1 and I received: I extracted the link:
http://www.grandforest.us/TheAxeBook.pdf from a2
Sender:I am a1 and I received: I downloaded a .pdf or .html file from the address:
http://www.grandforest.us/TheAxeBook.pdf   from a2
```

After the information searching agents populate the corpus repository users may access this repository and start annotation through the annotator's interface.

The annotator's interface also enables users to access SemanticScuttle, Annotation Pilot and ShapeAnnotator that are open source software integrated into CREST collaborative learning environment. An example is given in Fig. 3. Users can create and attach annotations to resources retrieved via a web search interface.

Fig. 3. Example of annotation for Bronze Age Knifes

Further on, a 2D object can be loaded into the Annotation Pilot, the image being captured and annotated with tags describing the most important parts of the object. This is an important tool for learning classifications of ancient artefacts (Fig. 4).

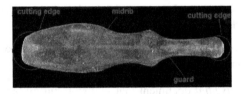

Fig. 4. Example of Annotator Pilot for images

The environment supports the annotation of 3D objects, as well as provides a user interface for browsing and searching annotations. Users can search across annotation attributes that include: creator, date, keywords or free-text searching over the description. In the ontology-driven annotation, the tags are defined by the ontology. The coupling of segmentation and knowledge formalization fosters the development of totally new approaches to shape retrieval. An example is provided for the bronze axe presented in Fig. 5. For this artefact it is possible to address queries such as: "*find a shape containing a loop and socketed body*", or more specifically to refer directly to subparts, e.g. "*find a socketed axe with rectangular mouth*", or "*rope moulding around the mouth*" obtaining as results, proper subparts of shapes. Semantics can be associated to the content itself, thus providing an enriched representation of the content.

Fig. 5. Bronze axe

The bronze axe presented in Fig.5 has an associated ontology developed in Protégé 3.4, which is an open source ontology editor and knowledge-base framework [16]. The ontology is loaded into the ShapeAnnotator tool [17]. After loading of the model and ontology, the first step of the annotation pipeline is execution of the segmentation algorithms to build the multi-segmented mesh. Once done, from the resulting multi-segmented mesh interesting features can be easily selected by simple mouse-clicks. Each interesting feature can then be annotated by creating an instance of a concept described in the ontology.

5   Results and Discussions

The result of the annotation process is a set of instances that, together with the domain ontology form a knowledge base (Fig. 6).

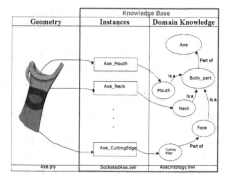

Fig. 6. Instances with specific relations representing a bridge between the geometry and semantics

All instances produced during the annotation pipeline are automatically assigned values for the above properties, so that the link between semantics and geometry is maintained within the resulting knowledge base. The ontology is stored in the *AxeOntology.owl* file (Fig.6). The knowledge base can be further exploited in educational activities. CREST was evaluated by more than thirty students and two professors in the field of Archaeology. Users have assembled data storage and retrieved structure for the archaeological field that allowed long-term data storage, (re)annotation, free-text searching, and dynamic record assembly. The system proved to be effective because students became active and interactive learners.

6   Conclusions

The presented collaborative annotation environment enables extensible and flexible storage of a specific domain data through collaborative tagging and use of specific domain ontologies. Focusing on the needs of virtual learning environments, this paper has presented how new approaches for data representation address changes due to the inevitable growth (both in diversity and volume) of data stores. The main components of the model and implementation architecture have been discussed with examples. Further developments will focus on models for melting different indexing solutions: automatic or by humans, including experts or simple users, based on structured models (e.g. ontologies) or on flexible metadata (e.g. tags). Structurable tags prove the technical possibility to make inferences on tags while keeping their spontaneous and flexible aspect. The contribution of domain experts and students in the design of such learning materials will increase the usability of structured tags created within CREST by the community of students and teachers and will extend the test bed to other fields of interest.

References

1. SemanticScuttle, http://semanticscuttle.sourceforge.net/
2. Annotation Pilot, http://www.colorpilot.com/annotation.html
3. Aim&Shape repository, http://shapes.aimatshape.net/
4. WG 12: Learning Object Metadata, http://ltsc.ieee.org/wg12/index.html
5. Brooks, C., Bateman, S., Liu, W., McCalla, G., Greer, J., Gasevic, D., Eap, T., Richards, G., Hammouda, K., Shehata, S., Kamel, M., Karray, F., Jovanovic, J.: Issues and directions with educational metadata. In: Third Annual International Scientific Conference of the Learning Object Repository Research Network, Montreal (2006)
6. Downes, S.: Resource profiles. Journal of Interactive Media in Education 5 (2004)
7. SMW, http://semantic-mediawiki.org/wiki/Semantic_MediaWiki
8. Klett, F.: A Design Framework for Interaction in 3D Real-Time Learning Environments. In: Second IEEE International Conference on Advanced Learning Technologies (2001)
9. Aracena-Pizzaro, D., Mamani-Castro, J.: Museum Guide Through Annotations Using Augmented Reality. In: Poster Papers Proceedings of the 18th International Conference on Computer Graphics, Visualization and Computer Vision 2010, pp. 35–38 (2010)
10. Jung, T., Gross, M.D., Do, E.I.Y.: Annotating and sketching on 3D Web models. In: Proceedings of the 7th International Conference on Intelligent User Interfaces, San Francisco, California, USA, pp. 95–102 (2002)
11. Hunter, J., Henderson, M., Khan, I.: Collaborative annotation of 3D crystallographic models. J. Chem. Inf. Model. 47(5), 2475–2484 (2007)
12. Bilasco, I.M., Gensel, J., Villanova-Oliver, M., Martin, H.: 3DSEAM: a model for annotating 3D scenes using MPEG-7. In: Proceedings of the 7th IEEE International Symposium on Multimedia (ISM 2005), pp. 310–319 (2005)
13. Koivunen, M.R., Kahan, J.: Annotea: an open RDF infrastructure for shared Web annotations. In: Proceedings of the 10th International Conference on World Wide Web, Hong Kong. ACM Press (2001)
14. Dublin Core Metadata Initiative, http://dublincore.org
15. Hunter, J., Gerber, A.: Harvesting community annotations on 3D models of museum artefacts to enhance knowledge, discovery and re-use. Journal of Cultural Heritage 11, 81–90 (2010)
16. Protégé, http://protege.stanford.edu/
17. Attene, M., Robbiano, F., Spagnuolo, M., Facidieno, B.: Part-based Annotation of Virtual 3D Shapes. In: Proceedings of the International Conference of Cyberworlds, pp. 427–436 (2007)

GC-MAS – A Multiagent System for Building Creative Groups Used in Computer Supported Collaborative Learning

Gabriela Moise, Monica Vladoiu, and Zoran Constantinescu

UPG University of Ploiesti, Romania
gmoise@upg-ploiesti.ro, {monica,zoran}@unde.ro

Abstract. Group creativity is a hot topic in the creativity literature, yet no method to obtain the most creative teams given a group of individuals is available. We introduce here a method for building creative teams, based on unsupervised learning and implemented with support from a multiagent system. Our first experiments with using this method for grouping learners involved in online brainstorming are presented as well.

Keywords: creativity, creative group, multi-agent system, unsupervised learning, computer supported collaborative learning.

1 Introduction

The concept of group creativity has been lately in the attention of educational institutions and companies alike. However, it is quite challenging to determine in which way the interactions that take place inside a group result in either increases or decreases in creative group performances. *Creative learning* is concerned with instructional processes that focus on the development of creative abilities of individuals. *Collaborative creative learning* approaches learning that results from interactions and collaborations that take place between learners and that aspires to enhance creativity at both individual level and group level. Group creativity may be improved by providing appropriate contextual instructional environments and by organizing the individuals in suitable groups [1]. *Computer Supported Collaborative Learning (CSCL)* has appeared as a reaction to software used previously in learning, which have been forcing learners to study and learn as isolated individuals [2]. In CSCL, learning is obtained by computer-supported interactions both between learners and between learners and teachers. Thus CSCL is defined as *a field of study centrally concerned with meaning and the practices of meaning-making in the context of joint activity and the ways in which these practices are mediated through designed artifacts* [3].

In this paper, we introduce a method of grouping team members in creative groups whose creativity is increased iteratively during the process. Our method is based on an adapted version of the unsupervised learning algorithm introduced by Watkins in [4] and it is under implementation with support from a multiagent system. We have experimented with this method for grouping learners involved in a CSCL process by

G. Jezic et al. (eds.), *Agent and Multi-Agent Systems: Technologies and Applications,*
Advances in Intelligent Systems and Computing 296,
DOI: 10.1007/978-3-319-07650-8_31, © Springer International Publishing Switzerland 2014

building up on the results obtained in our previous works [5, 6], which have approached the triggers that influence creativity in learning groups.

The structure of the paper is as follows: the next section presents the related work, the third one introduces our multi-agent system for building creative groups within CSCL processes, with which we have done some preliminary experiments presented in Section 4, and the last section include some conclusions and future work ideas.

2   Related Work

In this section we overview the related work, and point out some ideas that we have based our work on.

2.1   Creativity in Groups

Creativity is a concept highly debated in the psychological literature. Sternberg and his co-authors view *creativity as the ability to produce work that is novel (i.e., original, unexpected), high in quality, and appropriate* [7]. The challenge of understanding creativity has lead to the elaboration of many theories, for instance *the investment theory of creativity* proposed in [8, 9]. According to it, creative people are the ones *who are willing and able to, metaphorically, buy low and sell high in the realm of ideas.* Buying low refers to work on ideas that are unknown or unpopular, which have, however, built-in potential for growth. It is quite common that when such ideas are introduced for the first time they may encounter resistance. Nevertheless, a creative person would persist resisting to this opposition, and s/he will, eventually "sell" high, a new, powerful, or popular idea, achieving this way a *creativity habit* [9]. The creativity is multifaceted and it can be assessed by measuring *fluency (creative production of nonredundant ideas, insights, problem solutions, or products), originality (uncommonness or rarity of these outcomes)*, and *flexibility (how creativity manifests itself when using comprehensive cognitive categories and perspectives)* [10].

Nevertheless, group creativity is a recent topic in the literature, and it is seen as one of the expression of *the social nature of the creative act* [11]. However, group creativity means more that summing up the individual creativities of the members, as the interactions that take place between them within the group, the diversity of their backgrounds, abilities, and knowledge generate added value in creative processes. Baruah and Paulus approach the importance of interactions between the group members and their role in stimulating creative processes and point out that synergy refers to the added gain of collaboration within the group, which is obtained as a result of the stimulation, both cognitive and motivational, that results from these interactions. Further, based on the theoretical bases of synergy, the authors identify the cognitive, social, and motivational factors that influence the increase of group creativity: exchange of ideas, potential for competitiveness that allow individuals to compare their performances with the ones of their teammates, concept, product and perspective sharing, intrinsic motivation, openness to new experiences, etc. [12].

2.2   Modeling Group Creativity

The work of Amabile introduces the *componential theory of creativity,* along with the elements that influence creativity [13]. Three of them concern the individual level: *domain-relevant skills, creativity-relevant processes,* and *task motivation.* The fourth component is external to the individual: *the social environment* in which the work takes place. Domain-relevant skills refer to knowledge and expertise of the individual in a specific field. *Creativity-relevant processes* include individual characteristics that favor creativity: cognitive style, personality traits etc. Internal motivation of the individual is captured in the *task-motivation* component. Moreover, the author points out that *a central tenet of the componential theory is the intrinsic motivation principle of creativity.* In his model of group creativity, Sawyer sees creativity as a synergy between *synchronic interactions* and *diachronic exchanges* [14]. While developing his *multilevel model of group creativity,* Taggar highlights that besides including creative members, team creativity is significantly influenced by *relevant processes that emerge as part of group interaction* [15]. In their theoretical multilevel model of group creativity, Pirolla-Merlo and Mann explain how creativity evolve over time within teams and how it is influenced by the "climate" of creativity [16]. The contextual factors that influence creativity presented in [17] are divided in three categories: (1) factors that facilitate team creativity *(supervisory and co-workers support, psychological safety, group process),* (2) factors that obstruct the generation of creative ideas *(conformity, insufficient resources, bureaucratic structure),* and uncertain factors *(team diversity, conflicts in teams, group cohesion).*

   The interactionist model of creative behavior at the individual level of Woodman et al. provides an *interactionist perspective on organizational creativity.* Thus, group creativity is seen as *a function of individual creative behavior "inputs", the interaction of the individuals involved (e.g. group composition), group characteristics (e.g., norms, size, degree of cohesiveness), group processes (e.g., approaches to problem solving), and contextual influences (e.g. the larger organization, characteristics of group task).* Further, organizational creativity is considered to be *a function of the creative outputs of its component groups and contextual influences (organizational culture, reward systems, resource constraints, the larger environment and so on).* This multifaceted mix boosts *the gestalt of creative output (new products, services, ideas, procedures, and processes).* When building creative groups several features may be considered, at various levels: *individual (cognitive abilities/style, personality, intrinsic motivation, knowledge), group (cohesiveness, size, diversity, role, task, problem-solving approaches),* and *organizational (culture, structure, strategy, technology, resources, rewards etc.)* [18, 19, 20].

2.3   Similar Approaches of Building Creative Groups

Limited experiments with grouping students in creative teams are available in the literature. In [21], the authors present their work on using learning styles for grouping students involved in collaborative learning. A research project that investigates empirically whether knowledge sharing in community contexts can result in group knowledge that exceeds the individual knowledge of the group's members is done in [22].

The authors see that as *the hallmark of collaborative learning, understood in an emphatic sense*. An experimental study that evaluated the assumption that *shared cognition influences the effectiveness of collaborative learning* and it is crucial for cognitive construction and reconstruction of meaning is presented in [23]. A model of collaborative learning that aimed to build an intelligent collaborative learning system able to identify and target group interaction problem areas is available in [24]. Intense social interaction and collaboration is proven to contribute to the creation of a community of learning that nurtures *a space for fostering higher order thinking through co-creation of knowledge processes* in a case study presented in [25]. In [26], groups are classified and guided toward the optimal class that is a *high performing cooperative group with positive interdependence*. The issue of identifying peers and checking their fittingness for collaboration, as an essential pre-collaboration task, is approached in [27], where is shown that a more personalized cooperation can take place provided that individual tastes and styles of the peers are taken into consideration. In [28], the authors are concerned with the *liberating role of conflict in group creativity*, as a possible approach for weaknesses of group creativity, such as social loafing, production blocking, and evaluation apprehension. They have carried out an experiment in two countries to prove that brainstorming may benefit significantly from dissent, debate, and competing views, stimulating this way divergent and creative thought.

3   GC-MAS – A Multiagent System for Building Creative Groups

In this section we introduce our multi-agent system for building creative groups that we have experimented with in CSCL processes. Our approach is similar to the ones presented in Subsection 2.3, being concerned with teaming up individuals in the most appropriate teams with respect to creativity, but it is innovative in the sense that grouping students in creative teams in an iterative, semi-automated process has not been performed in our country or worldwide, up to our knowledge. Moreover, our first experiments are focused on online brainstorming to address some of the shortcomings of the face-to-face one revealed in the literature. Within our current stage of our work we focus on individual components of creativity when building the learning groups likely to be creative. The architecture of the system is presented in Fig. 1 and it includes the following agents (except for CommGC, all the other are *task agents*):

- *The Communication Agent (CommGC)* that has a dual role, being responsible with interfacing with the users (both students and instructors) and with the agents, along with managing the activities of the other agents;
- *The Creative Groups' Builder (BuildGC)* that is an agent that assists the instructor in the construction of the creative groups based on an unsupervised learning algorithm and various classification techniques;
- *The Creativity Evaluation Agent (EvalGC)* that has a support role in assessment of group creativity;
- *The Creativity Booster (EnvrGC)* that stimulates the development and the maintenance of creative contextual environments that provide for increasing group creativity;

- *The Glue Role Agent (GlueGC)* that supports the instructor in seeking out and taking on otherwise neglected tasks that have potential to facilitate creative group performances;
- *The Facilitator Agent (FCL-GC)* that supports the facilitator in helping groups to interact more efficiently;
- *The Team Relational Support Agent (TRS-GC)* that supports the team members in providing support for the other group members.

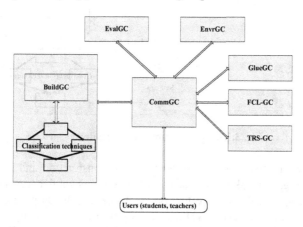

Fig. 1. GC-MAS - the bird's eye view architecture of the system

CommGC has a horizontally stratified structure, in which each level is connected directly to both the input sensors and the output actors (software entities that perform particular actions). Each level acts as an individual agent that provides the expected action. CommGC has two levels as follows: *(1) the social level* that ensures the communication with the other agents, the users, and with the external environment, as a true *personal/interface agent*, and *(2) the administrative level* that coordinates the actions of all the agents, so CommGC acts as a *middle agent* as well (see Fig. 2).

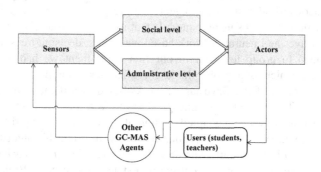

Fig. 2. CommGC– the agent's architecture

The agents **BuildGC, EvalGC, EnvrGC, GlueGC, TRS-GC** are execution agents that perform precise actions in the process of construction of the creative groups. They have a very simple structure, are goal-oriented, and they use plans libraries or classification techniques to perform their duties, as it can be seen in Fig. 3.

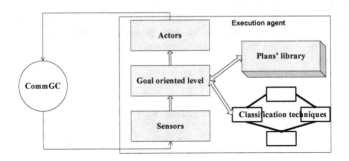

Fig. 3. The architecture of an execution agent

BuildGC - The Creative Groups' Builder aims at construction and iterative refinement of creative groups taking into account the components that generate creativity, their interdependencies that have effect on creativity and the purpose of building of creative groups (because, generally, the creativity of the group is sought for a specific goal - to solve a problem, to complete a task etc.). The data inputs for BuildCG are:

- *Student data* that include the individual characteristics that influence (both positively and negatively) the group creativity;
- *Group data* that contain the purpose of constructing creative groups (the problem to be solved, the task to be completed, the research to be undertaken etc.), the group size, the diversity of group members and so on;
- *Support data* that is generated by both users and other agents autonomously or as a result to the queries addressed by BuildGC.

The output data of BuildGC consists of both the most creative learning groups buildable and the queries to other users and agents with respect to the process of the group construction. BuildGC works using a module that includes various classification techniques (naïve Bayes and neural network based classifiers, decision trees, and support vector machines) to group learners. In our first experiment we had used a Naïve Bayes classifier, which is a probabilistic classifier based on Bayes theorem [29]. A detailed description of the Bayesian networks-based classification techniques can be found in [30]. The current reasoning process of the BuildCG agent is based on a combined approach between an adapted version of the Q-learning algorithm [4] and a classification technique based on Bayesian networks. In brief, this algorithm is a reward learning algorithm that starts with an initial estimate $Q(s, a)$ for each pair *<state, action>*. When a certain action a is chosen in a state s, the system (BuildCG) gets a reward $R(s, a)$ and the next state of the system is acknowledged. The Q-learning algorithm estimates the function *value-state-action* as follows:

$$Q(s,a) := Q(s,a) + \alpha(R(s,a) + \gamma \max_{a'} Q(s',a') - Q(s,a)) \quad (1)$$

Where $\alpha \in (0,1)$ is the instruction rate, $\gamma \in (0,1)$ is the discount factor, and s' is the state reached after executing the action a in the state s.

The way in which the values for the instruction rate and for the discount factor are selected is presented in [31]. Value 0 for the instruction rate means that the value for Q is never updated, and that the system never learns. Selection of a high value for this rate means that learning is faster. When the instruction rate equals 1 it means that the immediate reward is much more important than a past reward. For dynamic environments a balance between the immediate rewards and the past rewards is sought for. In our first experiments we had used a 0.5 instruction rate. The discount factor has values between 0 and 1. Closeness to 1 means that a future reward is more important to the system than an immediate reward.

In our case, we tackle n students. For each student, a characteristic vector that includes m individual features is constructed, namely $(c_1, c_2, ..., c_m)$. A *state* consists of this vector and the group number, while *an action* refers to moving a student to another group. Q expresses the quality of association between a state and an action. Our goal is to build the most creative k groups (k being given). To fulfill this goal we use the *GC-Q-learning adapted algorithm*, which is presented below:

1. Build a bi-dimensional matrix Q for all the possible pairs *<state, action>*. The columns of this matrix consists of $(c_1, c_2, ..., c_m, no\_group, action\_number, q)$. The action 1 corresponds to the selection for a particular student (given by the tuple of his individual characteristics) of the group number 1 to which he pertains. The action number 2 corresponds to the selection of group 2, and so on. All the elements in the q column are initialized with the value 0 or with a random low value;

2. Initialize the optim_policy (in our case is the optimal grouping) with a guided policy, and $Q\_optimal$ with Q;

3. Group the students and undertake working sessions (in our first experiments, brainstorming), in which the group creativity is assessed and its score is assigned to $R(s,a)$. For each such working session, the matrix Q is calculated.

   ```
   procedure working_session_computation
   select action of (optimal_policy)/*student grouping*/
   compute R(s,a)/* using agent EvalCG*/
   compute table Q /* following the formula (1) */
   ```

4. Analyze matrix Q. The optimal policy is given by the action for which Q_optimal gets the maximum value.

Once the optimal policy consisting in tuples $(c_1, c_2, ..., c_m, group\ number)$ is obtained, predictions for each set of data can be made based on advanced classification techniques (Bayesian networks, neural networks etc.). The Q values are the same for all the members of a group.

EvalGC. The Creativity Evaluation Agent supports the instructor in assessing the group creativity. This agent evaluates the group creativity based on the criteria for measuring ideation, namely novelty, variety, quantity, and quality introduced in [32]. It uses a plan library to achieve its goals of (1) recording the ideas generated by the group and classifying them, (2) calculating the frequency of ideas' production (as the number of ideas per time unit), and (3) keeping the creativity score and ensuring the conversation with the instructor via CommGC. **EnvrGC,** The Creativity Booster aims to enhance group creativity by providing for contextual environments that provide for creativeness. The agent works by "pushing on" the creativity triggers identified in our previous works to obtain a better creativity score for each group [5, 6]. This action is performed using a fuzzy controller with which we have worked previously. More details can be found in [6]. **GlueGC,** The Glue Role Agent is concerned with the coordination of group members' contributions and the management of group conflict. It pro-actively prevents situations in which group members focus entirely on coming up with their own ideas and ignore completely (to build on) the ideas of others, which is an essential added value of working together in a group, as it is shown in [33].

4   Experimenting with GC-MAS

In this section we present briefly our first experiments with our system. After clarifying the conceptual aspects of GC-MAS, we have been concerned with investigating the viability of our approach and therefore we have undertaken a pedagogical experiment with our undergraduates and graduate students in Computer Science. The core of the experiment consists of brainstorming sessions concerned with the issues that regard the improvement of the curricula and of the syllabuses of the courses for our Computer Science programs, both at undergraduate and graduate level. In order to avoid some of the shortcomings of the face-to-face brainstorming sessions, we have undertaken these sessions online. This experiment consists in several stages:

- Assessing the individual student creativity with several evaluation tools. For the time being we have worked with the Gough Creative Personality Scale [34] and an extended version of Creative Achievement Questionnaire [35] that we have adapted for Computer Science students. We have chosen to start with Gough because is simple to use it and interpret it. Within our 27 students, the maximum score is 10 and the minimum one is -3. The average score is 2.9. In the Gough Scale the values are between -12 and 18. The student motivation can be low (having value 0), middle (1), or high (2);

- Activating BuildCG for the pool of 27 students based on the next procedure:
  0.   Build matrix Q;
  1.   Group them, let them have a brainstorming session, and obtain a reward R;
  2.   Update column q of matrix Q;
  3.   Iterate step 1,2 for the initial pool of students;
  4.   Consider (randomly for now) other student pools to undertake step 1,2, 3;
  5.   Analyze table Q. The optimal policy is given by the action for which Q_optimal gets the maximum value;

- Analyzing the preliminary results and improving of the multi-agent system.

Following this simple procedure, BuildCG undergoes a process of unsupervised learning based on the CG-Q-Algorithm, which associates an action to a state aiming at increasing the reward. The data from our first experiments are available at http://www.unde.ro/GC-MAS.zip. We will continue to update this archive.

5   Conclusions and Future Work

We introduced here our semi-automated method of grouping team members in increasingly creative groups, which is put to practice by a multiagent system prototype. Moreover, we had performed some experiments for grouping learners involved in online brainstorming, the results being encouraging. Future work ideas regard the improvement of both the method and the working prototype in several directions: corroborating the results obtained with several creativity evaluation scales, assessment of creativity before and after activities assumed to help trigger creativity, inclusion of contextual and organizational factors, testing the method in other activities, improving of the algorithm, offering the method as an online open service etc.

References

1. Mannix, E.A., Goncalo, J.A., Neale, M.A. (eds.): Creativity in Groups, Research on Managing Groups and Teams Series, vol. 12. Emerald Group Publishing Ltd. (2009)
2. Stahl, G., Koschmann, T., Suthers, D.: Computer-Supported Collaborative learning: An historical perspective. In: Sawyer, R.K. (ed.) Cambridge Handbook of the Learning Sciences, pp. 409–426. Cambridge University Press, Cambridge (2006)
3. Koschmann, T.: Dewey's Contribution to the Foundations of CSCL Research. In: Conference on Computer Support for Collaborative Learning: Foundations for a CSCL Community, pp. 17–22. International Society of the Learning Sciences (2002)
4. Watkins, C.: Learning from Delayed Rewards. PhD Thesis,University of Cambridge, England (1989), http://www.cs.rhul.ac.uk/home/chrisw/thesis.html
5. Moise, G.: Triggers for Creativity in CSCL. In: 9th International Scientific Conference eLearning and Software for Education, Bucharest, pp. 326–331 (2013)
6. Moise, G.: Fuzzy Enhancement of Creativity in Collaborative Online Learning. In: Chiu, D.K.W., Wang, M., Popescu, E., Li, Q., Lau, R. (eds.) ICWL 2011 and ICWL 2012. LNCS, vol. 7697, pp. 290–299. Springer, Heidelberg (2014)
7. Sternberg, R.J., Lubart, T.I., Kaufman, J.C., Pretz, J.E.: Creativity. In: Holyoak, K.J., Morrison, R.G. (eds.) The Cambridge Handbook of Thinking and Reasoning, pp. 351–369. Cambridge University Press, New York (2005)
8. Sternberg, R.J., Lubart, T.I.: An Investment Theory of Creativity and its Development. Human Development 34(1), 1–31 (1991)
9. Sternberg, R.J.: The Assessment of Creativity: An Investment-Based Approach. Creativity Research Journal 24(1), 3–12 (2012)

10. Rietzschel, E.F., De Dreu, C.K.W., Nijstad, B.A.: What are we talking about, when we talk about creativity? Group creativity as a multifaceted, multistage phenomenon. In: Mannix, E.A., Goncalo, J.A., Neale, M.A. (eds.) Creativity in Groups, Research on Managing Groups and Teams Series, vol. 12. Emerald Group Publishing Ltd. (2009)

11. Gorny, E. (ed.): Group creativity. Dictionary of Creativity: Terms, Concepts, Theories & Findings in Creativity Research (2007),
    `http://creativity.netslova.ru/Group_creativity.html`

12. Baruah, J., Paulus, P.B.: Enhancing Group Creativity: The Search for Synergy. In: Mannix, E.A., Goncalo, J.A., Neale, M.A. (eds.) Creativity in Groups, Research on Managing Groups and Teams Series, vol. 12. Emerald Group Publishing Ltd. (2009)

13. Amabile, T.M.: Componential Theory of Creativity. In: Kessler, E.H. (ed.) Encyclopedia of Management Theory. SAGE Publications Inc. (2013)

14. Sawyer, R.K.: Group Creativity. Lawrence Erlbaum, Mahwah (2003)

15. Taggar, S.: Individual Creativity and Group Ability to Utilize Individual Creative Resources. The Academy of Management Journal 45(2), 315–330 (2002)

16. Pirola-Merlo, A., Mann, L.: The Relationship Between Individual Creativity and Team Creativity: Aggregating Across People and Time. Journal of Organizational Behavior 25, 235–257 (2004)

17. Yeh, Y.C.: The Effects of Contextual Characteristics on Team Creativity: Positive, Negative, or still Undecided. Working paper, Lund University, Sweden (2012)

18. Woodman, R.W., Schoenfeldt, L.F.: Individual Differences in Creativity: An Interactionist Perspective. In: Glover, J.A., Ronning, R.R., Reynolds, C.R. (eds.) Handbook of Creativity, pp. 77–92. Plenum Press, New York (1989)

19. Woodman, R.W., Schoenfeldt, L.F.: An Interactionist Model of Creative Behaviour. Journal of Creative Behavior 24, 279–290 (1990)

20. Woodman, R.W., Sawyer, J.E., Griffin, R.W.: Toward a Theory of Organizational Creativity. Academy of Management Review 18(2), 293–321 (1993)

21. Martin, E., Paredes, P.: Using Learning Styles for Dynamic Group Formation in Adaptive Collaborative Hypermedia Systems. In: 1st International Workshop on Adaptive Hypermedia and Collaborative Web-Based Systems, AHCW 2004, pp. 188–198 (2004)

22. Stahl, G.: Cognition in Computer Assisted Collaborative Learning. Journal of Computer Assisted Learning 21, 79–90 (2005)

23. Stoyanova, N., Kommers, P.: Concept Mapping as a medium of shared cognition in Computer-Supported Collaborative Learning. Journal of Interactive Learning Research 13(1), 111–133 (2012)

24. Soller, A.: Supporting Social Interaction in an Intelligent Collaborative Learning System. International Journal Artificial Intelligence in Education 12, 40–62 (2001)

25. Ma, A.W.W.: Computer Supported Collaborative Learning and Higher Order Thinking Skills: A Case Study of Textile Studies. The Interdisciplinary Journal of E-Learning and Learning Objects 5, 145–167 (2009)

26. Israel, J., Aiken, R.: Supporting Collaborative Learning with an Intelligent Web-based System. International Journal of Artificial Intelligence and Education 17(1), 3–40 (2007)

27. Kumar, V.: Computer Supported Collaborative Learning - Issues for Research. University of Saskatchewan (1996),
    `http://www.cos.ufrj.br/~jano/CSCW2008/Papers/Kumar_.pdf`

28. Nemeth, C.J., et al.: The Liberating Role of Conflict in Group Creativity: A Study in Two Countries. European Journal of Social Psychology 34, 365–374 (2004)

29. Joyce, J.: Bayes' Theorem. The Stanford Encyclopedia of Philosophy. In: Zalta, E.N. (ed.) (2008), http://plato.stanford.edu/archives/fall2008/entries/bayes-theorem/
30. Friedman, N., Geiger, D., Goldszmidt, M.: Bayesian Network Classifiers. Machine Learning 29, 131–163 (1997)
31. Leon, F., Şova, I., Gâlea, D.: Reinforcement Learning Strategies for Intelligent Agents in Knowledge-Based Information Systems. In: 8th International Symposium on Automatic Control and Computer Science, Iaşi, Romania (2004)
32. Shah, J.J., Vargas-Hernandez, N.: Metrics for measuring ideation effectiveness. Design Studies 24(2), 111–134 (2003)
33. Bolinger, A.R., Bonner, B.L., Okhuysen, G.A.: Sticking together: the glue role and group creativity. In: Mannix, E.A., Goncalo, J.A., Neale, M.A. (eds.) Creativity in Groups, Research on Managing Groups and Teams Series, vol. 12, pp. 267–291. Emerald Group Publishing Ltd. (2009)
34. Gough, H.G.: A Creative Personality Scale for the Adjective Check List. Journal of Personality and Social Psychology 37, 1398–1405 (1979)
35. Carson, S., Peterson, J.B., Higgins, D.M.: Reliability, Validity, and Factor Structure of the Creative Achievement Questionnaire. Creativity Research Journal 17(1), 37–50 (2005)

Author Index